AF181716

ALLE ZEIT WACH
1842

D. A. Carlson A. B. Haurie A. Leizarowitz

Infinite Horizon Optimal Control

Deterministic and Stochastic Systems

Second, Revised and Enlarged Edition

Springer-Verlag
Berlin Heidelberg New York
London Paris Tokyo
Hong Kong Barcelona
Budapest

Dean A. Carlson
Department of Mathematics,
University of Toledo,
2801 W. Bancroft Street, Toledo, OH 43606, USA

Alain B. Haurie
Département d'Economie Commerciale et Industrielle,
Université de Genève,
2, rue de Candolle, CH-1211 Genève, Suisse

Arie Leizarowitz
Department of Mathematics,
Technion-Israel Institute of Technology,
32000 Haifa, Israel

The first edition was published in 1987 as Volume 290 of the series "Lecture Notes in Economics and Mathematical Systems".

ISBN-13:978-3-642-76757-9 e-ISBN-13:978-3-642-76755-5
DOI: 10.1007/978-3-642-76755-5

Library of Congress Cataloging-in-Publication Data
Carlson, D. A. (Dean A.) 1955-
Infinite horizon optimal control: deterministic and stochastic systems / D. A. Carlson, A. B. Haurie, A. Leizarowitz. - 2nd, rev. and enl. ed. p. cm.
Includes bibliographical references and index.
ISBN-13:978-3-642-76757-9

1. Control theory. 2. Mathematical optimization. I. Haurie, A. (Alain) 1940-. II. Leizarowitz, A. (Arie) III. Title.
QA402.3.C35 1991 629.8'312-dc20 91-23886

This work is subject to copyright. All rights are reserved, whether the whole or part of the material is concerned, specifically the rights of translation, reprinting, reuse of illustrations, recitation, broadcasting, reproduction on microfilms or in other ways, and storage in data banks. Duplication of this publication or parts thereof is only permitted under the provisions of the German Copyright Law of September 9, 1965, in its current version, and a copyright fee must always be paid. Violations fall under the prosecution act of the German Copyright Law.

© Springer-Verlag Berlin Heidelberg 1991
Softcover reprint of the hardcover 2nd edition 1991

Typesetting: Camera ready by author
42/3130-543210 - Printed on acid-free paper

to Jadzia

to Dominique

to Natty

Preface

This monograph deals with various classes of deterministic and stochastic continuous time optimal control problems that are defined over unbounded time intervals. For these problems the performance criterion is described by an improper integral and it is possible that, when evaluated at a given admissible element, this criterion is unbounded. To cope with this divergence new optimality concepts, referred to here as *overtaking optimality, weakly overtaking optimality, agreeable plans, etc.*, have been proposed. The motivation for studying these problems arises primarily from the economic and biological sciences where models of this type arise naturally. Indeed, any bound placed on the time horizon is artificial when one considers the evolution of the state of an economy or species. The responsibility for the introduction of this interesting class of problems rests with the economists who first studied them in the modeling of capital accumulation processes. Perhaps the earliest of these was F. Ramsey [152] who, in his seminal work on the theory of saving in 1928, considered a dynamic optimization model defined on an infinite time horizon. Briefly, this problem can be described as a *Lagrange problem with unbounded time interval.* The advent of modern control theory, particularly the formulation of the famous *Maximum Principle* of Pontryagin, has had a considerable impact on the treatment of these models as well as optimization theory in general. During the late sixties and seventies, applications of the optimal control formalism to the modeling and analysis of dynamic economic systems began to flourish. In 1965 Kurz [113] and Stoleru [171] simultaneously initiated this research by applying Pontryagin's Maximum Principle (see [150]) to finite horizon optimal economic growth problems. Shortly afterward, Albouy and Breton [1] in 1968, and Dorfman [64] in 1969, presented an economic interpretation of Pontryagin's Theorem. An equally significant development, for the purposes of this monograph, occurred in 1965 when von Weizäcker [180] introduced the overtaking optimality concept. This concept provided a means for dealing with a general class of infinite horizon optimal control problems with unbounded cost criterion for the first time. Another development that has played an important role in this theory also occurred in 1965 when Cass [36] established the first continuous time *Turnpike Theorem* for optimal control problems. The turnpike property was first coined by Samuelson in 1958 (see [164]) where he showed that an efficient expanding economy would spend most of the time in the vicinity of a balanced equilibrium path (also called a *Von Neumann path*).

In conjunction with these developments, a similar property was systematically studied in the realm of discrete time multi-sector optimal capital accumulation models. The works of Radner [151] in 1961 and McKenzie [141] in 1963, present a rather complete turnpike theory for these models. Their results are based on duality theory and convex analysis. With regards to infinite horizon optimization, Brock [27] in 1970 presented existence results for both overtaking and weakly overtaking optimal solutions for the models studied by Radner [151] and McKenzie [141]. At the time of their development, Brock's results had no counterpart in the continuous time framework. The first systematic treatment of continuous time infinite horizon optimal control problems is presented in the now classic book of Arrow and Kurz [7], 1970. In that work, they present a sufficient maximum principle and study various asymptotic properties of optimal trajectories. Further, at approximately the same time period, Shell [169] had begun to investigate the nature of the *Hamiltonian systems* that arise from these economic systems.

All of the developments discussed above are treated within the realm of economics and occurred in parallel with the widening theory of dynamic optimization which took place in the areas of automatic control and applied mathematics. With regard to these matters we remark that the maximum principle for infinite horizon optimal control problems is discussed only briefly in Pontryagin's book, but is not considered at all in the complete text of Lee and Markus [134] published in 1967. Further, in 1966 Mangasarian [134] provides a set of sufficient conditions for optimality in optimal control problems which is in the spirit of convex mathematical programming. Also, at the same time, the classical existence results of the calculus of variation were extended to the realm of optimal control in the work of Cesari [48] in 1966, as well as other authors. All these contributions to the theory of optimal control were primarily concerned with finite horizon problems. The exception to this case was the *regulator problem.* In this problem, the linear structure of the dynamics coupled with a quadratic performance criterion permitted the development of a link between infinite horizon optimization and Lyapunov stability analysis. We further mention that in 1964 Bellman and Bucy [18] briefly mention the area of asymptotic control theory.

From the above we see that the investigation of infinite horizon optimal control was largely undeveloped in the late sixties. This changed in 1974 when Halkin [89] presented the first correct extension of Pontryagin's Maximum Principle to the case of weakly overtaking optimality. Since Halkin's paper, research in this area of optimal control increased dramatically, and much is known concerning these models. Throughout the development of this the theory it is possible to distinguish three major categories of research that constitute the theory of infinite horizon optimal control:

(i) The extension of the necessary conditions for optimality either in the form of a maximum principle or as the classical Euler-Lagrange equations;

(ii) The extension of various finite horizon sufficient conditions and verification theorems that are known either in optimal control theory or the calculus of variations; and

(iii) The discussion of existence theorems for the various notions of optimality that have been introduced.

A common link between these three categories is played by the *Turnpike Property.* Indeed once it can be established that the optimal trajectories emanating from different initial states, ultimately "bunch together" it is possible to investigate (i), (ii), and (iii). In particular we will see that it is possible to cope with the fact that there is no transversality condition in Halkin's extension of the maximum principle, to extend the results of Mangasarian [134], as well as to provide conditions which insure the existence of an optimal solution.

The monograph which you are about to read is an extended version of the lecture note entitled **Infinite Horizon Optimal Control; Theory and applications**, Lecture Notes in Economic and Mathematical Systems, **290**, published in 1987. As before we have attempted to present a systematic account of the development of continuous time infinite horizon optimal control theory beginning primarily with the early seventies to the present. Consequently, in addition to the topics treated in our last effort we have included many more recent results which serve to unify as well as extend the existing theory so that many more classes of systems may be dealt with. In particular we mention the addition of the new chapter on *reduction to finite rewards*, the extension of the results of Chapter 4 to some models with time delay, the treatment of the overtaking criterion for controlled diffusion processes as well as the discussion of the turnpike property to optimal control problems with random modal jumps. We hope that previous readers will welcome these new additions and find them useful. From the significant amount of new material which we have included it is clear that this area is still growing rapidly and as a consequence our presentation is not all inclusive. We apologize for any results which we may have omitted.

Acknowledgements

In undertaking a task such as this there are many people along the way who deserve our sincerest gratitude and appreciation. Unfortunately it is impossible for us to mention all of these wonderful people, although there are some who we feel deserve particular mention. These include T. S. Angell, A. Blaquière, W. A. Brock, G. Leitmann, for their encouragement and inspiration. We further thank T. Eirola, R. Hamalaïnen, R.F. Hartl, M.L. Petit, G. Ricci, for their many suggestions and comments on the earlier edition of this work. We have attempted to include their input in this new edition. In addition we want to acknowledge the significant contributions of A. Jabrane and C. van Delft to the results presented in Chapters 9 and 11 respectively. We also extend our sincerest gratitude to our TEXpert Nicole Paradis for the truly excellent and competent composition of the countless revisions of the work that follows. Her patience and cooperation are without measure. Finally we wish to acknowledge the cooperation of École des Hautes Études Commerciales, the stimulating environment found at GERAD, and the financial support provided by NSERC-Canada to A. Haurie and A. Leizarowitz.

Contents

Chapter 1

Dynamical Systems with Unbounded Time Interval in Engineering, Ecology and Economics

1.1 Introduction

The aim of this first chapter is to provide the reader with some basic motivation for the study of optimal control problems with an infinite time horizon. Three examples borrowed respectively from the fields of engineering, ecology and economics are presented. The first example is the well known regulator problem, the second one deals with a pest-control problem and finally the third one concerns optimal economic growth. From the consideration of these three typical problems it will be possible to introduce the basic optimality concepts for infinite horizon problems.

1.2 The regulator problem

Consider a linear system

$$\dot{x}(t) = Ax(t) + Bu(t) \tag{1.1}$$

$$y(t) = Cx(t) \tag{1.2}$$

$$x(0) = x_0 \tag{1.3}$$

where $x \in \mathbb{R}^n$ is the state variable, $u \in \mathbb{R}^m$ is the control variable and $y \in \mathbb{R}^p$ is the output variable. A, B and C are given matrices of dimensions $n \times n$, $n \times m$ and $p \times n$ respectively.

Given a measurable control $u(\cdot) : [0, T] \to \mathbb{R}^n$, with functions $x(\cdot) : [0,T] \to \mathbb{R}^n$ and $y(\cdot) : [0,T] \to \mathbb{R}^p$ satisfying (1.1)–(1.3), a performance criterion is defined as follows on the time interval $[0,T]$

$$J_T(x_0, u(\cdot)) = \frac{1}{2} \int_0^T [y'(t)Qy(t) + u'(t)Ru(t)]\, dt \tag{1.4}$$

where Q and R are given $p \times p$ and $m \times m$ positive definite matrices respectively, and the transpose of a matrix A is denoted A'.

The problem of finding the control $u^*(\cdot)$ which minimizes the performance criterion (1.4) under the constraints (1.1)–(1.3) has received considerable attention in the control literature. The optimal control seeks to maintain the output close to zero while avoiding excessive expenditure of control energy. Practical applications of this control scheme are countless. The solution of the problem is well known, see Athans and Falb [4].

Proposition 1.1 *Assume that the system* (1.1)–(1.3) *is observable*[1] *and controllable*[2]. *Then the optimal control minimizing the performance criterion* (1.4) *exists, is unique and is given by*

$$u^*(t) = -R^{-1}B'K(t)x^*(t) \tag{1.5}$$

where the $n \times n$ *symmetric and positive definite matrix* $K(t)$ *is the solution of the matrix Riccati equation*

$$\dot{K}(t) = -K(t)A - A'K(t) + K(t)BR^{-1}B'K(t) - C'QC \tag{1.6}$$

with the boundary condition

$$K(T) = 0. \tag{1.7}$$

The optimal trajectory is then the solution of the linear differential equation

$$\begin{aligned} \dot{x}(t) &= [A - BR^{-1}B'K(t)]x(t) \\ x(0) &= x_0. \end{aligned} \tag{1.8}$$

The minimal cost is given by

$$J_T(x_0, u^*(\cdot)) = \frac{1}{2}x_0'K(0)x_0. \tag{1.9}$$

The optimal control has the form of a linear feedback with a time varying gain matrix $K(t)$. A time-invariant gain matrix is much easier to implement from the engineering point of view. Thus the following problem was formulated: *under what circumstances is the matrix* $K(t)$ *constant*? The answer was obtained by considering the asymptotic behavior of the solution of the Riccati equation (1.6)–(1.7) as T tends to infinity. Kalman [110] has shown that under the assumption of controllability $\lim_{T\to\infty} K(t)$ exists, is unique and is a constant matrix $\hat{K}$.

Thus it was possible to prove the following:

Proposition 1.2 *Assume that the system (1.1)–(1.3) is observable and controllable. Then there exists a unique control which minimizes the cost functional*

$$J(x_0, u(\cdot)) = \frac{1}{2} \int_0^\infty [y'(t)Qy(t) + u'(t)Ru(t)]\, dt. \tag{1.10}$$

The optimal control is given by

$$u^*(t) = -R^{-1}B'\hat{K}x(t) \tag{1.11}$$

where $\hat{K}$ is the positive definite solution of the matrix algebraic Riccati equation

$$-\hat{K}A - A'\hat{K} + \hat{K}BR^{-1}B'\hat{K} - C'QC = 0. \tag{1.12}$$

The minimal cost is given by

$$J(x_0, u^*(\cdot)) = \frac{1}{2}\, x_0'\hat{K}x_0. \tag{1.13}$$

Moreover the eigenvalues of the matrix

$$G = A - BR^{-1}B'\hat{K} \tag{1.14}$$

have negative real parts.

It is important to notice that the optimal control is stabilizing. The condition (1.14) implies the asymptotic stability of the optimally controlled system. Notice also that the boundary condition (1.7), present in the finite horizon optimality conditions, is not satisfied, even in an asymptotic way, when T tends to infinity. Thus two important features of asymptotic optimal control appear in this specific example:

(i) The transversality conditions obtained in the maximum principle for the finite horizon problem do not extend directly to the infinite horizon case, and

(ii) the optimal control over the infinite time horizon is stabilizing.

The assumption of controllability assures that the infimum of the integral (1.10) exists and is finite.

1.3 The pest control problem and other problems of optimal control of interacting species

The Volterra system describing the dynamics of two interacting species (x, y) is based on the following assumptions:

H1. At any instant the x-population, in the absence of interaction, increases at a rate proportional to its size at that instant.

H2. At any instant, the size of the x-population is reduced at a rate proportional both to its own size and the size of the y-population at that instant.

H3. At any instant, the size of the y-population decreases at a rate proportional to its size at that instant.

H4. At any instant, the size of the y-population is increased at a rate proportional both to its own size and to the size of the x-population of that instant.

We call the x-population the prey, and the y-population the predator, and let x and y be the state variables giving the size of the prey and predator populations respectively. The Volterra system is described, in its simplest form, by the two equations

$$\dot{x}(t) = x(t)(1 - y(t)) \tag{1.15}$$

$$\dot{y}(t) = y(t)(x(t) - 1). \tag{1.16}$$

Suppose that the populations x and y are harvested at intensities u and v respectively. Then (1.15) and (1.16) become

$$\dot{x}(t) = x(t)(1 - y(t) - u(t)) \tag{1.17}$$

$$\dot{y}(t) = y(t)(x(t) - 1 - v(t)). \tag{1.18}$$

This is a bilinear control system with state and control in $\mathbb{R}^2$. At any instant t the harvesting intensities are constrained to stay inside given bounds

$$u_m \leq u(t) \leq u_M \tag{1.19}$$

$$v_m \leq v(t) \leq v_M. \tag{1.20}$$

Given measurable controls $u(\cdot) : [0,T] \to [u_m, u_M]$, $v(\cdot) : [0, T] \to [v_m, v_M]$, and an initial population state x_0, y_0, there exists a unique solution to (1.17)–(1.18) and there is a cost associated with the trajectory and the control, defined by

$$J(x_0, y_0, u(\cdot), v(\cdot)) = \int_0^T \mathcal{L}(x(t), y(t), u(t), v(t))\, dt \tag{1.21}$$

where $\mathcal{L}(\cdot) : \mathbb{R}^4 \to \mathbb{R}$ is a given function, C^1 in x, y, continuous in u and v.

The problem of finding the optimal harvesting effort which minimizes (1.21) under the constraints (1.17)–(1.20) has been considered by several authors with various specifications for the constraints (1.19)–(1.20) and the cost function $\mathcal{L}(\cdot)$.

Goh, Leitmann & Vincent [88], and Vincent [179] have considered a pest-control problem where

$$u_m = 0 \tag{1.22}$$

while an additional constraint is introduced

$$v = \delta u \tag{1.23}$$

for a given parameter $\delta \geq 0$. The cost function $\mathcal{L}(\cdot)$ is defined as

$$\mathcal{L}(x, y, u, v) = x + \alpha u \tag{1.24}$$

with $\alpha > 0$, a given parameter.

This optimal control problem may be interpreted as follows. The x-species is a nuisance for man who can control its growth through chemical spraying at a rate u. As a consequence the y-species which is a predator living on the x-species is also destroyed. The cost (1.24) represents a balance between the cost of the nuisance and the cost of spraying.

Similar control problems, based on the assumption that both species, beneficial to man, are exploited concurrently, have been considered by Gilbert and Smith [87] and Getz [85].

One shortcoming in the formulation of the model (1.17)–(1.21) is the consideration of a given terminal time T. There is no natural reason for bounding the time interval on which the system has to be controlled. Some authors ([88],[179]) have considered terminal constraints on the state of the population with a free terminal time T. The problem here is that the time at which the terminal constraint is satisfied can become infinite. If one bounds the time interval, the choice of the terminal constraint can always be contested.

As in the regulator problem of section 1.2 it is desirable to be able to define the optimization problem over an infinite time interval. However, in the present case the consideration of the criterion

$$J(x_0, y_0, u(\cdot), v(\cdot)) = \int_0^\infty \mathcal{L}(x(t), y(t), u(t), v(t))\, dt \tag{1.25}$$

could be meaningless since nothing guarantees the existence (convergence) of the integral in (1.25). In fact, in most problems, the criterion (1.25) would be infinite for any admissible control. There is here a definite need for a theory of optimal control of nonlinear systems over an infinite time horizon.

Another interesting problem concerns the asymptotic behavior of optimal trajectories. It is well known [162] that the Volterra system (1.15)–(1.16) admits a nontrivial stable stationary point

$$\hat{x} = \hat{y} = 1.$$

However, the trajectories are not converging toward this stationary point, they are cycling around the equilibrium in a kind of conservative motion.

The question is the following: *Under which meaningful interpretation of the criterion* (1.25) *is the (suitably defined) asymptotic optimal control such that the optimal trajectory is asymptotically stable?*

1.4 The optimal economic growth problem

The researchers in mathematical economics seem to have been the first to include systematically an infinite time horizon in their modeling of dynamical economic systems. As stated by Arrow and Kurz in [7]:

> *The infinite horizon is an idealization of the fundamental point that the consequences of investment are very long-lived; any short horizon requires some methods of evaluating end-of-period capital stocks, and the only proper evaluation is their value in use in the subsequent future.*

The earliest consideration of an economic optimization problem with a dynamical system observed on an unbounded time interval is due to Ramsey [152]. In his seminal work this author considered a problem of Lagrange

$$\max_{k(\cdot)} \int_0^\infty e^{-\rho t} \mathcal{U}\big(\varphi(k(t)) - \dot{k}(t)\big)\, dt \tag{1.26}$$

where $k(\cdot)\ :\ [0,\infty) \to \mathbb{R}$ is C^1, $k(0) = k_0$ is given, $\varphi(\cdot)\ :\ \mathbb{R} \to \mathbb{R}$, $\mathcal{U}(\cdot)\ :\ \mathbb{R} \to \mathbb{R}$ are both concave and C^1. The parameter ρ is a discount rate, which usually is positive.

Under the assumption that for $k > 0$ we have

$$\varphi(k) > 0,\ \varphi'(k) > 0,\ \varphi''(k) < 0, \tag{1.27}$$

and moreover, that

$$\sup_{k>0} \varphi(k) = B < \infty \tag{1.28}$$

it may be shown that the integral in (1.26) is bounded in the class of functions $k(\cdot)$ with positive value $k(t)$, provided that the discount rate ρ is positive and $k(\cdot)$ satisfies $\int_0^\infty e^{-\rho t}\, k(t)\, dt < \infty$.

In this model $k(t)$ is the amount of capital accumulated at time t by the economy and $\varphi(k(t)) - \dot{k}(t)$ is the amount of consumption possible, given the capital stock $k(t)$ and the net investment $\dot{k}(t)$. Ramsey derived optimality conditions for this economic model with a positive discount rate although he advocated an ethical preference for neutrality between the welfare of different generations, i.e. $\rho = 0$.

A more recent version of the preceding problem has been considered by Cass in [35] and [36]. His model is the following

$$y(t) = f(k(t)) \tag{1.29}$$

$$c(t) = y(t) - z(t) \tag{1.30}$$

$$\dot{k}(t) = z(t) - \mu k(t) \tag{1.31}$$

$$k(0) = k_0. \tag{1.32}$$

Here again $k(t)$ is the stock of capital accumulated at time t. The function $f(\cdot)$ is the production function which associates an output y with the capital stock k. The output $y(t)$ can be either consumed (at a rate $c(t)$) or invested (at a rate $z(t)$). The capital stock depreciates at a constant positive rate μ. The initial capital stock is k_0. Cass considered a concave utility function $\mathcal{U}(\cdot)$ and the welfare functional

$$W = \int_0^T e^{-\rho t}\mathcal{U}(c(t))\,dt \tag{1.33}$$

with $\rho > 0$, a given discount rate.

The maximization of W with respect to $y(\cdot)$, $c(\cdot)$, $z(\cdot)$ and $k(\cdot)$ subject to the constraints (1.29)–(1.33) is a standard optimal control problem, and Cass has been able to show that under the following assumptions:

$$f(k) > 0,\ f'(k) > 0,\ f''(k) < 0 \text{ for } k > 0 \tag{1.34}$$

$$\lim_{k \to 0^+} f'(k) = \infty,\ \lim_{k \to \infty} f'(k) = 0 \tag{1.35}$$

$$\mathcal{U}'(c) > 0, \quad \mathcal{U}''(c) < 0 \text{ for } c > 0 \tag{1.36}$$

$$\lim_{c \to 0^+} \mathcal{U}''(c) = \infty,\ \lim_{c \to \infty} \mathcal{U}''(c) = 0 \tag{1.37}$$

there exists a sustainable constant capital level $\bar{k}$, such that for any T, the optimal capital accumulation path $k^*(\cdot)$ will be "attracted" by $\bar{k}$, i.e. $k^*(t)$ will be near $\bar{k}$ for most of the time. Figure 1.1 illustrates this property.

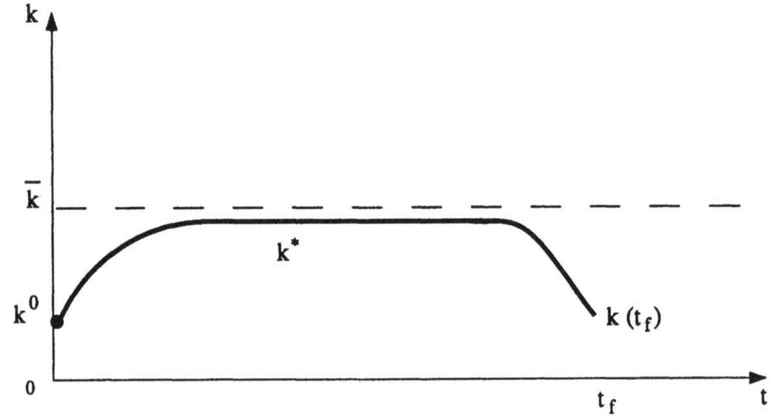

Figure 1.1: The "Turnpike" property

The constant trajectory $\tilde{k}(t) \equiv \bar{k}$ for all $t \geq 0$ acts like a "turnpikw", as the optimal path consists in getting on the turnpike and leaving it only near the end of the journey.

The important fact about this property is that $\bar{k}$ is defined independently of the initial state k_0 and the terminal time T. Furthermore the total time spent by the optimal trajectory at a distance greater than $\varepsilon > 0$ from $\bar{k}$ is bounded by a constant whatever be the value of T.

This property is related to the finite-horizon control probem (1.29)–(1.33) but it will be seen later that the state $\bar{k}$ plays also a central role in the characterization of the optimal control when T becomes infinite. Actually it will be shown that, for this particular model, for any nonnegative value of ρ the optimal trajectory over an infinite time horizon exists and converges toward $\bar{k}$, and this is true for any initial state k_0.

Of course this global asymptotic stability property is reminiscent of the solution of the regulator problem when T becomes infinite. Notice however, that when ρ is equal to zero the welfare functional is unbounded when T tends to infinity.

In more general economic growth models the global asymptotic stability of optimal trajectories will be guaranteed only for nonnegative and not too large values of ρ, as a positive discount rate may induce instability.

1.5 Definition of optimality on an unbounded time interval

The three preceding examples have shown the necessity of a good definition of optimality for a controlled system observed during an infinite time. Although the definition of optimality is unique when T is finite, several nonequivalent definitions of optimality are available when T is infinite.

Consider the system

$$\dot{x}(t) = f(x(t),\, u(t),\, t) \tag{1.38}$$

where $x \in \mathbb{R}^n$ is the state, $u \in \mathbb{R}^m$ is the control and $t \in [0, \infty)$ is the time. The function $f(\cdot,\cdot,\cdot) : \mathbb{R}^n \times \mathbb{R}^m \times \mathbb{R} \to \mathbb{R}^n$ is continuous, and with each $(x,t) \in \mathbb{R}^n \times [0, \infty)$ a compact subset $U(x,t)$ is associated such that the mapping

$$(x,t) \to U(x,t)$$

is upper semicontinuous [(3)].

We assume that there exists a finite number M such that

$$\|f(x,u,t)\| \leq M(1 + \|x\|) \tag{1.39}$$

for all $(x,t) \in \mathbb{R}^n \times [0, \infty)$ and $u \in U(x,t)$.

Definition 1.1 *A pair of functions* $(x(\cdot)\,, u(\cdot)\,) \,:\, [0, \infty) \to \times\mathbb{R}^m$ *is called admissible if* $x(\cdot)$ *is absolutely continuous,* $u(\cdot)$ *is measurable,*

$$\dot{x}(t) \;=\; f(x(t),\, u(t),\, t) \;\; a.e. \;in\; [0, \infty), \tag{1.40}$$

and

$$x(0) \;=\; x_0, \tag{1.41}$$

$$u(t) \;\in\; U(x(t),\, t), \;\; a.e. \;\; t \geq 0. \tag{1.42}$$

Following standard terminology we call $x(\cdot)$ a *trajectory* emanating from x_0 and $u(\cdot)$ an *admissible control.* We denote $\mathcal{A}_\infty$ the set of all admissible pairs $(x(\cdot)\,, u(\cdot)\,)$.

Given a trajectory $x(\cdot)$ emanating from x_0 and generated by an admissible control $u(\cdot)$ we consider a performance criterion for any $T \geq 0$ by

$$J_T(x_0, u(\cdot)\,) = \int_0^T f_0(x(t),\, u(t),\, t)\, dt \tag{1.43}$$

where f_0 is a continuous function from $\mathbb{R}^n \times \mathbb{R}^m \times [0, \infty)$ into $\mathbb{R}$.

Loosely speaking the controller's task is to maximize $J_T(\cdot,\cdot)$ as T grows to infinity, and the four following definitions of optimality can be proposed

Definition 1.2 *A trajectory $x^*(\cdot)$ emanating from x_0 is said to be:*

(i) **Strongly optimal** *at x_0 if it is generated by $u^*(\cdot)$ such that*

$$J_\infty(x_0, u^*(\cdot)) = \lim_{T\to\infty} J_T(x_0, u^*(\cdot)) < \infty \tag{1.44}$$

and for any other trajectory $x(\cdot)$ emanating from x_0 and generated by $u(\cdot)$ the following holds

$$J_\infty(x_0, u^*(\cdot)) \geq \limsup_{T\to\infty} J_T(x_0, u(\cdot)). \tag{1.45}$$

(ii) **Overtaking optimal** *at x_0 if* (1.44) *does not necessarily hold while condition* (1.45) *is replaced by (see [4])*

$$\liminf_{T\to\infty} \{J_T(x_0, u^*(\cdot)) - J_T(x_0, u(\cdot))\} \geq 0 \tag{1.46}$$

which can also be written

$$\forall \varepsilon > 0 \ \exists T(\varepsilon, u(\cdot)) \ s.t. \ T > T(\varepsilon, u(\cdot)) \Rightarrow J_T(x_0, u^*(\cdot)) > J_T(x_0, u(\cdot)) - \varepsilon. \tag{1.47}$$

(iii) **Weakly overtaking optimal** *at x_0 if condition* (1.46) *is replaced by*

$$\limsup_{T\to\infty} \{J_T(x_0, u^*(\cdot)) - J_T(x_0, u^*(\cdot))\} \geq 0 \tag{1.48}$$

which can also be written

$$\forall \varepsilon > 0, \forall T^* \ \exists T = T(\varepsilon, T^*, u(\cdot)) > T^* \ s.t. \ J_T(x_0, u^*(\cdot)) > J_T(x_0, u(\cdot)) - \varepsilon. \tag{1.49}$$

(iv) **Finitely optimal** *at x_0 if for every $T > 0$ and every trajectory $x(\cdot)$ emanating from x_0 and generated by an admissible control $u(\cdot)$ on $[0, T]$ with $x(T) = x^*(T)$ we have*

$$J_T(x_0, u^*(\cdot)) \geq J_T(x_0, u(\cdot)). \tag{1.50}$$

Notice that the optimality has been defined from the point of view of maximization of the performance index.

Clearly finite optimality should be satisfied by any reasonably defined optimal trajectories. It will be shown in the next chapter that weak overtaking optimality implies finite optimality, so that we have

Strong optimality⇒overtaking optimality⇒weak overtaking optimality⇒finite optimality.

Loosely speaking a trajectory is overtaking optimal if the performance index "catches up to" the performance index of any other trajectory emanating from x_0. Figure 1.2 illustrates this behavior.

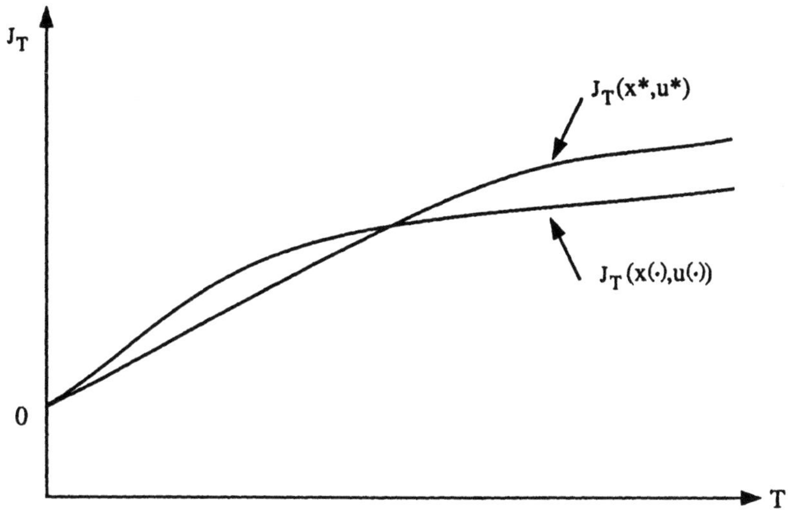

Figure 1.2: Overtaking optimality

A trajectory is weakly overtaking optimal if the performance index for any other trajectory and admissible control $J_T(x_0, \tilde{u}(\cdot))$ can never "catch up to" the performance index $J(x_0, \tilde{u}^*(\cdot))$. Figure 1.3 depicts this behavior.

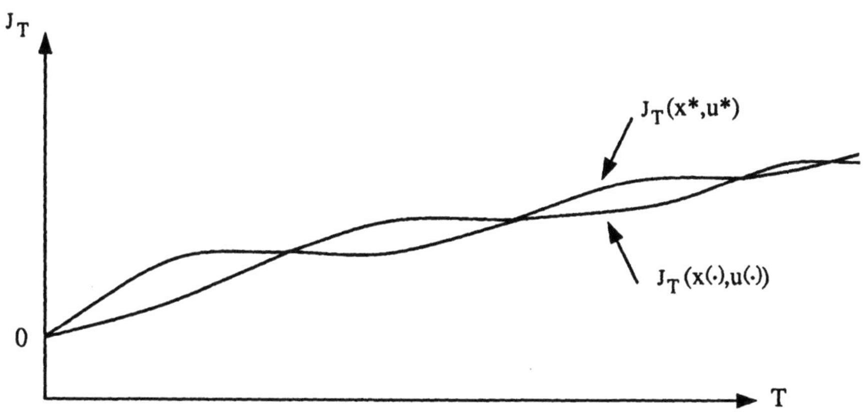

Figure 1.3: Weak overtaking optimality

Strong optimality is the simplest definition, and it is meaningful only when the performance index remains bounded above for all trajectories. Notice that, for the regulator problem, the controllability assumption permits the use of the strong optimality concept. For the two other examples either overtaking optimality or weak overtaking optimality can be used.

We now define another class of optimal solutions which stems from a completely different premise. These notions originate in the 1973 work of Hammond and Mirrlees [92](see also [39],[103], and [93]). To define these notions we introduce the following notation. Let $x(\cdot)$ be an admissible trajectory emanating from x_0 generated by a control $u(\cdot)$. For $\theta > 0$ we let $(x^\theta(\cdot), u^\theta(\cdot))$ denote the restriction of the pair $(x(\cdot), u(\cdot))$ to the interval $[0, \theta]$. We further let $\mathcal{A}(x(\cdot), \theta)$ be the set of admissible trajectory-control pairs $(y(\cdot), v(\cdot))$ emanating from x_0 whose trajectories coincide with $x(\cdot)$ on $[0, \theta]$. That is,

$$\mathcal{A}(x(\cdot), \theta) \triangleq \left\{(y(\cdot), v(\cdot)) \text{ admissible} | y^\theta(\cdot) = x^\theta(\cdot)\right\}.$$

We now define, for any $T > 0$

$$W(T) \triangleq \sup \left[J_T(x_0, u(\cdot)) | (x(\cdot), u(\cdot)) \text{ is admissible}\right]$$

and

$$W(T, x(\cdot), \theta) \triangleq \sup \left[J_T(x_0, v(\cdot)) | (y(\cdot), v(\cdot)) \in \mathcal{A}(x(\cdot), \theta)\right].$$

Clearly since $\mathcal{A}(x(\cdot), \theta)$ is a subset of all admissible trajectory-control pairs we have

$$W(T) \geq W(T, x(\cdot), \theta)$$

for all $T \geq 0$ and $\theta > 0$. This gives rise to the following definition.

Definition 1.3 *A trajectory $x^*(\cdot)$ emanating from x_0 is said to be*

(i) **Decision Horizon Optimal** *if for all $\theta > 0$, there exists $\hat{T} = \hat{T}(\theta) \geq 0$ such that for all $T \geq \hat{T}$,*

$$W(T, x^*(\cdot), \theta) = W(T).$$

(ii) **Agreeable** *if for every $\theta > 0$, one has*

$$\lim_{T \to \infty} \left[W(T) - W(T, x^*(\cdot), \theta)\right] = 0.$$

(iii) **Weakly Agreeable** *if for every $\theta > 0$ one has*

$$\liminf_{T \to \infty} \left[W(T) - W(T, x^*(\cdot), \theta)\right] = 0.$$

Clearly one has

Decision Horizon Optimal $\Longrightarrow$ *Agreeable* $\Longrightarrow$ *Weakly Agreeable.*

What is less clear is the relationship between the concepts of optimality given in Definition 1.2 and those given above. This will be considered in the next section. The basic idea for this class of optimal solutions is to seek a comparison based on corresponding finite horizon problems. For example, when one optimizes over a finite horizon, say T, following the trajectory $x^*(\cdot)$ over the interval $[0, \theta]$ constitutes a "wrong start". The trajectory $x^*(\cdot)$ is agreeable if the effect of this "wrong start" becomes negligible when $T \to \infty$.

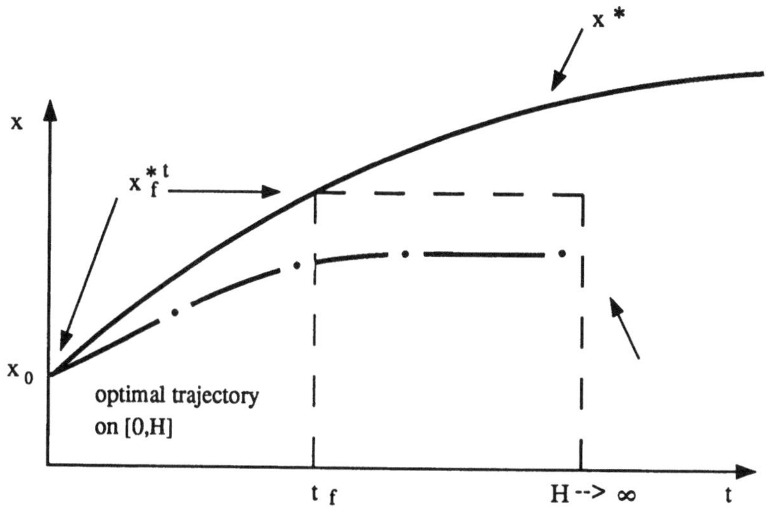

Figure 1.4: Illustration of an agreeable trajectory

We conclude this section by presenting several examples which show that the definitions of optimality presented in Definition 1.2 are distinct.

Example 1.1 A finitely optimal solution need not be a weakly overtaking solution.

The example we consider is known as "Gale's cake-eating example". This is the case of a society which has a piece of cake and may consume a certain amount at each instant until the cake is gone (i.e. a non-renewable resource problem). Specifically, this model consists of the functionals

$$J_T(x_0, u(\cdot)) = \int_0^T g(u(t))\, dt,$$

where g is a strictly concave function which satisfies

$$\max \{g(z) : 0 \leq z \leq b\} > g(0),$$

and the constraints

$$\dot{x}(t) = u(t), \text{ a.e. } t \geq 0,$$

$$x(0) = x_0, \quad 0 \le x_0 < a,$$
$$0 \le x(t) \le a, \quad t \ge 0,$$
$$0 \le u(t) \le b, \quad \text{a.e. } t \ge 0.$$

We now show by contradiction that the above problem does not have a weakly overtaking optimal solution and begin by assuming $(x^*(\cdot), u^*(\cdot))$ is such a solution. We divide our considerations into two cases as follows

Case 1: $u^*(t) \neq 0$ everywhere except on a set of measure zero.

In this case there exists a nonnegative $T > 0$ and $\delta > 0$ such that $u^*(t) \neq 0$ a.e. on $[T, T+\delta)$. Define the control $u(\cdot) : [0, \infty) \to [0, b]$ as follows

$$u(t) = \begin{cases} u^*(t) & \text{on } [0,T) \cup [T+\delta, \infty) \\ \dfrac{1}{\delta} \displaystyle\int_T^{T+\delta} u^*(s)\, ds & \text{on } [T, T+\delta). \end{cases}$$

It is easy to see that $u(\cdot)$ is an admissible control and that for δ sufficiently small, it generates an admissible trajectory $x(\cdot)$, defined by

$$x(t) = \begin{cases} x^*(t) & \text{on } [0,T) \cup [T+\delta, \infty) \\ x^*(T) + \dfrac{t-T}{\delta} \displaystyle\int_T^{T+\delta} u^*(s)\, ds & \text{on } [T, T+\delta). \end{cases}$$

We now show that for all $t \ge T + \delta$

$$\int_0^t g(u(s))\, ds > \int_0^t g(u^*(s))\, ds,$$

implying $(x^*(\cdot), u^*(\cdot))$ is not weakly overtaking optimal. For $t \ge T + \delta$ we note that

$$\int_0^t \left[g(u(s)) - g(u^*(s))\right] ds = \delta g(u(T)) - \int_T^{T+\delta} g(u^*(s))\, ds.$$

From Jensen's inequality and the strict concavity of g we deduce

$$g(u(T)) = g\left[\frac{1}{\delta} \int_T^{T+\delta} u^*(s)\, ds\right] > \frac{1}{\delta} \int_T^{T+\delta} g(u^*(s))\, ds$$

and so,

$$\int_0^t \left[g(u(s)) - g(u^*(s))\right] ds = g(u(T)) - \frac{1}{\delta} \int_T^{T+\delta} g(u^*(s))\, ds > 0,$$

as desired. Consequently, the only possibility for a weakly overtaking optimal control is $u^*(t) = 0$ a.e. on $[0, \infty)$, which is our second case.

Case 2: $u^*(t) = 0$ *a.e. on* $[0, \infty)$.

Since $g(\cdot)$ is strictly concave on $[0, b]$ it has a unique maximizer, say v^*, which by hypothesis is nonzero. Choose $T = (a - x_0)/v^*$ and define the admissible control $u(\cdot)$: $[0, \infty) \to [0, b]$ by

$$u(t) = \begin{cases} v^* & \text{for } 0 \le t \le T \\ 0 & \text{for } T \le t. \end{cases}$$

This control generates the admissible trajectory $x(\cdot)$: $[0, \infty) \to [0, a]$ given by

$$x(t) = \begin{cases} v^* t + x_0 & \text{if } 0 \le t < T \\ a & \text{if } T \le t \end{cases}$$

Notice however, that for $t > T$ we have

$$\begin{aligned} \int_0^t g(u(s))\, ds &= g(v^*)T + g(0)(t - T) \\ &> g(0)T + g(0)(t - T) \\ &= g(0)t \\ &= \int_0^t g(u^*(s))\, ds. \end{aligned}$$

Thus as in case 1, $u^*(\cdot)$ cannot be a weakly overtaking optimal control.

The above discussion shows that this example does not have a weakly overtaking optimal solution. On the other hand it is an easy matter to show that the control pair $(x(\cdot), u(\cdot))$ where $x(t) \equiv x_0, u(t) \equiv 0$ is finitely optimal and the desired conclusion is reached.

Example 1.2 A weakly overtaking optimal solution is not necessarily overtaking optimal.

In this example we consider the problem where

$$J_T(x_0, u(\cdot)) = \int_0^T y(s)\, ds$$

with admissible pairs $(x(\cdot), y(\cdot), u(\cdot))$: $[0, \infty) \to \mathbb{R}^2 \times [-1, 1]$ satisfying

$$\begin{aligned} \dot{x}(t) &= y(t), & \dot{y}(t) &= -x(t) + u(t) \quad \text{a.e. on } [0, \infty), \\ x(0) &= y(0) = 0, & -1 &\le u(t) \le 1 \qquad \text{a.e. on } [0, \infty). \end{aligned}$$

By appealing to the Pontryagin Maximum Principle (see Chapter 2) it can be shown that for each positive integer n, an optimal control on $[0, 2n\pi]$ is given by

$$u_1^*(t) = \begin{cases} -1 & \text{on } [2k\pi,\ (2k+1)\pi) \\ 1 & \text{on } [(2k+1)\pi,\ 2(k+1)\pi) \end{cases}$$

for $k = 0, 1, 2, \ldots, (n-1)$, which generates the optimal trajectory

$$\begin{bmatrix} x_1^*(t) \\ y_1^*(t) \end{bmatrix} = \int_0^t \begin{bmatrix} \sin(t-s) \\ \cos(t-s) \end{bmatrix} u_1^*(s)\, ds, \qquad 0 \leq t \leq 2n\pi$$

and optimal cost

$$J_{2n\pi} x_0, u_1^*(\cdot) = \int_0^{2n\pi} y_1^*(s)\, ds = 4n.$$

Letting $n \to \infty$ it is easy to see that $(x_1^*(\cdot), y_1^*(\cdot), u_1^*(\cdot))$ is weakly overtaking optimal.

Similarly if one considers the intervals $[0, (2n+1)\pi]$ for $n = 0, 1, 2, \ldots$ one obtains an optimal solution $(x_2^*(\cdot), y_2^*(\cdot), u_2^*(\cdot))$ given by

$$u_2^*(t) = \begin{cases} 1 & \text{on } [2k\pi,\ (2k+1)\pi] \\ -1 & \text{on } [(2k+1)\pi,\ 2(k+1)\pi], \end{cases}$$

for $k = 0, 1, 2, \ldots$, and

$$\begin{bmatrix} x_2^*(t) \\ y_2^*(t) \end{bmatrix} = \int_0^t \begin{bmatrix} \sin(t-s) \\ \cos(t-s) \end{bmatrix} u_2^*(s)\, ds$$

with optimal cost

$$J_{(2n+1)\pi} x_0, u_2^*(\cdot) = 4n + 2.$$

Once again it is easy to see that $(x_2^*(\cdot), y_2^*(\cdot), u_2^*(\cdot))$ is weakly overtaking optimal. Moreover, neither $(x_1^*(\cdot), y_1^*(\cdot), u_1^*(\cdot))$ or $(x_2^*(\cdot), y_2^*(\cdot), u_2^*(\cdot))$ is overtaking optimal since for any finite $T > 0$ we can find T_1 and T_2 greater than T such that the first is optimal on $[0, T_1]$ and the second on $[0, T_2]$.

Example 1.3 An overtaking optimal solution is not necessarily strongly optimal.

In this example we consider the functional

$$J_T(x_0, u(\cdot) = \int_0^T (1 - u(s))x(s)\, ds$$

with admissible pairs $(x(\cdot), u(\cdot)) : [0, \infty) \to [0, \infty) \times [0, 1]$ satisfying

$$
\begin{aligned}
\dot{x}(t) &= -x(t)u(t), \quad \text{a.e. on } [0, \infty), \\
x(0) &= x_0 > 0, \quad x(t) \geq 0 \text{ on } [0, \infty), \\
0 &\leq u(t) \leq 1, \quad \text{a.e. on } [0, \infty).
\end{aligned}
$$

It is easy to show that for each $T > 0$, the admissible pair $(x^*(t), u^*(t)) \equiv (x_0, 0)$ is optimal on $[0, T]$. Indeed for any admissible pair $(x(\cdot), u(\cdot))$ it can be shown that for all $t \geq 0$

$$x_0 \geq (1 - u(t))x(t),$$

so that

$$x_0 T = \int_0^T (1 - u^*(t))x^*(t)\, dt \geq \int_0^T (1 - u(t))x(t)\, dt.$$

This implies $(x^*(\cdot), u^*(\cdot))$ is overtaking optimal, as desired. However, letting $T \to \infty$ clearly shows

$$\lim_{T \to \infty} J_T(x^*(\cdot), u^*(\cdot)) = \infty,$$

and so $(x^*(\cdot), u^*(\cdot))$ is not strongly optimal.

1.6 Uniformly optimal solutions are agreeable

In this section we investigate the relationship between the concepts of optimality introduced in Definitions 1.2 and 1.3. First we observe that in Definition 1.3 a uniformity is imposed through the definitions of $W(T)$ and $W(T, x^*(\cdot), \theta)$. Consequently to establish the relationship we seek we introduce the following definitions which strengthen the concepts of optimality found in Definition 1.2

Definition 1.4 *A trajectory $x^*(\cdot)$ emanating from x_0 and generated by $u^*(\cdot)$ is said to be*

(i) **Uniformly strongly optimal** *at x_0 if* (1.44) *holds and if* (1.45) *holds uniformly over all trajectories $x(\cdot)$ emanating from x_0.*

(ii) **Uniformly overtaking optimal** *at x_0 if* (1.46) *holds with $T(\epsilon, u(\cdot)) = T(\epsilon)$ (i.e.,* (1.46) *holds uniformly over all trajectories emanating from x_0).*

(iii) **Uniformly weakly overtaking optimal** *if* (1.48) *holds with $T(\epsilon, T^*, u(\cdot)) = T(\epsilon, T^*)$ (i.e.,* (1.48) *holds uniformly over all trajectories emanating from x_0).*

Clearly the uniform optimality concepts imply their nonuniform counterparts. What is more interesting is that they are also related to the concepts of optimality presented in Definition *1.3.* We now establish this relationship in the following result.

Proposition 1.3 *Let $x^*(\cdot)$ be a trajectory emanating from x_0 and generated by $u^*(\cdot)$. Then if $x^*(\cdot)$ is*

(i) *uniformly strongly optimal it is decision horizon optimal.*

(ii) *uniformly overtaking optimal it is agreeable.*

(iii) *uniformly weakly overtaking optimal it is weakly agreeable.*

Proof: For brevity we prove only (ii) remarking that the others are established similarly. To begin let $x^*(\cdot)$ be uniformly overtaking optimal. This means that for every $\epsilon > 0$ there exists $T(\epsilon) > 0$ so that for all $T \geq T(\epsilon)$,

$$J_T(x_0, u^*(\cdot)) > W(T) - \epsilon.$$

This implies that for any $\theta \geq 0$

$$W(T, x^*(\cdot), \theta) \geq J_T(x_0, u^*(\cdot)) > W(T) - \epsilon$$

giving us

$$\epsilon > W(T) - W(T, x^*(\cdot), \theta) \geq 0$$

for all $T \geq T(\epsilon)$. That is,

$$\lim_{T \to \infty} [W(T) - W(T, x^*(\cdot), \theta)] = 0$$

as desired. ∎

Remark 1.1 *The above result was given in* [39]. *As we shall see in the next chapter, this observation permits us to strengthen the hypotheses of Theorem* 2.6 *so that we have conditions under which and admissible trajectory $x^*(\cdot)$ is, for example, uniformly overtaking optimal. Once this is done we know that $x^*(\cdot)$ is both overtaking optimal and agreeable.*

Notes and Comments

(1) A linear system (1.1)–(1.3) is observable if any state x_0 is observable, i.e. if the knowledge of the control $u(\cdot)$ and of the output $y(\cdot)$ on $[0,T]$ is sufficient to determine x_0.

A necessary and sufficient condition for observability for autonomous linear systems is that the matrix

$$H \triangleq \left[C' \vdots A'C' \vdots \cdots \vdots (A')^{n-1}C'\right]$$

be of full rank n.

(2) A linear system (1.1)–(1.3) is controllable if any initial state x_0 is controllable, i.e. if there exists a control $u(\cdot) : [0,T] \to \mathbb{R}^m$ such that $x(T) = 0$, where $x(\cdot) : t \to x(t)$ is the trajectory emanating from x_0 and generated by $u(\cdot)$.

A necessary and sufficient condition for controllability for autonomous linear systems is that the matrix

$$G \triangleq \left[B \vdots AB \vdots A^2B \vdots \cdots, \vdots A^{n-1}B\right]$$

be of full rank n.

(3) The definition of upper semicontinuity we refer to is given by C. Berge, *Espace Topologiques*, Dunod 1959, p. 114.

The mapping $U(\cdot,\cdot) : (x,t) \to U(x,t) \subset \mathbb{R}^m$ is upper semi continuous if for any open subset G such that $G \supset U(x_0,t_0)$ there exists a neighborhood $V(x_0,t_0) \subset \mathbb{R}^n \times [0,\infty)$ such that

$$(x,t) \in V(x_0,t_0) \Rightarrow U(x,t) \subset G.$$

(4) If $g(\cdot)$ is a real valued function we write $S = \limsup_{x \to x_0} g(x)$ if:

(i) for every $\varepsilon > 0$ there is $\delta > 0$ such that $\|x - x_0\| < \delta \Rightarrow g(x) < S + \varepsilon$ and

(ii) for every $\varepsilon > 0$ and $\delta > 0$ there is an x such that $\|x - x_0\| < \delta$ and $g(x) > S - \varepsilon$.

The limit inferior of $g(x)$ is defined as

$$\liminf_{x \to x_0} g(x) = -\limsup_{x \to x_0} (-g(x)).$$

Chapter 2

Necessary Conditions and Sufficient Conditions for Optimality

2.1 Introduction

In this chapter we present both necessary conditions and sufficient conditions for optimality. These conditions are obtained by generalizing the appropriate finite horizon results. In particular we present the Pontryagin Maximum Principle, and its extension to the infinite horizon.

When considering optimization problems with infinite horizon, one is not allowed to assume that all the results which are known for the finite horizon case can be carried to the infinite horizon case by replacing evaluations of quantities at the terminal time with evaluations of the limit of the same quantities as the time tends to infinity. The first correct formulation of necessary conditions for the infinite horizon optimal control problem has been obtained by Halkin [89]. He also provided examples where the transversality conditions of the infinite horizon case cannot be obtained as the limit of the transversality conditions in the finite horizon case. In Halkin's maximum principle for infinite horizon optimal control problems there do not appear transversality conditions. This could be a disturbing result since the absence of transversality conditions makes the system of necessary conditions incomplete and seems to preclude the computation of extremal trajectories as candidates for optimality.

2.2 The maximum principle with a finite horizon

Consider a dynamical system described by the state equation

$$\dot{x}(t) = f(x(t), u(t), t) \tag{2.1}$$

$$x(0) = x_0, \tag{2.2}$$

where $x \in \mathbb{R}^n$ and $u \in \mathbb{R}^m$ are the state and control as usually defined, and $t \in [0,T]$ for some given fixed terminal time T. The function $f(\cdot,\cdot,\cdot) : \mathbb{R}^n \times \mathbb{R}^m \times [0, T] \to \mathbb{R}^n$ is continuous and its first partial derivatives with respect to x exist and are continuous.

Consider also a fixed constraint set on the control

$$u(t) \in U \subset \mathbb{R}^m \text{ a.e. } t \in [0,T) \tag{2.3}$$

and terminal constraints, defining the target set,

$$\begin{aligned} g_j(x(T)) &\geq 0 \qquad 1 \leq j \leq p \\ g_j(x(T)) &= 0 \qquad p+1 \leq j \leq q \end{aligned} \tag{2.4}$$

for continuously differentiable functions $g_j(\cdot) : \mathbb{R}^n \to \mathbb{R}$, $1 \leq j \leq q$.

Finally, let $f_0(\cdot,\cdot,\cdot) : \mathbb{R}^n \times \mathbb{R}^m \times [0, T] \to \mathbb{R}$ be continuous with continuous first partial derivatives with respect to x. The optimization problem we consider consists of maximizing the integral functional

$$J_T(x_0, u(\cdot)) = \int_0^T f_0(x(t), u(t), t)\, dt \tag{2.5}$$

over all admissible pairs $(x(\cdot), u(\cdot))$ satisfying the relations (2.1) to (2.4). We further impose the following constraint qualification:

The system

$$\left.\begin{array}{ll} c_j \geq 0, & \text{for } 0 \leq j \leq p \\ c_j g_j(x^*(T)) = 0 & \text{for } 1 \leq j \leq p \\ \sum_{j=1}^{q} c_j \nabla g_j(x^*(T)) = 0 & \end{array}\right\} \qquad \text{(CQ)}$$

admits only the trivial solution $c_j = 0$, $1 \leq j \leq q$, where here x^* denotes an optimal trajectory.

With this notation we now state the classical first order necessary conditions for optimality known as Pontryagin's Theorem.

Theorem 2.1 (The Pontryagin maximum principle with fixed terminal time and integral performance criterion). *If $(x^*(\cdot), u^*(\cdot))$ is an optimal solution of the optimal control problem described above, and if the constraint qualification* (CQ) *holds, then there exist multipliers μ_j, $0 \leq j \leq q$ and an absolutely continuous function $\lambda(\cdot)$: $[0, T] \to \mathbb{R}^n$ such that if one defines the Hamiltonian*

$$H(x, u, t, \lambda, \mu_0) = \mu_0 f_0(x, u, t) + \lambda' f(x, u, t), \tag{2.6}$$

are satisfied

$$\text{(i)} \quad (\mu_0, \mu_1, \ldots, \mu_q, \lambda(0)) \neq 0 \tag{2.7}$$

$$\text{(ii)} \quad \dot{\lambda}(t) = -\nabla_x H(x, u^*(t), t, \lambda(t), \mu_0)|_{x=x^*(t)}, \text{a.e. on } [0, T] \tag{2.8}$$

$$\text{(iii)} \quad H(x^*(t), u^*(t), t, \lambda(t), \mu_0) = \max_{u \in U} H(x^*(t), u, t, \lambda(t), \mu_o) \tag{2.9}$$

$$a.e. \ on \ [0, T]$$

$$\text{(iv)} \quad \lambda(T) = \sum_{i=1}^{q} \mu_i \nabla g_i(x^*(T)). \tag{2.10}$$

The proof of the above Theorem is given in a variety of places (e.g. Pontryagin *et al.* [150], Halkin [91], Blaquière and Leitmann [20], Leitmann [116], etc.). Due to the technical nature of these arguments we have chosen not to present a proof within the text. We further remark that the smoothness hypotheses given above, while sufficiently general for the problems considered here, have been significantly weakened. In particular, we refer the reader to the monograph of F. Clarke [56] for a discussion of these matters.

2.3 The optimality principle

In this section we prove a result which can be viewed as a reformulation of the Bellman optimality principle. In Chapter 1 we denoted $\mathcal{A}_\infty$ the set of all admissible pairs $(x(\cdot), u(\cdot))$ over an infinite time horizon. We show that if $(x^*(\cdot), u^*(\cdot))$ is an optimal solution for the infinite horizon optimal control problem in any of the senses described in Definition 1.2 then $(x^*(\cdot), u^*(\cdot))$ is also finitely optimal.

Theorem 2.2 Optimality principle. *If the pair $(x^*(\cdot), u^*(\cdot)) \in \mathcal{A}_\infty$ is optimal according to any optimality concept given in Definition 1.2, then for any $T \geq 0$ the restriction $x^{*^T}(\cdot)$ of $x^*(\cdot)$ associated with the restriction $u^{*^T}(\cdot)$ of $u^*(\cdot)$ maximizes $J_T(x_0, u(\cdot))$ given by (2.5) on the set $\mathcal{A}_T^*$ of all pairs $(x(\cdot), u(\cdot))$ for which*

$$x(0) = x_0 \qquad x(T) = x^*(T), \tag{2.11}$$

and thus $(x_0, u^*(\cdot))$ is finitely optimal.

Proof: If the result is not true for some $\hat{T} > 0$, then for some $(\hat{x}(\cdot), \hat{u}(\cdot)) \in \mathcal{A}_{\hat{T}}^*$ one has

$$\int_0^{\hat{T}} f_0(\hat{x}(t), \hat{u}(t), t)\, dt > \int_0^{\hat{T}} f_0(x^*(t), u^*(t), t)\, dt \tag{2.12a}$$

and

$$\hat{x}(\hat{T}) = x^*(\hat{T}) \tag{2.12b}$$

From the above it is clear that there exists $\varepsilon_0 > 0$ so that

$$\int_0^{\hat{T}} f_0(\hat{x}(t), \hat{u}(t), t)\, dt > \int_0^{\hat{T}} f_o(x^*(t), u^*(t), t)\, dt + \varepsilon_0. \tag{2.13}$$

Defining

$$(\tilde{x}(t), \tilde{u}(t)) = \begin{cases} (\hat{x}(t), \hat{u}(t)) & \text{for } t \in [0, \hat{T}) \\ (x^*(t), u^*(t)) & \text{for } t \in [\hat{T}, \infty) \end{cases} \tag{2.14}$$

we obtain

$$\int_0^{T} f_0(\hat{x}_0(t), \hat{u}_0(t))\, dt > \int_0^{T} f_0(x^*(t), u^*(t))\, dt + \varepsilon \tag{2.15}$$

for every $T \geq \hat{T}$ implying that $x^*(\cdot)$ is not optimal in any of the definitions given in Definition 1.2. ∎

Figure 2.1 illustrates this theorem. Generally speaking, the theorem shows that if a trajectory is optimal then any section of this trajectory, emanating from x_0 is optimal, in the classical sense, in the class of all trajectories which have the same fixed end-point.

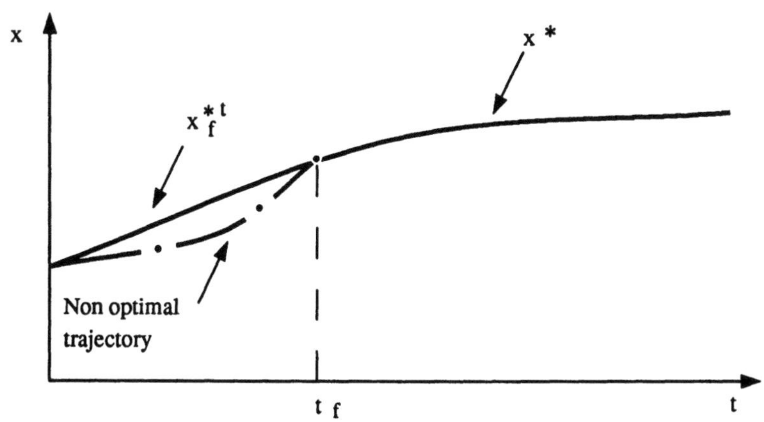

Figure 2.1: Principle of optimality for infinite horizon optimal trajectories

2.4 A maximum principle for an infinite horizon control problem

Now the *Infinite Horizon Maximum Principle* can be proved.

Theorem 2.3 *If* $(x^*(\cdot), u^*(\cdot)) \in \mathcal{A}_\infty$ *is optimal according to any definition of optimality given in Definition* 1.2, *then there exists a non-negative number* μ *and a continuous piecewise differentiable function* $\lambda(\cdot) : [0, \infty) \to \mathbb{R}^n$ *such that*

$$\text{(i)} \quad \|(\mu, \lambda(0))\| = 1$$

$$\text{(ii)} \quad \dot{\lambda}(t) = -\frac{\partial}{\partial x} H(x^*(t), u^*(t), t, \lambda(t), \mu) \quad \textit{a.e. on } [0, \infty) \tag{2.16}$$

$$\text{(iii)} \quad H(x^*(t), u^*(t), t, \lambda(t), \mu) \geq H(x^*(t), u, t, \lambda(t), \mu)$$
$$\textit{for every } t \in [0, \infty), \textit{ and for all } u \in U. \tag{2.17}$$

Proof: Consider a strictly increasing sequence $\tau_1, \tau_2, \ldots$ in $[0, \infty)$ with $\tau_i \to \infty$. The τ_i restriction of $(x^*(\cdot), u^*(\cdot))$ is optimal in the class $\mathcal{A}^*_{\tau_i}$ of pairs $(x(\cdot), u(\cdot))$ satisfying $x(\tau_i) = x^*(\tau_i)$. From the maximum principle for finite horizon problems (Theorem 2.1) one knows that there exist $\mu^i \geq 0$, $\lambda^i(\cdot) : [0, \tau_i] \to \mathbb{R}^n$ such that:

$$\text{(i)} \quad \|(\mu^i, \lambda^i(0))\| \neq 0 \tag{2.18}$$

$$\text{(ii)} \quad \dot{\lambda}^i(t) = -\frac{\partial}{\partial x} H(x, u^*(t), t, \lambda^i(t), \mu^i)\Big|_{x=x^*(t)} \tag{2.19}$$
$$\text{a.e. on } [0, \tau_i]$$

$$\text{(iii)} \quad H(x^*(t), u^*(t), t, \lambda^i(t), \mu^i) \geq H(x^*(t), u, t, \lambda^i(t), \mu^i)$$
$$\forall\, t \in [0, \tau_i], \ \forall\, u \in U. \tag{2.20}$$

By normalization, the condition (2.18) could be equivalently replaced by

$$\text{(i}'\text{)} \quad \|(\mu^i, \lambda^i(0))\| = 1. \tag{2.21}$$

By going to an appropriate subsequence one may assume that $\lim_{i\to\infty} \mu^i = \mu$ and $\lim_{i\to\infty} \lambda^i(0) = \lambda(0)$ exist. One has then $\|(\mu, \lambda(0))\| = 1$.

Let $\lambda(\cdot) : [0, \infty) \to \mathbb{R}^n$ be the unique continuous, piecewise C^1 solution of the linear time varying differential system

$$\dot{\lambda}(t) = -\frac{\partial}{\partial x} H(x^*(t), u^*(t), t, \lambda(t), \mu) \tag{2.22}$$

with the initial condition

$$\lambda(0) = \lim_{i\to\infty} \lambda^i(0). \tag{2.23}$$

We have then that

$$H(x^*(t),\, u^*(t),\, t,\, \lambda(t),\, \mu) \geq H(x^*(t),\, u,\, t,\, \lambda(t),\, \mu)$$

for every $t \in [0, \infty)$ and every $u \in U$, since H is linear in λ and μ and since for any t one has

$$\lambda(t) = \lim_{i \to \infty} \lambda^i(t)$$

due to the continuous dependence of the solutions of a differentiable system with respect to the initial data. This proves the theorem. ∎

Extensions of Halkin's result to the case when the control constraint set depends explicitly on both x and t requires more restrictive hypothesis (see e.g. Arrow and Kurz [7], Sethi and Thompson [168], and note (1)). As a consequence of the free endpoint at infinity for the infinite horizon optimal control problem, one might expect that the transversality condition $\lim_{t \to \infty} \lambda(t) = 0$ holds and that the multiplier μ be nontrivial (i.e. $\mu > 0$). Halkin [89] gives the following two examples to show that this need not be the case.

Example 2.1 This first example shows that the transversality condition does not necessarily hold. We consider the problem of finding optimal solutions for

$$J_T(x_0, u(\cdot)\,) = \int_0^T (1 - x(t))u(t)\, dt$$

over all admissible pairs $(x(\cdot)\,, u(\cdot)\,)$ satisfying

$$\begin{aligned} \dot{x}(t) &= (1 - x(t))u(t) && \text{for a.e. } t \geq 0 \\ x(0) &= 0 && \\ u(t) &\in [0, 1] && \text{for a.e. } t \geq 0. \end{aligned}$$

To exhibit an optimal solution we observe that for any $T > 0$,

$$J_T(x(\cdot)\,, u(\cdot)\,) = x(T)$$

for each admissible pair $(x(\cdot)\,, u(\cdot)\,)$. Moreover, it is easy to see that

$$J_T(x_0, u(\cdot)\,) = x(T) = \Big(1 - exp\Big(-\int_0^T u(s)\, ds\Big)\Big)$$

holds for each admissible pair. This leads us to observe that $u^*(\cdot) \; : \; [0, \infty) \to \mathbb{R}$ defined by

$$u^*(t) \equiv 1$$

is optimal, since for any admissible control u we have $0 \le u(t) \le 1$, implying

$$J_T(x_0, u(\cdot)) = 1 - exp\Big(-\int_0^T u(s)\,ds\Big) \le 1 - e^{-T} = J_T(x^*(\cdot), u^*(\cdot)).$$

Therefore, by the maximum principle, there exists $\mu \ge 0$ and $\lambda(\cdot) : [0, \infty) \to \mathbb{R}$ such that

$$\dot{\lambda}(t) = (\lambda(t) + \mu)$$

with

$$H(x^*(t), u, t, \lambda(t), \mu) \le H(x^*(t), 1, t, \lambda(t), \mu)$$

a.e. on $[0, \infty)$. From the above differential equations we have

$$\lambda(t) = (\lambda(0) + \mu)e^t - \mu$$

and

$$x^*(t) = 1 - e^{-t}$$

so that

$$H(x^*(t), u, t, \lambda(t), \mu) = (\lambda(0) + \mu)u,$$

which is maximized by $u = 1$ if and only if $\lambda(0) \ge -\mu$. For these initial values $\lambda(t)$ is either constant (for $\lambda(0) = -\mu$), or diverges, and so the transversality condition does not hold.

Example 2.2 The second example shows that it is possible to have $\mu = 0$. Consider the problem of finding the optimal solutions of

$$J_T(x_0, \mu(\cdot)) = \int_0^T (u(t) - x(t))\,dt$$

over all admissible pairs $(x(\cdot), u(\cdot))$ satisfying

$$\begin{aligned} \dot{x}(t) &= u^2(t) + x(t) && \text{for a.e. } 0 \ge 0 \\ x(0) &= 0 \\ u(t) &\in [0, 1] && \text{for a.e. } 0 \ge 0. \end{aligned}$$

For $u^*(t) \equiv 0$ we observe that for all $T \ge 0$,

$$J_T(x_0^*, u^*(\cdot)) = 0,$$

while for any other choice of control $u(\cdot)$, we get for $t \ge 0$

$$x(t) = \int_0^t u^2(s)e^{t-s}\,ds,$$

which implies

$$J_T(x_0, u(\cdot)) = \int_0^T \Big(u(t) - \int_0^t u^2(s)e^{t-s}\, ds\Big)\, dt.$$

Therefore $J_T(x(\cdot), u(\cdot)) \to -\infty$ as $T \to \infty$, since $\int_0^T u(t)\, dt$ grows at most linearly, while $\int_0^T \int_0^t u^2 e^{t-s} ds\, dt$ grows exponentially. Thus $(x^*(\cdot), u^*(\cdot)) \equiv (0,0)$ is optimal, and by the maximum principle there exists $\mu \geq 0$ and $\lambda(\cdot) : [0, \infty) \to \mathbb{R}$ such that

$$\dot{\lambda}(t) = -\lambda(t) + \mu$$

and

$$H(x^*(t), u, t, \lambda(t), \mu) = \mu u + \lambda(t)u^2$$

is maximized at $u^*(\cdot) \equiv 0$. From the above differential equation

$$\lambda(t) = (\lambda(0) - \mu)e^{-t} + \mu.$$

If $\mu > 0$, then $u^* \equiv 0$ is not a maximizer of $\mu u + \lambda(t)u^2$, therefore we must have $\mu = 0$.

2.5 Sufficient conditions for overtaking optimality

Mangasarian [134] has given a set of sufficient conditions for optimality in a finite horizon control problem which can easily be extended to infinite horizon overtaking optimality (See [7]).

Theorem 2.4 *Suppose that:*

(i) *The control set U is compact and there exists a compact set X such that any trajectory emanating from x_0 and generated by an admissible control, stays in the interior, X_0, of X.*

(ii) *The function:* $H^*(x, t, \lambda, \mu) = \max_{u \in U} H(x, u, t, \lambda, \mu)$ *is well defined for $x \in X$ and every t, λ, μ, and it is concave function of x for every fixed t, λ, μ.*

(iii) *There exists a trajectory $x^*(\cdot)$ emanating from x_0 and generated by an admissible control $u^*(\cdot)$ which satisfies the necessary conditions of Theorem 2.3 with $\mu > 0$.*

(iv) *The adjoint function $\lambda(\cdot)$ satisfies the asymptotic transversality condition* $\lim_{t \to \infty} \|\lambda(t)\| = 0$.

Then the trajectory $x^*(\cdot)$ is overtaking optimal at x_0.

Proof: Since $H^*(\cdot, t, \lambda, \mu)$ is concave in x it satisfies

$$H^*(x, t, \lambda(t), \mu) \leq H^*(x^*(t), t, \lambda(t), \mu) + \left.\frac{\partial H^*}{\partial x}\right|'_{x=x^*(t)} (x - x^*(t)) \tag{2.24}$$

for all $t \in [0, \infty)$ and $x \in X$.

Making use of the adjoint equation in the necessary condition (2.16) and of (2.18) it is easily shown (see [7] for details) that

$$\mu\Big[f_0(x^*(t), u^*(t), t) - f_0(x(t), u(t), t)\Big] \geq \frac{d}{dt}\Big[\lambda(t)'(x(t) - x^*(t))\Big] \tag{2.25}$$

for any other trajectory $x(\cdot)$ emanating from x_0 and generated by $u(\cdot)$.

Integrating (2.25) over $[0, T]$ one obtains

$$\mu\Big[J_T(x_0, u^*(\cdot)) - J_T(x_0, u(\cdot))\Big] \geq \lambda(T)'\Big[x(T) - x^*(T)\Big]. \tag{2.26}$$

As μ is positive, the asymptotic transversality condition (iv), the compactness of X and (2.26) yield

$$\liminf_{T\to\infty} (J_T(x_0, u^*(\cdot)) - J_T(x_0, u(\cdot)) \geq 0,$$

hence $x^*(\cdot)$ is overtaking optimal at x_0. ∎

When dealing with optimal economic growth problems, the concavity of H^* w.r.t. x is a direct consequence of the concavity assumptions on the production and utility functions. Furthermore when a positive discount rate is considered it is possible to show that the asymptotic transversality condition (iv) can be guaranteed.

When the discount rate ρ is equal to zero, the asymptotic transversality conditions (iv) will not be satisfied in general. A more general set of sufficient conditions for overtaking optimality has to be obtained.

The concavity of H^* with respect to x is a condition which can be relaxed. The following theorem is an extension of a result established by Leitmann & Stalford [117] and Peterson [148]. This proof has been given by Feinstein and Luenberger [75].

Theorem 2.5 *Let $x^*(\cdot)$ be a trajectory emanating from x_0 and generated by an admissible control $u^*(\cdot)$. Let $p^*(\cdot)$ be a continuous, piecewise continuously differentiable function from $[0, \infty)$ into $\mathbb{R}^n$. Define the Hamiltonian*

$$H(x, t, u, p) = f_0(x, u, t) + p'f(x, u, t).$$

Suppose that

(i) *For all* $x \in X$, $u \in U$ *and a.e.* $t \geq 0$

$$H(x^*(t), t, u^*(t), p^*(t)) + \dot{p}^*(t)'x^*(t) \geq H(x, t, u, p^*(t)) + \dot{p}^*(t)'x$$

(ii) $\lim_{t\to\infty} p^*(t)'x^*(t)$ *exists, and the relation*

$$-\infty < \lim_{t\to\infty} p^*(t)'x^*(t) \leq \liminf_{t\to\infty} p^*(t)'x(t)$$

holds for any admissible trajectory $x(\cdot)$.

Then $(x^*(\cdot), u^*(\cdot))$ *is overtaking optimal at* x_0.

Proof: Let $(x(\cdot), u(\cdot)) \in \mathcal{A}_\infty$. By assumption (i), the following holds for every $T \geq 0$:

$$\int_0^T \left[f_0(x^*(t), u^*(t), t) + \frac{d}{dt}(p^*(t)'x^*(t))\right] dt$$
$$\geq \int_0^T \left[f_0(x(t), u(t), t) + \frac{d}{dt}(p^*(t)'x(t))\right] dt.$$

Then

$$\int_0^T \left[f_0(x^*(t), u^*(t), t) - f_0(x(t), u(t), t)\right] dt \geq p^*(T)'\left[x(T) - x^*(T)\right].$$

Hence, in view of (ii)

$$\liminf_{t\to\infty} \Big(J_T(x_0, u^*(\cdot)) - J_T(x_0, u(\cdot))\Big) \geq \liminf_{t\to\infty} p^*(T)'\left[x(T) - x^*(T)\right] \geq 0$$

■

From the proof of the above theorem it is clear that condition (ii) could be replaced by

(ii)′ $\liminf_{t\to\infty} p^*(t)'\left[x(t) - x^*(t)\right] \geq 0$
for all admissible trajectories $x(\cdot) : [0, \infty) \to \mathbb{R}^n$.

Further, if (ii)′ is replaced by

(ii)″ $\limsup_{t\to\infty} p^*(t)'\left[x(t) - x^*(t)\right] \geq 0$,
for all admissible trajectories $x(\cdot) : [0, \infty) \to \mathbb{R}^n$,

the above theorem becomes a sufficiency theorem for weak overtaking optimality. These results are due to Seierstad and Sydsæter [166].

A strengthening of the conditions (ii) gives us sufficient conditions for uniformly overtaking and uniformly weakly overtaking optimal solutions. In this way we obtain conditions under which a trajectory $x^*(\cdot)$ is both overtaking (resp. weakly overtaking) optimal and agreeable (resp. weakly agreeable). We state this result without proof.

Theorem 2.6 *Let $x^*(\cdot)$, $u^*(\cdot)$, $p^*(\cdot)$ and* H *as in Theorem* 2.5. *If* (ii)′ (resp. (ii)″) *holds uniformly over all admissible trajectories $x(\cdot) : [0,\infty) \to \mathbb{R}^n$, then $x^*(\cdot)$ is uniformly overtaking (resp. uniformly weakly overtaking) optimal.*

We conclude this chapter by rexamining Example 1.2.

Example (1.2)′ *From Halkin's Maximum Principle, the adjoint variable, $(p_1^*(\cdot), p_2^*(\cdot))$, corresponding to the weakly overtaking optimal trajectory $(x_1^*(\cdot), y_1^*(\cdot))$ is given by*

$$\begin{pmatrix} p_1^*(t) \\ p_2^*(t) \end{pmatrix} = \begin{pmatrix} \cos(t) - 1 \\ -\sin(t) \end{pmatrix}.$$

Further, any admissible trajectory $(x(\cdot), y(\cdot))$ generated by a control $u(\cdot)$ is given by

$$\begin{pmatrix} x(t) \\ y(t) \end{pmatrix} = \int_0^t \begin{pmatrix} \sin(t-s) \\ \cos(t-s) \end{pmatrix} u(s)\, ds.$$

Thus for any $t \geq 0$ we have

$$\begin{aligned} p(t)'\big[x(t) - x^*(t)\big] &= p_1^*(t)\big[x_1(t) - x_1^*(t)\big] + p_2^*(t)\big[y_1(t) - y_1^*(t)\big] \\ &= p_1^*(t)\Big[\textstyle\int_0^t \sin(t-s)\big(u(s) - u_1^*(s)\big)\, ds\Big] \\ &\quad + p_2^*(t)\Big[\textstyle\int_0^t \cos(t-s)(u(s) - u_1^*(s))\, ds\Big] \\ &= \int_0^t \Big[(\cos(t) - 1)\sin(t-s) - \sin(t)\cos(t-s)\Big]\Big(u(s) - u_1^*(s)\Big)\, ds \\ &= \int_0^t \Big[-\sin(s) - \sin(t-s)\Big]\Big(u(s) - u_1^*(s)\Big)\, ds \end{aligned}$$

Choosing $t = 2k\pi$, $k = 0, 1, 2, \ldots$, implies that

$$p(t)'(x(t) - x^*(t)) = 0.$$

Thus for any $\epsilon > 0$ and $T > 0$ we can find an integer k, such that $2k\pi > T$ and

$$p(t)'(x(t) - x^*(t)) = -\epsilon.$$

This shows that (ii)″ holds uniformly over all admissible trajectories $x(\cdot)$, so that we can assert that $(x_1^*(\cdot), y_1^*(\cdot))$ is uniformly overtaking optimal and therefore also weakly agreeable.

Notes and Comments

1. The following set of necessary conditions can be obtained when $f : \mathbb{R}^n \times \mathbb{R}^m \times \mathbb{R} \to \mathbb{R}^n$ and $f_0 : \mathbb{R}^n \times \mathbb{R}^m \times \mathbb{R} \to \mathbb{R}$ are C^1 w.r.t. x, u, t, while the control constraint set is given by

$$U(x,t) \triangleq \{u : h(x,u,t) \geq 0\},$$

where

$$h(\cdot,\cdot,\cdot) : \mathbb{R}^n \times \mathbb{R}^m \times \mathbb{R} \to \mathbb{R}^p,$$

for $p < m$, and h is C^1 on its domain. Let

$$H \triangleq \mu f_0(x,u,t) + \lambda' f(x,u,t) + \nu' h(x,u,t).$$

Then an optimal pair $(x^*(\cdot), u^*(\cdot))$ satisfies the following equations

$$\begin{aligned} x(0) &= x_0 \\ \dot{x} &= \frac{\partial H}{\partial \lambda} \\ \dot{\lambda} &= -\frac{\partial H}{\partial x} \\ \frac{\partial H}{\partial u} &= 0 \\ h(x,u,t) &\geq 0 \\ \nu &\geq 0 \\ \nu' h(x,u,t) &= 0 \end{aligned}$$

for some $\lambda(\cdot)$ and ν, provided constraint qualification conditions are satisfied. (See Arrow and Kurz, [7].)

Chapter 3

Asymptotic Stability and the Turnpike Property in Some Simple Control Problems

3.1 Introduction

This chapter is devoted to the presentation of the fundamental ideas concerning the asymptotic stability of optimal trajectories and the related *turnpike theorems*, as they first appeared in the economic literature. In order to avoid the use of too much economic jargon, only two simple examples will be considered. The one in section 3.3 concerns the optimal exploitation of a fish population: on any fixed, sufficiently large horizon with given initial and terminal states, the optimal trajectory will first reach an "optimal steady state", stay on this arc as long as possible and leave it only when the terminal state will have to be reached. This is an exact "turnpike" property. The example of section 3.4 is one of the simplest optimal economic growth models. Using a phase diagram technique it is possible to show that an optimal trajectory over an infinite time horizon will have to converge toward an asymptotic equilibrium and that the costate variable also has to converge.

These two examples are treated by ad hoc techniques which are effective due to the low dimensionality of the state space. It is generally impossible to generalize these approaches to the study of larger systems.

The chapter will begin by a short presentation of a general result due to Samuelson [163] which concerns the saddle point property of the Hamiltonian in a convex problem of Lagrange. Samuelson seems to have been the first to recognize that the turnpike property as well as the asymptotic stability of optimal trajectories in most economic models were deeply related to this property of the Hamiltonian.

The next two chapters will provide a general framework reuniting all these separate fragments of theory.

3.2 Saddle point property of the Hamiltonian in a convex problem of Lagrange. Implications on local asymptotic stability of optimally controlled systems

Consider a function $\mathcal{L}(\cdot,\cdot) : \mathbb{R}^n \times \mathbb{R}^n \to \mathbb{R}$ of class C^2 and the class $D[0, T]$ of continuous, differentiable functions $x(\cdot) : [0, T] \to \mathbb{R}^n$ with piecewise continuous derivative, a given initial value $x(0)$ and terminal value $x(T)$. The problem of Lagrange is to characterize the function $x^*(\cdot)$ in $D[0,T]$ which solves

$$\max_{x(\cdot)\in D[0,T]} \int_0^T \mathcal{L}(x(t), \dot{x}(t))\, dt. \tag{3.1}$$

The well known necessary conditions for optimality to be satisfied by $x^*(\cdot)$ are the Euler-Lagrange equations

$$\frac{d}{dt}\frac{\partial}{\partial \dot{x}}\mathcal{L}(x(t), \dot{x}(t)) - \frac{\partial}{\partial x}\mathcal{L}(x(t), \dot{x}(t)) = 0 \tag{3.2}$$

(see [116] for a presentation of the calculus of variations).
The following canonical relations of Hamilton (see Goldstein, [167]), are equivalent to (3.2)

$$\left.\begin{aligned} \dot{p}_i(t) &= -\frac{\partial}{\partial x_i} H(x(t), p(t)) \\ \dot{x}_i(t) &= \frac{\partial}{\partial p_i} H(x(t), p(t)) \end{aligned}\right\} (i = 1, 2, \ldots, n) \tag{3.3}$$

where the "generalized momenta" $p = (p_1, p_2, \ldots, p_n)$ and the "Hamiltonian" H are defined in terms of $\mathcal{L}$ as follows. For a fixed x consider the mapping $v \to \mathcal{L}(x, v)$ and let v be such that $p = -\frac{\partial}{\partial v}\mathcal{L}(x, v)$. Then $H(x,p) = \mathcal{L}(x,v) + p'v$. It then turns out that $\frac{\partial}{\partial x} H(x,p) = \frac{\partial \mathcal{L}}{\partial x}(x,v)$. We summarize this in the relation

$$\left.\begin{aligned} & H(x,p) = p'v + \mathcal{L}(x,v) \\ & p = -\frac{\partial}{\partial v}\mathcal{L}(x,v) \\ & -\frac{\partial}{\partial x}\mathcal{L}(x,v) + \frac{\partial}{\partial x} H(x,p) = 0. \end{aligned}\right\} \tag{3.4}$$

The Hamiltonian system (3.3) is a dynamical system in the (x,p) phase space. Along a trajectory $(x(\cdot), p(\cdot))$ which satisfies (3.3), the Hamiltonian H is constant since by (3.3)

$$\dot{H} = \frac{\partial H}{\partial x}\dot{x} + \frac{\partial H}{\partial p}\dot{p} = 0.$$

Let $(\overline{x}, \overline{p})$ be a critical point, i.e. a solution of the algebraic equations

$$\begin{aligned} 0 &= -\frac{\partial}{\partial x_i} H(\overline{x}, \overline{p}) \\ 0 &= \frac{\partial}{\partial p_i} H(\overline{x}, \overline{p}). \end{aligned} \tag{3.5}$$

The following result will be used repeatedly in our analysis of the local asymptotic behavior of optimally controlled systems.

Lemma 3.1 *If the function $(x, v) \rightarrow \mathcal{L}(x, v)$ is concave in both of its arguments, then the Hamiltonian H could be also defined as*

$$H(x, p) = \max \{p'v + \mathcal{L}(x, v) \, : \, v \in \mathbb{R}^n\}. \tag{3.6}$$

Furthermore H is concave *with respect to x and* convex *with respect to p.*

Proof: See Rockafellar [160].

Corollary 3.1 *Assume that $(x, v) \rightarrow \mathcal{L}(x, v)$ is a strictly concave function, and that the Hessian of H exists at $(\overline{x}, \overline{p})$ a stationary point for the Hamiltonian system* (3.4). *Then $(\overline{x}, \overline{p})$ is a saddle point of the Hamiltonian as defined in* (3.6) *and there cannot be closed trajectories with periodic motion as solutions of the system* (3.3) *in the neighborhood of $(\overline{x}, \overline{p})$.*

Proof: See Samuelson [163] and Levhari and Liviatan [126] ∎

This result have far reaching consequences. An optimal trajectory when the horizon is infinite has to satisfy the Euler-Lagrange equation, if the problem can be formulated as a problem of Lagrange with unbounded time interval. The preceding result suggests that the extremal trajectory will either converge toward $\overline{x}$ or diverge depending on the correct choice of the initial value $p(0)$ for the costate variable. A typical example of this behavior will be shown in section 3.4.

3.3 An exact turnpike property: Optimal fish harvest

The following model has been introduced by Cliff and Vincent [57].

Let x be the density of a single species of fish in a pond. Assume that the population dynamics obeys the following law

$$\dot{x}(t) = x(t)(x_s - x(t)),\ 0 < x(0) < x_s \tag{3.7}$$

where x_s is a given parameter representing the maximal sustainable density level.

One has the possibility of controlling x by catching an amount of fish proportional to the total population. The state equation of the controlled system becomes

$$\dot{x}(t) = x(t)(x_s - x(t) - u(t)) \tag{3.8}$$

where $u(t)$ is the "fishing effort" control taking its value in a given bounded interval

$$0 \le u(t) \le M. \tag{3.9}$$

Given an initial state x_0, a terminal time T and a terminal state x_T, the problem is to find a piecewise continuous control $u(\cdot)\ :\ [0,\,T] \to [0, M]$ bringing x from x_0 to x_T while minimizing the following functional

$$J(x_0, u(\cdot)\,) = \int_0^T (ax(t) + bu(t) - cx(t)u(t))\, dt \tag{3.10}$$

with

$$x(0) = x_0 \qquad x(T) = x_T$$

and where $a > 0$ is the unit cost rate for maintaining the population, $b > 0$ the unit cost rate for the fishing effort and $c > 0$ the unit selling price of fish.

This problem is characterized by the presence of singular arcs where the control is not well defined by the first order necessary conditions. We will use the approach originally proposed by Miele [143], and utilized by Sanders [165]. As each admissible control $u(\cdot)$ (satisfying (3.9)) generates a unique admissible trajectory, by (3.8), the functional (3.10) is a function of $u(\cdot)$ only and we may denote it by $\tilde{J}(u(\cdot)\,)$. Given two controls $u(\cdot)$ and $v(\cdot)$, one can write the difference between $\tilde{J}(u(\cdot)\,)$ and $\tilde{J}(v(\cdot)\,)$ as a contour integral in the (t, x)-plane and use Stokes' theorem (1). Thus

$$\begin{aligned} \tilde{J}(u(\cdot)\,) - \tilde{J}(v(\cdot)\,) &= \oint_{\pi} \left\{ [ax + (b - cx)(x_s - x)]\, dt - \left[\frac{b - cx}{x}\right] dx \right\} \\ &= \iint_{\mathcal{R}} (-2cx - a + b + cx_s)\, dt\, dx \end{aligned} \tag{3.11}$$

The closed contour is composed of two parts corresponding to the two trajectories generated by $u(\cdot)$ and $v(\cdot)$ respectively. $\mathcal{R}$ is the region bounded by π. The integral in (3.11) is zero if x is kept at a constant value

$$\overline{x} = \frac{1}{2} x_s + \frac{b-a}{2c}. \tag{3.12}$$

It is positive if x is smaller than $\overline{x}$, negative if x is larger than $\overline{x}$. One assumes in the sequel that a, b and c are such that

$$0 < \overline{x} < x_s. \tag{3.13}$$

Notice that this value $\overline{x}$ has an interesting interpretation. It is the solution of the static optimization problem

minimize

$$ax + bu - cxu$$

subject to

$$x_s - x - u = 0 \tag{3.14}$$

which reduces to

$$\min_x \{ax + (b - cx)(x_s - x)\},$$

a minimization problem in x only. By differentiation one obtains the first order necessary condition,

$$a - (b - cx) - c(x_s - x) = a - b - cx_s + 2cx = 0$$

whose only solution is $\overline{x}$. Hence $\overline{x}$ is a sustainable density level, such that, among all sustainable levels, this particular one minimizes the instantaneous cost $ax + bu - cxu$.

Assume first that

$$\overline{x} \leq x_0 \leq x_T \leq x_s. \tag{3.15}$$

The intuitive idea in trying to construct a "good" control policy $\tilde{u}(\cdot)$ is to get on the stationary path $x \equiv \overline{x}$ as quickly as possible, stay on this path for most of the time period and leave it as late as possible to reach the assigned terminal state x_T at time T. Figure 3.1 illustrates this kind of control and the trajectory thus generated.

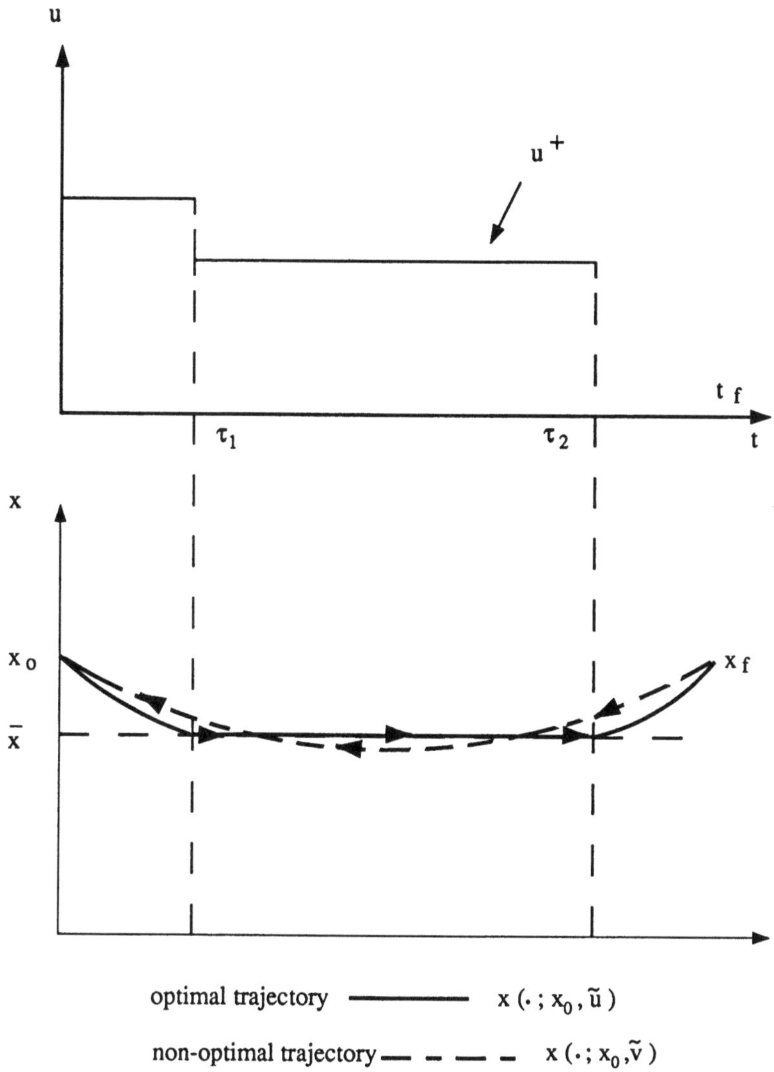

Figure 3.1: Optimal and non-optimal fish harvest trajectories

The control policy considered is defined by

$$\tilde{u}(t) = \begin{cases} M & \text{for } t \in [0, \tau_1) \\ \overline{u} & \text{for } t \in [\tau_1, \tau_2) \\ 0 & \text{for } t \in [\tau_2, T] \end{cases}$$

(One implicitly assumes that T is large enough) and the trajectory generated, denoted

$$x(\,\cdot\,; x_0, \tilde{u}(\cdot)\,) \,:\, [0,\, T] \to x(t\,;\, x_0, \tilde{u}(\cdot)\,)$$

satisfies

$$x(0\,;\,x_0,\tilde{u}(\cdot)\,) \;=\; x_0,$$
$$x(t\,;\,x_0,\tilde{u}(\cdot)\,) \;\equiv\; \overline{x}, \quad \forall\, t \in [\tau_1, \tau_2],$$
$$x(T\,;\,x_0,\tilde{u}(\cdot)\,) \;=\; x_T.$$

Consider any other control $\tilde{v}(\cdot)$ which brings x from x_0 to x_T. Refering to Figure 3.1 one sees that it is impossible to have, for any t

$$x(t\,;\,x_0,\tilde{u}(\cdot)\,) < x(t\,;\,x_0,\tilde{v}(\cdot)\,)$$

and

$$x(t\,;\,x_0,\tilde{v}(\cdot) \;\leq \overline{x}. \tag{3.16}$$

Thus, for any admissible $\tilde{v}(\cdot)$ one has either

$$x(t\,;\,x_0\,;\,\tilde{u}(\cdot)\,) \geq x(t\,;\,x_0,\tilde{v}(\cdot)\,)$$

or

$$x(t\,;\,x_0,\tilde{v}(\cdot)\,) > \overline{x}, \tag{3.17}$$

and this last result with (3.11) implies[(12)] that

$$J(\tilde{u}(\cdot)\,) \leq J(\tilde{v}(\cdot)\,).$$

The other cases corresponding to other relative positions of $\overline{x}, x_0, x_T$ and x_s are treated similarly. Therefore an exact turnpike property holds for the optimal fish harvest problem. Any trajectory optimal over a sufficiently large horizon will contain an arc along which the fish density has the steady state value $\overline{x}$. This is an optimal steady state which minimizes the instantaneous cost. This is also what we call an *exact turnpike.*

3.4 Use of a phase diagram for a one-state-variable control problem: The simple optimal economic growth model

Consider again the model of optimal economic growth proposed by Cass [36] and briefly introduced in section 1.4, Eqs. (1.29)–(1.37).

In a compact formulation the model (1.29)–(1.33) can be written

$$\dot{k}(t) = f(k(t)) - \mu k(t) - c(t) \tag{3.18}$$

$$k(0) = k_0 \tag{3.19}$$

$$W_T = \int_0^T e^{-\rho t}\mathcal{U}(c(t))\,dt \tag{3.20}$$

where k is the one-dimensional state variable, c is the one-dimensional control variable and W_T is the accrued (discounted) utility up to time T.

Let $k^*(\cdot) : [0, \infty) \to \mathbb{R}$ be a trajectory emanating from k_0 and generated by an admissible control $c^*(\cdot) : [0, \infty) \to \mathbb{R}$ which is optimal over an infinite time horizon. The Hamiltonian of the system is

$$H(k, c, \lambda, t) \triangleq e^{-\rho t}\mathcal{U}(c) + \lambda(f(k) - \mu k - c),$$

and there exists $\lambda(\cdot) : [0, \infty) \to \mathbb{R}$, a solution to the adjoint system

$$\dot{\lambda}(t) = -\frac{\partial}{\partial k} H(k^*(t), c^*(t), \lambda(t), t) = -\lambda(t)(f'(k^*(t)) - \mu) \tag{3.21}$$

such that for any $t > 0$

$$\frac{\partial}{\partial c} H(k^*(t), c^*(t), \lambda(t), t) = e^{-\rho t}\mathcal{U}'(c^*(t)) - \lambda(t) = 0. \tag{3.22}$$

The condition (3.22) permits the computation of $c^*(t)$ as a function of $\lambda(t)$ if $\mathcal{U}'(\cdot)$ is monotone, (a condition assumed in (1.36)). Therefore Equations (3.18), (3.21) and (3.22) constitute a complete system which would characterize $k^*(\cdot)$, $c^*(\cdot)$, $\lambda(\cdot)$ if a complete set (k_0, λ_0) of initial conditions were specified. The difficulty is that only k_0 is known. Due to the absence of transversality conditions in the maximum principle for infinite horizon control problem, one has to find an "ad hoc" method for completing the boundary conditions and then obtain extremal trajectories.

A convenient transformation of Eqs. (3.18), (3.21), occurs if one defines a new adjoint variable

$$q(t) = e^{\rho t}\lambda(t) \tag{3.23}$$

which will be called the *current value costate variable.* Introducing also the *current value Hamiltonian*

$$\hat{H}(k, c, q) \triangleq \mathcal{U}(c) + q(f(k) - \mu k - c) \tag{3.24}$$

the following autonomous dynamic system in (q, k) is obtained from (3.18)–(3.22)

$$\dot{k} = f(k) - \mu k - c = \frac{\partial \hat{H}}{\partial q} \tag{3.25}$$

$$\dot{q} = -\left[f'(k) - (\mu + \rho)\right]q = -\frac{\partial \hat{H}}{\partial k} + \rho q \tag{3.26}$$

$$0 = \mathcal{U}'(c) - q = \frac{\partial \hat{H}}{\partial c} \tag{3.27}$$

where, for convenience, we have omitted the arguments $k^*(t)$, $c^*(t)$, $q(t)$.

A stationary equilibrium for the sytem (3.25)–(3.27) is a solution of the three algebraic equations

$$0 = f(k) - \mu k - c \tag{3.28}$$

$$0 = -q[f'(k) - (\mu + \rho)] \tag{3.29}$$

$$0 = \mathcal{U}'(c) - q \tag{3.30}$$

Excluding for the moment $q = 0$ as an acceptable equilibrium there is a unique pair $(\overline{k}, \overline{q})$, solution of (3.28)–(3.30). Recall that the condition (1.35) assures that there exists $\overline{k}$ such that

$$f'(\overline{k}) = \mu + \rho \tag{3.31}$$

then $\overline{c}$ is determined by

$$\overline{c} = f(\overline{k}) - \mu\overline{k} \tag{3.32}$$

and finally $\overline{q}$ is uniquely defined as

$$\overline{q} = \mathcal{U}'(\overline{c}). \tag{3.33}$$

The pair $(\overline{k}, \overline{q})$ is such that if the initial state k_0 coincides with $\overline{k}$ and if the current-value costate variable is given the initial value $\overline{q}$, then the necessary conditions of the maximum principle will yield a steady-state trajectory

$$k^*(t) \equiv \overline{k}$$

with

$$\lambda(t) = \overline{q}e^{-\rho t}, \quad t \geq 0$$

and a constant control (i.e. consumption)

$$c^*(t) \equiv \overline{c}.$$

With the help of a phase diagram it is now possible to show that for any initial state $k_0 \neq \overline{k}$, if there exists an optimal trajectory then it must converge to $\overline{k}$ and the associated costate trajectory must also converge to $\overline{q}$.

The analysis proceeds as follows: In the (k, q) space, draw the loci of points for which $\dot{k} = 0$ and $\dot{q} = 0$ respectively, as according to Eqs. (3.28)–(3.29).

The line $\dot{q} = 0$ is a vertical line passing through $k = \overline{k}$.
The line $\dot{k} = 0$ has the equation

$$f(k) - \mu k - c(q) = 0$$

where $c(q)$ is the unique solution of (3.30). According to the conditions (1.36), $c(q)$ is a decreasing function of q and according to (1.34), $f(k) - \mu k$ is a concave function. The figure 3.2 illustrates the (k, q) phase space and the four quadrants delimited by the two lines $\dot{k} = 0$, $\dot{q} = 0$.

The two lines intersect at the stationary point $(\bar{k}, \bar{q})$. In each quadrant the arrows indicate the direction of change for k and q respectively, according to the equations (3.25)-(3.27).

A simple, though lengthy, analysis of the paths originating from points situated in quadrant I and III show that they cannot be optimal. This analysis is fully detailed in Arrow and Kurz [7].

Therefore, given k_0 the only possible values for q_0 are such that the solution of (3.25)-(3.27) converges to the steady-state equilibrium $(\bar{k}, \bar{q})$. This shows that the optimal k-trajectory is globally asymptotically stable. Notice that, as $q(t) \to \bar{q}$ when $t \to \infty$, then $\lambda(t) \to 0$ when $t \to \infty$ and the asymptotic transversality condition (iii) of Theorem 2.4 holds. Hence the necessary conditions for optimality will also be sufficient in this particular case.

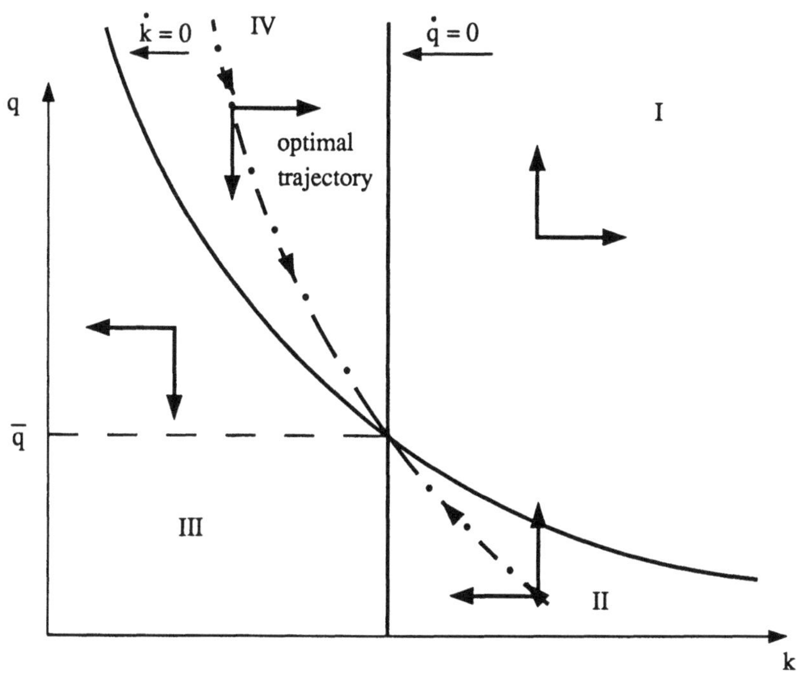

Figure 3.2 Phase diagram for the optimal economic growth model

As a final remark one can notice that the optimal economic growth problem could be reformulated as a convex problem of Lagrange by introducing the functional

$$J_T(k(\cdot)) = \int_0^T \mathcal{U}\Big[f(k(t)) - \mu k(t) - \dot{k}(t)\Big] e^{-\rho t}\, dt.$$

Also a turnpike property exists for the finite horizon optimal economic growth problem. As mentioned in section 1.4, Cass [36] has been able to show that the stationary value $\overline{k}$ will be an "attractor" for any finite-horizon optimal capital accumulation path k^*.

The results obtained in this particular example seem to be related to the saddle point property of Hamiltonians in convex problems of Lagrange, although one has to be careful in the treatment of nonzero discount rates ρ. Because of the term ρq in the right-hand side of Eq. (3.26), the state and costate trajectories $k(\cdot)$, $q(\cdot)$ are not exactly the solution of an Hamiltonian system, when ρ is nonzero. For more general models it will be shown in Chapter 6 that to assure asymptotic stability of optimal trajectories with positive discount rate one requires more stringent assumptions than concavity of the integrand $\mathcal{L}(x, \dot{x})$ in the Lagrange problem reformulation to hold.

Notes and Comments

(1) Use of Stokes' Theorem for the comparison of control policies. Formally, Stokes' Theorem can be written as

$$\oint_{\pi} \{F(t,x)\,dt + G(t,x)\,dx\} = \iint_{\mathcal{R}} \left[\frac{\partial G}{\partial t} - \frac{\partial F}{\partial x}\right] dt\,dx$$

where F and G are C^1 in x and t, π is a closed curve oriented so that the enclosed region $\mathcal{R}$ is on the left as one advances along π. A change of orientation for π implies a change of sign of the surface integral.

The comparison method is described in Hermes and LaSalle book [106]; or in Miele [144]. We further remark that Sethi [167] utilizes a similar approach to the optimal control of the Vidale-Wolfe advertising model to obtain an exact turnpike result.

We restrict the presentation of the method to the particular case of the system (3.8) with cost functional (3.10).

Let $u(\cdot)$ be a given control generating a trajectory $x(\cdot)$ such that $x(0) = x_0$, $x(T) = x_T$. Along this trajectory one has

$$dx = x(x_s - x - u)\,dt$$

and thus

$$udt = -\frac{dx}{x} + (x_s - x)\,dt.$$

Bringing this expression of udt into the cost functional one gets

$$\int_{(0,x_0)}^{(T,x_T)} \left\{[ax + (b - cx)(x_s - x)]\,dt - \frac{b - cx}{x}\,dx\right\}$$

where the integral is now a line integral taken along the trajectory x joining x_0 and x_T.

Let $v(\cdot)$ be another control generating another trajectory joining x_0 and x_T. Assume that x_0 and x_T are the only points which are common to both trajectories. Let π be the closed curve formed by these trajectory arcs. One follows first the arc generated by u from x^T to x_0. One thus finds

$$\tilde{J}(u(\cdot)) - \tilde{J}(v(\cdot)) = \oint_{\pi} \left\{(ax + (b - cx)(x_s - x))\,dt - \frac{b - cx}{x}\,dx\right\}$$

The bounding curve is a Jordan arc, thus applying Stokes' Theorem (or Green's Theorem) one has

$$\begin{aligned}\tilde{J}(u(\cdot)) - \tilde{J}(v(\cdot)) &= \iint_{\mathcal{R}} \left\{\frac{-\partial}{\partial x}(ax + (b - cx)(x_s - x)) - \frac{\partial}{\partial t}\left[\frac{b - cx}{x}\right]\right\} dt\,dx \\ &= \iint_{\mathcal{R}} (-2cx - a + b + cx_s)\,dt\,dx\end{aligned}$$

where $\mathcal{R}$ is the domain enclosed by π.

Chapter 4

Global Asymptotic Stability and Existence of Optimal Trajectories for Infinite Horizon Autonomous Convex Systems

4.1 Introduction

In this chapter the optimal control problem with infinite time horizon will be considered for a class of autonomous systems obeying specific convexity assumptions. For this class of systems it will be possible to demonstrate that:

(i) If there exists a unique extremal steady state $\bar{x}$, then an overtaking trajectory must converge toward $\bar{x}$.

(ii) The Mangasarian sufficient conditions can be extended to the case where the adjoint function $\lambda(\cdot)$ does not tend to zero at infinity.

(iii) For any finite horizon, the turnpike property holds, with $\bar{x}$ being the attractor.

(iv) Under strict convexity assumptions the existence of overtaking trajectories is asserted for any initial state.

(v) Under less restrictive assumptions the existence of weak-overtaking trajectories is asserted.

With the maximum principle established in Theorem 2.3, this set of results constitute a rather complete theory of optimal control of autonomous convex systems over an infinite time horizon. Although this theory was initially developed in the realm of control theory (Haurie [99], Brock and Haurie [28]), it has been very much influenced by previous studies in Mathematical Economics dealing with multisector models of optimal capital accumulation. The turnpike theory, presented in this chapter, is merely a translation into the continuous time optimal control formalism of a similar property obtained by McKenzie [140] for discrete time economic growth models. The existence result is a generalization of similar results obtained by Brock [27], also for discrete time economic growth models.

It is felt that the optimal control framework is much more general than the rather specialized formalism used by economists when dealing with efficient capital accumulation paths, and thus the global theory presented here could find a wider domain of applications.

4.2 The class of systems considered

Consider the autonomous system

$$\begin{aligned} \dot{x}(t) &= f(x(t), u(t)) \\ x(0) &= x_0 \end{aligned} \tag{4.1}$$

where $f(\cdot,\cdot) : \mathbb{R}^n \times \mathbb{R}^m \to \mathbb{R}^n$ is a continuous mapping with respect to x and u. For each x, $U(x)$ is a compact subset of $\mathbb{R}^m$ such that the mapping $U(\cdot) : x \to U(x)$ is upper semicontinuous.

The function $f(\cdot,\cdot)$ is also assumed to satisfy: there exist finite numbers K, M, N such that for all x, $y \in \mathbb{R}^n$ and $u \in U \triangleq \cup_x U(x)$ one has:

$$\|f(x,u) - f(y,u)\| \leq K\|x-y\| \tag{4.2}$$

$$\|f(x,u)\| \leq K(M + N\|x\|). \tag{4.3}$$

The performance criterion considered is also analogous to the one defined in (1.43), and is given by

$$J_T(x_0, u(\cdot)) = \int_0^T f_0(x(t), u(t))\, dt, \tag{4.4}$$

where $f_0(\cdot,\cdot)$ is an upper semicontinuous map of $\mathbb{R}^n \times \mathbb{R}^m$ into $\mathbb{R}$.

We now turn to a description of the basic convexity assumptions characterizing the class of systems studied in this chapter.

Assumption 4.1 *Let Ω be the subset of $\mathbb{R}^n \times \mathbb{R}^n \times \mathbb{R}$ defined as*

$$\Omega \triangleq \left\{(\xi,\eta,\gamma) : \xi \in \mathbb{R}^n,\ \exists\, u \in U(\xi)\ \text{s.t.}\ \eta \leq f(\xi,u), \gamma \leq f_0(\xi,u)\right\}. \tag{4.5}$$

It is assumed that Ω is a convex set.

A sufficient condition for Assumption 4.1 to hold is that $f(\cdot,\cdot)$ and $f_0(\cdot,\cdot)$ be concave functions of x, u and $U(\cdot)$ be a *biconvex* mapping as defined below.

Definition 4.1 *The mapping* $U(\cdot)\ : x \to U(x) \subset \mathbb{R}^m$ *is* biconvex *if for any* x, *the graph of* $U(\cdot)$ *is convex, that is the set*

$$G = \left\{(x,u)\ :\ x \in \mathbb{R}^n, \quad u \in U(x)\right\}$$

is convex in $\mathbb{R}^n \times \mathbb{R}^m$.

An example of a biconvex mapping $U(\cdot)$ is obtained if for any x, $U(x)$ is defined as

$$U(x) \triangleq \left\{u \in \mathbb{R}^m\ :\ h(x,u) \geq 0\right\} \tag{4.6}$$

where $h(\cdot,\cdot)\ :\ \mathbb{R}^n \times \mathbb{R}^m \to \mathbb{R}^p$ is a concave function of both x and u.

It will be useful for later developments to introduce the following set valued map $V(\cdot,\cdot)$ and function $\mathcal{L}(\cdot,\cdot)$.

$$\forall\,(x,v) \in \mathbb{R}^n \times \mathbb{R}^n \quad V(x,v) \triangleq \{u \in U(x)\ :\ v = f(x,u)\} \subset \mathbb{R}^m \tag{4.7}$$

$$\mathcal{L}(x,v) \triangleq \begin{cases} \sup\{f_0(x(\cdot)\,,u(\cdot))\ :\ u \in V(x,v)\} & \text{if } V(x,v) \neq \emptyset \\ -\infty & \text{if } V(x,v) = \emptyset. \end{cases} \tag{4.8}$$

A subclass of the class of systems that satisfy Assumption 4.1 is the one for which the following holds.

Assumption 4.1′ *The function* $\mathcal{L}(\cdot,\cdot) = \mathbb{R}^n \times \mathbb{R}^n \to \overline{\mathbb{R}} \triangleq \mathbb{R} \cup \{-\infty\}$ *defined by* (4.7) *and* (4.8) *is concave.*

Lemma 4.1 *Assumption* 4.1′ *implies Assumption* 4.1.

Proof: Let $y = (\xi,\eta,\gamma)$ and $y' = (\xi',\eta',\gamma')$ be two points in Ω. One has to show that for any α in $(0,1)$ the convex combination

$$\alpha y + (1-\alpha)y' \text{ is in } \Omega.$$

By definition of Ω there exist u and u' respectively in $U(\xi)$ and $U(\xi')$ such that

$$\begin{aligned} &\eta \leq f(\xi,u), \qquad \gamma \leq f_0(\xi,u) \\ &\eta' \leq f(\xi',u'), \qquad \gamma' \leq f_0(\xi',u'). \end{aligned} \tag{4.9}$$

Let's call

$$v \triangleq f(\xi,u), \qquad v' \triangleq f(\xi',u') \tag{4.10}$$

one has therefore, by definition of $\mathcal{L}(\cdot,\cdot)$:

$$\gamma \leq \mathcal{L}(\xi, v), \qquad \gamma' \leq \mathcal{L}(\xi', u'). \tag{4.11}$$

Since $\mathcal{L}(\cdot,\cdot)$ is concave one has also

$$\alpha\gamma + (1-\alpha')\gamma \leq \mathcal{L}(\alpha\xi + (1-\alpha)\xi',\ \alpha v + (1-\alpha)v'). \tag{4.12}$$

As $V(x,v)$ is a closed subset of a compact set $U(x)$, the supremum in Eq. (4.8) can be replaced by a maximum, hence there exists u^α in $V(\alpha\xi + (1-\alpha)\xi',\ \alpha v + (1-\alpha)v')$ such that, according to (4.9)–(4.12)

$$f_0(\alpha\xi + (1-\alpha)\xi',\ u^\alpha) \geq \alpha\gamma + (1-\alpha)\gamma'$$

and

$$f(\alpha\xi + (1-\alpha)\xi',\ u^\alpha) = \alpha v + (1-\alpha)v' \geq \alpha\eta + (1-\alpha)\eta'.$$

This establishes the desired result. ∎

The transformation (4.7)–(4.8) will be useful when the framework of a problem of Lagrange in the calculus of variations will be preferred to the standard optimal control framework. In particular this will be the case while proving the existence theorems.

The following lemma gives a sufficient condition for Assumption 4.1′ to hold.

Lemma 4.2 *Assume that $f(\cdot,\cdot)$ and $f_0(\cdot,\cdot)$ are both concave functions of x and u, assume that $U(\cdot)$ is biconvex and that $\mathcal{L}(x,v)$ as defined by (4.7), (4.8) satisfies the following:*

$$\forall x \in \mathbb{R}^n, \quad v \geq v' \Rightarrow \mathcal{L}(x,v) \leq \mathcal{L}(x,v'). \tag{4.13}$$

Then Assumption 4.1′ holds.

Proof: Consider (x,v) and (x',v') such that $\mathcal{L}(x,v)$ and $\mathcal{L}(x',v')$ are both finite. There exists $u \in U(x)$ and $u' \in U(x')$ such that

$$\begin{aligned} \mathcal{L}(x,v) &= f_0(x,u), \quad v = f(x,u) \\ \mathcal{L}(x',v') &= f_0(x',u'), \quad v' = f(x',u'). \end{aligned} \tag{4.14}$$

For any α in $(0,1)$, by the concavity of $f(\cdot,\cdot)$ one has

$$\alpha v + (1-\alpha)v' \leq f(\alpha x(1-\alpha)x',\ \alpha u + (1-\alpha)u') \tag{4.15}$$

where, by the biconvexity of $U(\cdot)$

$$\alpha u + (1-\alpha)u' \in U(\alpha x + (1-\alpha)x'). \tag{4.16}$$

Therefore, if one defines

$$v^\alpha \triangleq f(\alpha x + (1-\alpha)x',\ \alpha u + (1-\alpha)u'), \tag{4.17}$$

by the concavity of $f_0(\cdot,\cdot)$ one has

$$\begin{aligned}\alpha\mathcal{L}(x,v) + (1-\alpha)\mathcal{L}(x',v') &\le f_0(\alpha x + (1-\alpha)x',\ \alpha u + (1-\alpha)u') \\ &\le \mathcal{L}(\alpha x + (1-\alpha)x',\ v^\alpha).\end{aligned} \tag{4.18}$$

Now, following (4.13), (4.15), (4.17), and (4.18) one has

$$\alpha\mathcal{L}(x,v) + (1-\alpha)\mathcal{L}(x',v') \le \mathcal{L}(\alpha x + (1-\alpha)x',\ \alpha v + (1-\alpha)v') \tag{4.19}$$

and the result is established for this case. The result is also trivially true if for (x,v), $\mathcal{L}(x,v)$ is equal to $-\infty$. ∎

Another sufficient condition for Assumption 4.1′ to hold is the following.

Lemma 4.3 *If $f(\cdot,\cdot)$ is linear in x and u, if $U(\cdot)$ is biconvex and $f_0(\cdot,\cdot)$ is concave, the Assumption 4.1′ is satisfied.*

Proof: The proof follows closely the preceding one, except that (4.15) is replaced by the equality:

$$\alpha v + (1-\alpha)v' = f[\alpha x + (1-\alpha)x', \alpha u + (1-\alpha)u']$$

because of the linearity of $f(\cdot,\cdot)$. This implies immediately that (4.19) holds and hence the result follows. ∎

Remark 4.1 *When $f(\cdot,\cdot)$ is linear and $f_0(\cdot,\cdot)$ is quadratic, the concavity of $\mathcal{L}(x,v)$ will be assured even if $U(x)$ is not a compact set (e.g. $U(x) \equiv \mathbb{R}^m$). Therefore the linear-quadratic systems described in Chapter 1 satisfy Assumption 4.1′, although they were technically excluded by the assumption of compactness for $U(x)$.*

Example 4.1 *Multisector capital accumulation model:* This is an adaptation of Uzawa's two-sector economy model [176]. Consider an economy with n capital goods. Let $k_i(t), i = 1,2,\ldots,n$ be the stocks of capital goods accumulated at time t. Let μ_i be the depreciation rate of capital i, and $z_i(t)$ the investment in capital i. The state equations describing the capital accumulation are thus

$$\dot{k}_i(t) = z_i(t) - \mu_i k(t), \quad 1,2,\ldots,n. \tag{4.20}$$

A production function $F(\cdot) : \mathbb{R}^n \to \mathbb{R}$ associates with a vector $k \triangleq (k_1, k_2, \ldots, k_n)$ of the n capital stocks a potential production of a composite commodity which can either be consumed or invested. Let $\mathcal{C}(t)$ be the consumption flow at t, thus the following constraint has to be satisfied

$$\mathcal{C}(t) + \sum_{i=1}^{n} z_i(t) \leq f(k(t)). \tag{4.21}$$

Furthermore all the variables have to be nonnegative

$$k(t) \geq 0, \quad z(t) \geq 0, \quad C(t) \geq 0. \tag{4.22}$$

The state variables are the capital stocks, the control variables are the investment rates and consumption flow. Consider a performance criterion associated with any horizon T, where $\mathcal{U}(\cdot) \; : \mathbb{R} \rightarrow \mathbb{R}$ is a utility function.

$$J_T = \int_0^T \mathcal{U}(\mathcal{C}(t))\, dt. \tag{4.23}$$

If the production function $F(\cdot)$ and the utility function $\mathcal{U}(\cdot)$ are both concave, then the system (4.20)–(4.23) clearly satisfies Assumption 4.1. The control constraint set defined by (4.21), (4.22) gives rise to a biconvex mapping since it has the form referred to in (4.6). Furthermore, as the state equations (4.21) are linear, the system also satisfies Assumption 4.1′.

Example 4.2 *Optimal exploitation of a renewable resource.* This is an economic growth model with exploitation of a natural resource proposed by Haurie and Hung in [100].

The system is described by the following state equations

$$\begin{aligned} \dot{K}(t) &= I(t) - \mu K(t), \quad \mu > 0 \text{ given} && (4.24)\\ \dot{R}(t) &= G(R(t)) - E(t) && (4.25)\\ K(0) &= K_0, \quad R(0) = R_0 \end{aligned}$$

with the following constraints on state and control

$$\begin{aligned} I(t) + C(t) &\leq F_1(K_1(t), E(t)) && (4.26)\\ E(t) &\leq F_2(K_2(t), R(t)) && (4.27)\\ K_1(t) + K_2(t) &\leq K(t) && (4.28)\\ 0 \leq I(t),\ 0 &\leq C(t),\ 0 \leq E(t),\ 0 \leq K_1(t),\ 0 \leq K_2(t),\ 0 \leq R(t) && (4.29) \end{aligned}$$

and performance criterion

$$J_T(K_0, R_0; C(\cdot), I(\cdot), E(\cdot)) = \int_0^T \mathcal{U}(C(t))\, dt. \tag{4.30}$$

Here the *state variables* are the stock of a reproducible composite commodity K (equipment or capital) and the stock of the natural resource R (e.g. a forest). The *control variables* are the consumption flow C, the investment I, the flow E of natural resource into production, the part K_1 of the stock K devoted to producing the composite commodity and the part K_2 devoted to extracting the natural resource. The functions $\mathcal{U}(\cdot)$, $G(\cdot)$, $F_1(\cdot,\cdot)$ and $F_2(\cdot,\cdot)$ are concave in all their arguments. The constraint set on the control satisfies the property of biconvexity since it has the form (4.6). Therefore, $\mathcal{U}(\cdot)$ and $G(\cdot)$ being concave functions, one can assert that Assumption 4.1′ is satisfied.

The model does not assume a specialization of capital, as the total capital stock can be reallocated at any time between the two concurrent uses. With this specification one cannot use Lemma 4.2 for proving that Assumption 4.1′ holds. If there is capital specialization the equations (4.24) and (4.28) are replaced by the two state equations

$$\dot{K}_1(t) = I_1(t) - \mu K_1(t) \tag{4.31}$$
$$\dot{K}_2(t) = I_2(t) - \mu K_2(t) \tag{4.32}$$

when $I_i(t)$ is the investment rate in capital of type i and the constraint (4.26) becomes

$$I_1(t) + I_2(t) + C(t) \leq F_1(K_1(t),\, E(t)). \tag{4.33}$$

Therefore, the system has now three state variables, K_1, K_2, R and four control variables, I_1, I_2, E, C. For this system it can be easily shown that the conditions of Lemma 4.2 are satisfied. Given K_1, K_2, R there is a one-to-one correspondence between the velocity triples $(\dot{K}_1,\, \dot{K}_2,\, \dot{R})$ and the control triples $(I_1,\, I_2,\, E)$. In order to increase simultaneously the three velocities one has thus to increase I_1, I_2 and to decrease E. If the production function F_1 has a positive partial derivative with respect to E, then by (4.33) the consumption flow must necessarily decrease. If $\mathcal{U}(\cdot)$ is a nondecreasing function of consumption this implies that the conditions of Lemma 4.2 hold and therefore Assumption 4.1′ is satisfied by this new system.

4.3 Convergence toward the Von Neumann Set for weakly overtaking trajectories

For the system defined in (4.1)–(4.4) consider the following associated stationary problem

$$\text{maximize } f_0(x,u) \text{ such that } f(x,u) = 0 \text{ and } u \in U(x). \tag{4.34}$$

A solution $(\overline{x},\overline{u})$ to the above static optimization problem defines an *optimal steady state* for the autonomous control system. The following assumption is called the *strict support property* for $\overline{x}$

Assumption 4.2 *Suppose that:*

(i) *there exists a unique solution, $\overline{x}$, to the stationary problem* (4.34),

(ii) *there exists $\overline{p}$ in $\mathbb{R}^n$ such that*

$$f_0(\overline{x},\overline{u}) = \max\{f_0(x,u) + \overline{p}'f(x,u) \,:\, x \in \mathbb{R}^n,\ u \in U(x)\}, \tag{4.35}$$

and,

(iii) *for any ε positive there exists $\delta > 0$ such that*

$$\|x - VN(\overline{x})\| > \varepsilon \Rightarrow \forall\, u \in U(x),\ f_0(\overline{x},\overline{u}) > (f_0(x,u) + \overline{p}'f(x,u)) + \delta. \tag{4.36}$$

where $VN(\overline{x})$ is the so-called Von Neumann set defined as

$$VN(\overline{x}) = \Big\{x \,:\, \exists\, u \in U(x) \text{ with } f_0(x,u) + p'f(x,u) = f_0(\overline{x},\overline{u})\Big\}$$

Theorem 4.1 *Consider an initial state x_0 for the system* (4.1) *such that $\overline{x}$ is reachable from x_0 in a finite time $T(x_0)$. Then, under Assumption* 4.2, *a bounded weakly overtaking trajectory x^* emanating from x_0 must satisfy*

$$\lim_{t\to\infty} x^*(t) \in VN(\overline{x}) \tag{4.37}$$

Proof: Assume that the asymptotic convergence property (4.37) does not hold. Then

$$\exists\, \varepsilon > 0,\ \forall\, t > 0,\ \exists\, T > t \text{ s.t. } \|x^*(T) - VN(\overline{x})\| > 2\varepsilon. \tag{4.38}$$

Introduce the subset of the time interval $[0, \infty)$ defined by

$$\Xi(\varepsilon, T) \triangleq \{t \in [0,T) \,:\, \|x^*(t) - VN(\overline{x})\| > \varepsilon\}.$$

We now show that for any positive ε, the Lebesgue measure of this set grows to infinity when T tends to infinity. To verify this consider t_n such that

$$\|x^*(t_n) - VN(\overline{x})\| > 2\varepsilon \tag{4.39}$$

and define t'_n by

$$t'_n = \inf\{t > t_n \,:\, \|x^*(t) - \overline{x}\| \le \varepsilon\}.$$

Since $U(x)$ is compact for all x and $f(\cdot,\cdot)$ is continuous, there exists a positive lower bound σ to the difference $t'_n - t_n$. It is therefore possible to construct a sequence $\{t_n\}_{n\in\mathbb{N}}$ such that (4.39) holds for all n so that one would have

$$\mu\Big[\{t \in [0, \infty) \,:\, \|x^*(t) - VN(\overline{x})\| > \varepsilon\}\Big] \ge \lim_{n\to\infty} n\sigma = \infty \tag{4.40}$$

where $\mu[\cdot]$ denotes the Lebesgue measure on the real line, thus proving our claim. By Assumption 4.2 one has

$$f_0(x^*(t), u^*(t)) + \overline{p}'(x^*(t), u^*(t)) \leq f_0(\overline{x}, \overline{u}), \quad \forall t \in [0, \infty). \tag{4.41}$$

Integrating (4.41) between 0 and T one obtains

$$J_T(x_0, u^*(\cdot)) - T f_0(\overline{x}, \overline{u}) \leq \overline{p}'(x^*(T) - x_0). \tag{4.42}$$

This inequality can be further restricted by using (4.36), i.e., property (iii) of Assumption 4.2. This yields

$$J_T(x_0, u^*(\cdot)) - T f_0(\overline{x}, \overline{u}) \leq -\overline{p}'(x^*(T) - x_0) - \delta\mu\Big[\Xi(\varepsilon, T)\Big] \tag{4.43}$$

and therefore

$$\lim_{T\to\infty} (J_T(x_0, u^*(\cdot)) - T f_0(\overline{x}, \overline{u})) = -\infty \tag{4.44}$$

since $x^*(T)$ stays bounded.

If one constructs another trajectory $\tilde{x}(\cdot)$ emanating from x_0 by first reaching $\overline{x}$ at time $T(x_0)$ and then maintaining the steady state $(\overline{x}, \overline{u})$, the following is true

$$\lim_{T\to\infty} (J_T(x_0, \tilde{u}(\cdot)) - T f_0(\overline{x}, \overline{u})) > -\infty \tag{4.45}$$

(here $\tilde{u}(\cdot)$ is the control generating $\tilde{x}(\cdot)$). From (4.44) and (4.45) one has

$$\lim_{T\to\infty} (J_T(x_0, u^*(\cdot)) - J_T(x_0, \tilde{u}(\cdot))) = -\infty \tag{4.46}$$

therefore the performance criterion $J(x_0, \tilde{u}(\cdot))$ catches up to $J_T(x_0, x^*(\cdot))$ and this contradicts the weak overtaking optimality of $x^*(\cdot)$. ■

Remark 4.2 *The convergence property of bounded weakly overtaking trajectories has been established for the general class of systems* (4.1)–(4.4). *The classical maximum principle does not hold necessarily in this class. In particular, it has not been assumed that $f(\cdot,\cdot)$ and $f_0(\cdot,\cdot)$ were differentiable with respect to x and so we have to consider the more general framework proposed by Clarke* [56] *or Halkin* [90].

We conclude this section by giving a set of conditions on $f(\cdot,\cdot)$, $f_0(\cdot,\cdot)$ and $U(\cdot)$ which assure that Assumption 4.2 is satisfied.

Lemma 4.4 *Let $f_0(\cdot,\cdot)$ and $f(\cdot,\cdot)$ be concave functions and $U(\cdot)$ be a biconvex mapping such that*

a) *There exists $x \in \mathbb{R}^n$, and $u \in U(x)$ such that $f(x,u) > 0$*

b) *If for some $x \in \mathbb{R}^n$ and $u \in U(x)$ $f(x,u) > 0$, then there exists $\hat{u} \in U(x)$ such that $f(x,\hat{u}) = 0$ and $f_0(x,\hat{u}) \geq f_0(x,u)$.*

Then Properties (i) *and* (ii) *of Assumption* 4.2 *are satisfied. If x is constrained to stay in a compact convex set X then Property* (iii) *of Assumption* 4.2 *is also satisfied.*

Proof: Under condition b) the associated stationary problem (4.34) can be replaced by the following *concave programming problem*

maximize

$$f_0(x(\cdot), u(\cdot))$$

subject to

$$0 \leq f(x(\cdot), u(\cdot)) \quad u \in U(x). \tag{4.47}$$

Condition a) is a constraint qualification condition. The properties (i) and (ii) of Assumption 4.2 are thus a direct consequence of the duality theory in convex programming (see Mangasarian [135]).

Now assume that property (iii) does not hold with x in a compact subset X of $\mathbb{R}^n$. Thus, it is possible to construct a sequence $\{x^n\}_{n\in\mathbb{N}}$ in X such that, for an associated sequence $\{u^n\}_{n\in\mathbb{N}} \in U(x)$, $\forall n \in \mathbb{N}$ $\|x^n - VN(\overline{x})\| > \varepsilon$ but

$$\lim_{n\to\infty} \{f_0(\overline{x}, \overline{u}) - f_0(x^n, u^n) - \overline{p}' f(x^n, u^n)\} = 0. \tag{4.48}$$

As X is compact one can extract converging subsequences from $\{x^n\}_{n\in\mathbb{N}}$ and $\{u^n\}_{n\in\mathbb{N}}$ with limits $\hat{x}$ and $\hat{u}$ respectively. For these points the following holds true

$$\|\hat{x} - VN(\overline{x})\| > \varepsilon \tag{4.49}$$

$$f_0(\overline{x}, \overline{u}) - f_0(\hat{x}, \hat{u}) - \overline{p}' f(\hat{x}, \hat{u}) = 0. \tag{4.50}$$

From (4.50) one can conclude that $\hat{x}$ is an element of the Von Neumann set $VN(\overline{x})$ which contradicts (4.49). Hence property (iii) of Assumption 4.2 is true. ∎

Remark 4.3 *Conditions* a) *and* b) *of the preceding lemma imply that the set*

$$\Omega_+ \triangleq \{(\xi, \eta, \gamma) \in \Omega \, : \, \eta \geq 0\} \tag{4.51}$$

has a nonempty interior, while Ω is convex. Condition (ii) *of Assumption* 4.2 *is then obtained as a consequence of the separation theorem for convex sets.*(3) *Notice also that the vector $\overline{p}$ is nonnegative in that case.*

4.4 The turnpike property

In this section the optimal control problem with fixed terminal time is considered and a general turnpike property is proved under the same assumptions which guarantee the asymptotic stability of weak overtaking trajectories.

Theorem 4.2 *Under Assumption 4.2, consider an initial state x_0 for the system* (4.1) *and a terminal state x_T such that $\overline{x}$ is reachable from x_0 in a finite time $T_1(x_0)$, x_T is reachable from $\overline{x}$ in a finite time $T_2(x_T)$. For any terminal time $T > T_1(x_0) + T_2(x_T)$, consider the control problem consisting of maximizing the criterion*

$$J_T(x_0, u(\cdot)) = \int_0^T f_0(x(t), u(t))\, dt$$

under the constraints (4.1). *Then along the optimal trajectory $x^*(\cdot)$ the following property holds*

$$\forall \varepsilon > 0\ \exists\, \nu(\varepsilon) > 0 \text{ s.t. } \mu\Big[\{t \in [0, T] : \|x^*(t) - VN(\overline{x})\| > \varepsilon\}\Big] < \nu(\varepsilon)$$

where $\nu(\varepsilon)$ does not depend on T.

Proof: For T larger than $T_1(x_0) + T_2(x_T)$ there exists a trajectory $x(\cdot)$ satisfying

$$\begin{aligned} &x(0) = x_0 \\ &x(t) \equiv \overline{x} \text{ for } T_1(x_0) \leq t \leq T - T_2(x_T) \\ &x(T) = x_T, \end{aligned}$$

generated by an admissible control $u(\cdot)$ satisfying

$$u(t) \equiv \overline{u} \text{ for } T_1(x_0) \leq t \leq T - T_2(x_T).$$

One has therefore,

$$\begin{aligned} J_T(x_0, u(\cdot)) \;=\; & \int_0^{T_1(x_0)} f_0(x(t), u(t))\, dt + (T - T_2(x_T) - T_1(x_0)) f_0(\overline{x}, \overline{u}) \\ & + \int_{T-T_2(x_T)}^{T} f_0(x(t), u(t))\, dt \leq J_T(x^*(\cdot), u^*(\cdot)). \end{aligned}$$

Call respectively I_1 and I_2 the two integrals appearing in the left hand side of the above expression, and get

$$(T - T_2(x_T) - T_1(x_0)) f_0(\overline{x}, \overline{u}) + I_1 + I_2 \leq J_T(x_0, u^*(\cdot)).$$

On the other hand, Assumption 4.2 gives us

(i) $f_0(\overline{x}, \overline{u}) \geq f_0(x^*(t), u^*(t)) + \overline{p}' f(x^*(t), u^*(t))$

and whenever $\|x^*(t) - \overline{x}\| > \varepsilon$

(ii) $f_0(\overline{x}, \overline{u}) \geq f_0(x^*(t), u^*(t)) + \overline{p}' f(x^*(t), u^*(t)) + \delta$

Combining (i) and (ii) and integrating from 0 to T gives us

$$T f_0(\overline{x},\overline{u}) - \overline{p}'(x_T - x_0) - \delta\mu(\varepsilon) > J_T(x_0, u^*(\cdot)),$$

where

$$\mu(\varepsilon) \triangleq \mu\Big[\{t \in [0,T] : \|x^*(t) - VN(\overline{x})\| > \varepsilon\}\Big].$$

Combining the above results gives us,

$$\mu(\varepsilon) < \frac{1}{\delta}\Big[(T_1(x_0) + T_2(x_T))f_0(\overline{x},\overline{u}) - \overline{p}'(x_T - x_0) - I_1 - I_2\Big].$$

Call $\nu(\varepsilon)$ the right hand side of the above relation and the theorem is proved. ∎

Since the measure of the set of times where $x^*(t)$ is at a distance from $VN(\overline{x})$ larger than ε, is bounded by a fixed number $\nu(\varepsilon)$, it appears that as T increases the trajectory $x^*(\cdot)$ spends most of the time period in the vicinity of the Von Neumann set $VN(\overline{x})$.

Remark 4.4 *If the Von Neumann set $VN(\overline{x})$ reduces to the singleton $\{\overline{x}\}$ then the preceding results imply that the optimal steady state is an attractor for optimal trajectories.*

The forthcoming sections deal more specifically with this global asymptotic stability property.

4.5 Global asymptotic stability for extremal trajectories

In this section the system (4.1)–(4.4) is specialized to the case where

$U(x) \triangleq \{u \in \mathbb{R}^n : h(x,u) \geq 0\}$

$h : \mathbb{R}^n \times \mathbb{R}^m \to \mathbb{R}^p$ continuous and C^1 in x.

Furthermore, we assume that $f(\cdot,\cdot)$ and $f_0(\cdot,\cdot)$ are C^1 in x.

A *strong maximum principle* has been obtained by Halkin [90] for systems governed by a biconvex differential inclusion

$$\dot{x}(t) \in F(x(t)),$$

with a finite time horizon and a performance criterion defined as a function of the terminal state only.

In this section we will be obtain a similar result for the infinite horizon control problem under Assumption 4.3 below.

With the help of this stronger maximum principle, it is possible to construct a Lyapunov-like function in p and x which will enable us to derive sufficient condition for any extremal trajectory $x^*(\cdot)$ to converge to a unique stationary state $\overline{x}$. We will then extend the asymptotic stability property to cover also the costate trajectory $p^*(\cdot)$ which must also converge to a unique stationary value $\overline{p}$.

Given the system (4.1)–(4.4), we associate with the performance criterion the state equation

$$\begin{aligned} \dot{\xi}(t) &= f_0(x(t), u(t)) \\ \xi(0) &= 0. \end{aligned} \tag{4.52}$$

A solution $(x(\cdot), \xi(\cdot))$ to the augmented system (4.1) and (4.52) which satisfies the constraints on the control is called an *extended trajectory.*

In the maximum principle, a trajectory for the adjoint variable, $\lambda(\cdot)$, is associated with an extremal trajectory $x^*(\cdot)$. In this section this associated trajectory will be called the *costate trajectory* and be denoted by $p^*(\cdot)$. Furthermore one assumes that in the maximum principle the coefficient μ is positive and thus can be taken equal to one. An extremal trajectory having this property will be called *regular.* It is also assumed that the constraint qualification conditions are satisfied along any optimal trajectory.

Assumption 4.3 *For all $t \geq 0$ the Hamiltonian*

$$H(x, p^*(t), u) = f_0(x, u) + p^*(t)' f(x, u)$$

is concave in x, u.

Lemma 4.5 *Under Assumption* 4.3 *any regular extremal trajectory $x^*(\cdot)$ and its associated costate trajectory $p^*(\cdot)$ are such that, if x and u satisfy $h(x,u) \geq 0$ then, for every $t \geq 0$,*

$$f_0(x^*(t), u^*(t)) + \frac{d}{dt}\left[p^*(t)' x^*(t)\right] \geq f_0(x, u) + p^*(t)' f(x, u) + \dot{p}^*(t)' x. \tag{4.53}$$

Proof: From the maximum principle, for every $t \geq 0$, if u satisfies $h(x^*, u) \geq 0$, then

$$f_0(x^*(t), u^*(t)) + p^*(t)' f(x^*(t), u^*(t)) \geq f_0(x^*(t), u) + p^*(t)' f(x^*(t), u). \tag{4.54}$$

and the associated adjoint variational equation

$$\dot{p}^*(t)' = -\frac{\partial}{\partial x} f_0(x^*(t), u^*(t))' - p^*(t)' \frac{\partial}{\partial x} f(x^*(t), u^*(t)) - \frac{\partial}{\partial x}\left[\nu(t)' h(x^*(t), u^*(t))\right] \tag{4.55}$$

where $\nu(t) \geq 0$ is the Lagrange multiplier associated with the constraint $h(x,u) \geq 0$, at time t.

However, Eqs. (4.54) and (4.55), with the complementary conditions

$$\nu(t)'h(x^*(t), u^*(t)) \geq 0$$
$$\nu(t) \geq 0$$
$$h(x^*(t), u^*(t)) \geq 0$$

give a sufficient condition for the joint maximization with respect to x and u satisfying $h(x,u) \geq 0$ of the extended Hamiltonian

$$\tilde{H}(x, p^*(t), u) = f_0(x,u) + p^*(t)'f(x,u) + \dot{p}^*(t)'x \tag{4.56}$$

and thus (4.53) holds. ∎

Remark 4.5 *Assumption* 4.3 *holds if* $p^*(t)$ *is nonnegative for all* t *and* $f_0(\cdot,\cdot)$ *and* $f(\cdot,\cdot)$ *are concave in* x *and* u.

Corollary 4.1 *Let* x_{01} *and* x_{02} *be two different initial states, let* $x_1(\cdot)$ *and* $x_2(\cdot)$ *be two regular extremal trajectories emanating from* x_{01} *and* x_{02} *respectively, generated by* $u_1(\cdot)$ *and* $u_2(\cdot)$ *respectively, and let their respective costate trajectories be* $p_1(\cdot)$ *and* $p_2(\cdot)$. *Then the following holds*

$$\frac{d}{dt}\left\{(p_1(t) - p_2(t))'(x_1(t) - x_2(t))\right\} \geq 0. \tag{4.57}$$

Proof: For each trajectory one can use (4.53) to get

$$f_0(x_1(t), u_1(t)) + \frac{d}{dt}[p_1(t)'x_1(t)] - f_0(x_2(t), u_2(t)) - \frac{d}{dt}[p_1(t)'x_2(t)] \geq 0 \tag{4.58}$$

and

$$f_0(x_2(t), u_2(t)) + \frac{d}{dt}[p_2(t)'x_2(t)] - f_0(x_1(t), u_1(t)) - \frac{d}{dt}[p_2(t)'x_1(t)] \geq 0. \tag{4.59}$$

Adding (4.59) and (4.58) yields the desired result. ∎

Now it is possible to establish a result similar to Theorem 4.1, which has been obtained in a discrete time setting by McKenzie [139] and in a continuous time setting by Cass and Shell [37].

Theorem 4.3 *Let* $x_1(\cdot)$ *be a regular extremal trajectory, generated by* $u_1(\cdot)$ *with the associated costate trajectory* $p_1(\cdot)$. *Suppose Assumption* 4.3 *holds true and that for every* $\varepsilon > 0$ *there exists a* $\delta > 0$ *so that if* $\|x - x_1(t)\| > \varepsilon$ *for some* $t \geq 0$ *then there is a* $u \in U(x)$ *such that*

$$\delta^1(t, x, u) \triangleq f_0(x_1(t), u_1(t)) + \frac{d}{dt}[p_1(t)'x_1(t)] - f_0(x,u) - p_1(t)'f(x,u) - \dot{p}(t)'x > \delta \tag{4.60}$$

Let $x_2(\cdot)$ be another regular extremal trajectory generated by $u_2(\cdot)$ with the associated costate trajectory $p_2(\cdot)$. Then

$$\lim_{t\to\infty} \|x_1(t) - x_2(t)\| = 0 \tag{4.61}$$

provided that

$$\limsup_{t\to\infty} \|p_i(t)\| < \infty, \qquad i = 1,2 \tag{4.62}$$

$$\limsup_{t\to\infty} \|x_i(t)\| < \infty, \qquad i = 1,2, \tag{4.63}$$

Proof: Given an $\varepsilon > 0$ one has

$$\mu\Big[\{t \geq 0 \;:\; \|x_1(t) - x_2(t)\| > \varepsilon\}\Big] = \infty \tag{4.64}$$

if (4.61) is not verified. Now (4.58)–(4.60) yield

$$\frac{d}{dt}\Big\{(p_1(t) - p_2(t))'(x_1(t) - x_2(t))\Big\} \geq \delta \tag{4.65}$$

if $\|x_1(t) - x_2(t)\|$ is greater than ε. It follows from (4.64) and (4.65) that if (4.61) is false then

$$\lim_{t\to\infty} (p_1(t) - p_2(t))'(x_1(t) - x_2(t)) = \infty$$

which is in contradiction with the boundedness assumptions (4.62) and (4.63). ∎

Corollary 4.2 *If Assumption 4.3 holds, and if the Von Neumann set $VN(\bar{x})$ reduces to the singleton $\{\bar{x}\}$ then any regular extremal trajectory $x^*(\cdot)$ which, with its associated costate trajectory $p^*(\cdot)$, satisfies*

$$\limsup_{t\to\infty} \|x^*(t)\| < \infty$$

$$\limsup_{t\to\infty} \|p^*(t)\| < \infty$$

must converge to the optimal steady state $\bar{x}$.

4.6 A Lyapunov function approach for GAS of optimal trajectories

At first glance Theorem 4.3 seems to be less general than Theorem 4.1, since it uses the apparatus of the maximum principle and the other theorem does not. However, the convergence property is now established for extremal trajectories rather than for optimal trajectories. This is a small difference. More important is the fact that the proof of Theorem 4.3 is actually based on the construction of a Lyapunov-like function for the

Hamiltonian system satisfied by x and p. This avenue of research for obtaining global asymptotic stability (GAS) results in mathematical economics has been fully explored by Brock and Sheinkman [30], [29] and Brock [26].

Introduce the maximized Hamiltonian

$$\mathcal{H}(x,p) \triangleq \max_{u \in U(x)} H(x,u,p) = \max_{u \in U(x)} \left(f_0(x,u) + p' f(x,u) \right)$$

and assume that $\mathcal{H}$ is differentiable with respect to both arguments y and p. Assume that the maximum principle for an extremal trajectory and its associated costate trajectory is equivalent to the Hamiltonian system[(1)].

$$\begin{aligned} \dot{x} = &= \frac{\partial \mathcal{H}}{\partial p}, \\ \dot{p} &= -\frac{\partial \mathcal{H}}{\partial x}. \end{aligned} \tag{4.66}$$

A stationary point $(\overline{x}, \overline{p})$ which satisfies

$$\begin{aligned} 0 &= \frac{\partial \mathcal{H}}{\partial p}(\overline{x}, \overline{p}), \\ 0 &= -\frac{\partial \mathcal{H}}{\partial x}(\overline{x}, \overline{p}) \end{aligned} \tag{4.67}$$

corresponds to an extremal steady state of the optimally controlled system. Assume that there exists a unique stationary point $(\overline{x}, \overline{p})$. For convenience in the writing of stability conditions it is preferable to change the origin in the phase space in order to have $(\overline{x}, \overline{p}) = 0$ in $\mathbb{R}^m \times \mathbb{R}^n$.

We are interested in obtaining a Lyapunov function $V(\cdot,\cdot) : (x,p) \rightarrow \mathbb{R}$ which will yield asymptotic stability conditions for the origin 0 in the system (4.66)–(4.67).

It is convenient to write

$$\begin{aligned} z &\triangleq (x,p) \\ \dot{x} = F_1(z) &\triangleq \frac{\partial \mathcal{H}}{\partial p} \\ \dot{p} = F_2(z) &\triangleq -\frac{\partial \mathcal{H}}{\partial x} \\ \dot{z} = F(z) &\triangleq (F_1(z), F_2(z)) \end{aligned} \tag{4.68}$$

Guided by the property shown in Corollary 4.1 we consider, as a candidate for a Lyapunov-like function,

$$V(z) = -p'x. \tag{4.69}$$

In order to show that 0 is globally asymptotically stable for uniformly bounded trajectories it suffices to show that

$$\text{a)} \quad \dot{V}(z) = \frac{\partial V(z)}{\partial z} F(z) \leq 0 \tag{4.70}$$

$$\text{b)} \quad E_0 \triangleq \{z : \frac{\partial V(z)}{\partial z} F(z) = 0\} = \{0\}, \tag{4.71}$$

since then the function $t \to V(z(t))$ converges to a limit which must be zero (otherwise $\frac{d}{dt} V(z(t))$ will be bounded away from zero). It is easy to see that if $z(t) \not\to 0$ then $V(z(t)) \to -\infty$, which is impossible for uniformly bounded trajectories.

Theorem 4.4 *Assume that the function $\mathcal{H}$ is C^2 in x and p and denote* $Q(z) \triangleq \begin{bmatrix} \mathcal{H}_{xx} & 0 \\ 0 & -\mathcal{H}_{pp} \end{bmatrix}$. *Assume that the following conditions are satisfied*

a) $0 = F(0)$ *is the unique critical point of* $\dot{z} = F(z)$.

b) *For all* $z \neq 0$

$$x'F_2(z) + p'F_1(z) = 0 \Rightarrow z'Q(z)z < 0. \tag{4.72}$$

c) *For all* $z \neq 0$

$$z'Q(0)z < 0. \tag{4.73}$$

Then the function $V(z) = -p'x$ *satisfies* (4.70)–(4.71) *and therefore, all state and costate trajectories that are bounded for* $t \geq 0$ *converge to 0 as* $t \to \infty$.

Proof: One has

$$V(z) = -p'x = -z'Az \tag{4.74}$$

with

$$A = \frac{1}{2} \begin{bmatrix} 0 & I \\ I & 0 \end{bmatrix}$$

and we have to prove that

$$\dot{V}(z) = \nabla V(z)F(z) < 0 \quad \text{for all} \quad z \neq 0. \tag{4.75}$$

For this purpose define, for a fixed $z \neq 0$

$$g(\lambda) \triangleq \nabla V(\lambda z)F(\lambda z) \tag{4.76}$$

(so that $g(\cdot)$ is a function of a single scalar variable λ) and show that $g(1)$ is negative. Once we have shown that

$$g(0) = 0, \; g'(0) = 0, \; g''(0) < 0 \tag{4.77}$$

and that

$$g(\overline{\lambda}) = 0 \text{ for some } \overline{\lambda} > 0 \text{ implies } g'(\overline{\lambda}) < 0, \tag{4.78}$$

then, (see Fig. 4.1 it is assured that $g(\lambda) < 0$ for every $\lambda > 0$ and in particular $g(1) < 0$. This is so, because (4.77) implies that $g(\lambda) < 0$ for an $\varepsilon > 0$ and every $0 < \lambda < \varepsilon$.

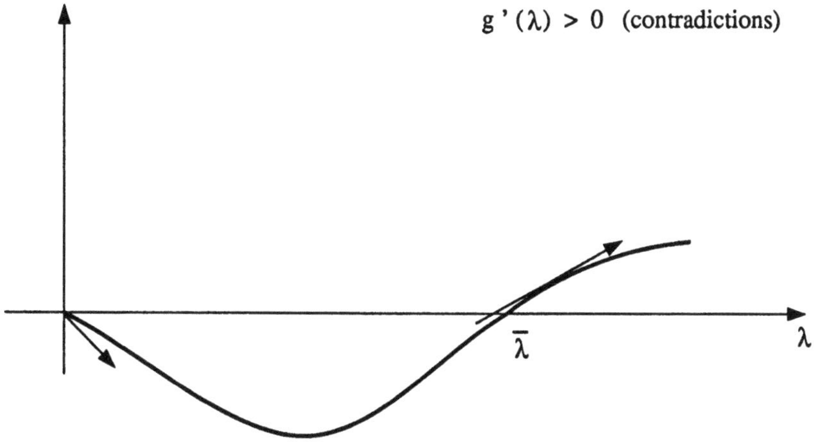

Figure 4.1: Shape of the function $g(\lambda)$ if $g(\overline{\lambda}) = 0$ for $\lambda > 0$.

If $g(\cdot)$ ever vanishes for a positive λ then let $\overline{\lambda} > 0$ be the smallest $\lambda > 0$ for which $g(\lambda) = 0$, and it follows that $g'(\overline{\lambda}) \geq 0$, contradicting (4.78). Differentiating (4.76) with respect to λ one has

$$g'(\lambda) = z'\nabla^2 V(\lambda z)F(\lambda z) + \nabla V(\lambda z)\nabla F(\lambda z)z \tag{4.79}$$

$$\begin{aligned} g''(\lambda) = z'\Big[\frac{d}{d\lambda}\nabla^2 V(\lambda z)\Big]F(\lambda z) + z'\nabla^2 V(\lambda z)\nabla F(\lambda z)z \\ +z'\nabla^2 V(\lambda z)\nabla F(\lambda z)z + \nabla V(\lambda z)\Big[\frac{d}{d\lambda}\nabla F(\lambda z)\Big]z. \end{aligned} \tag{4.80}$$

Now, using (4.74) and (4.68) one has

$$\nabla^2 V(\lambda z) = -2A$$

$$\nabla F(\lambda z) = \left[\begin{array}{cc} \mathcal{H}_{px} & \mathcal{H}_{pp} \\ -\mathcal{H}_{xx} & -\mathcal{H}_{xp} \end{array}\right]\Big|_{\lambda z}$$

$$\nabla V(\lambda z) = -2\lambda z'A = -\lambda[x'p']\left[\begin{array}{cc} 0 & I \\ I & 0 \end{array}\right].$$

Bringing these expressions into (4.79) and (4.80) one obtains

$$\begin{aligned} g'(\lambda) &= -2z'AF(\lambda z) - \lambda z'\left[\begin{array}{cc} -\mathcal{H}_{xx} & -\mathcal{H}_{xp} \\ \mathcal{H}_{xp} & \mathcal{H}_{pp} \end{array}\right]\Big|_{\lambda z} z \\ &= -\Big\{x'F_2(\lambda z) + p'F_1(\lambda z) + \lambda[x'p']\left[\begin{array}{cc} -\mathcal{H}_{xx} & 0 \\ 0 & \mathcal{H}_{pp} \end{array}\right]\Big|_{\lambda z}\left[\begin{array}{c} x \\ p \end{array}\right]\Big\} \\ &= -\{x'F_2(\lambda z) + p'F_1(\lambda z) - \lambda z'Q(\lambda z)z\} \end{aligned}$$

$$g''(\lambda) = 2z'Q(\lambda z)z - 2\lambda z'A\frac{d}{d\lambda}[\nabla F(\lambda z)]z.$$

Using the fact that $F(0) = 0$ and the condition (4.73) one easily obtains that

$$g(0) = g'(0) = 0 \quad g''(0) < 0.$$

Now consider $\lambda > 0$. From the definition we have

$$g(\overline{\lambda}) = 0 \Rightarrow x'F_2(\overline{\lambda}z) + p'F_1(\overline{\lambda}z) = 0$$

which, in view of (4.72) implies that $g'(\overline{\lambda})$ is negative. Hence, by the claim which follows (4.78), $g(1)$ is negative implying that

$$\begin{aligned} \dot{V}(z) &< 0 \quad \text{if} \quad z \neq 0 \\ \dot{V}(0) &= 0. \end{aligned}$$

Therefore, the stationary point is asymptotically stable for any bounded z-trajectory. ∎

Remark 4.6 *The proof of the preceding theorem follows closely the proof of Theorem* 3.1 *of* [30]. *Brock and Sheinkman give in that theorem a more general set of sufficient conditions for having a monotonically decreasing Lyapunov function for a Hamiltonian system. These ideas will be pursued in the next chapter while dealing with discounted performance index.*

Remark 4.7 *Theorem* 4.4 *establishes simultaneously the global asymptotic stability of both the state and the costate trajectories. Notice that, by* (4.72)–(4.73), *it is assumed that the Hamiltonian is, at least locally, strictly concave in x, strictly convex in p, and this establishes a link with Samuelson's remark on the concavity-convexity of Hamiltonians in convex problems of Lagrange* (*Lemma* 3.1). *Also Condition* (4.73) *gives more insight on the globally asymptotically stable condition of Theorem* 4.3. *However, both Theorems* (4.3 *and* 4.4) *are not completely equivalent since their respective regularity and convexity* (*concavity*) *assumptions are not exactly the same.*

4.7 Sufficient conditions for overtaking optimality

In Section 2.5 a set of sufficient conditions of the overtaking optimality of extremal trajectories has been given in Theorem 2.4. One of these conditions was that: $\lambda(t) \to 0$ when $t \to \infty$. Clearly the preceding results, obtained in Sections 4.5 and 4.6 show that one cannot expect such an asymptotic behavior of the adjoint (costate) variable when the discount rate ρ is equal to zero.

Fortunately, the convergence property of extremal trajectories will permit the extension of Mangasarian's sufficiency condition without assuming asymptotic transversality conditions.

Assumption 4.4 *Assume that Assumptions* 4.2 *and* 4.3 *hold and that the Von Neumann set* $VN(\overline{x})$ *reduces to the singleton* $\{\overline{x}\}$.

Theorem 4.5 *Suppose that Assumption* 4.4 *holds. Then a bounded regular extremal trajectory* $x^*(\cdot)$, *with a bounded associated trajectory* $p^*(\cdot)$ *is overtaking optimal.*

Proof: Let x_0 be the initial state. Consider another trajectory $\tilde{x}(\cdot)$ generated by an admissible control $\tilde{u}(\cdot)$ and emanating from x_0. Two cases are possible:

a) $\lim_{t\to\infty} \tilde{x}(t) = \lim_{t\to\infty} x^*(t) = \overline{x}$.

Then, by the same argument as in Theorem 2.4

$$f_0(x^*(t), u^*(t)) - f_0(\tilde{x}(t), \tilde{u}(t)) \geq \frac{d}{dt}\left[p^*(t)'(\tilde{x}(t) - x^*(t))\right]$$

and thus: $\forall T > 0$

$$J_T(x_0, u^*(t)) - J_T(x_0, \tilde{u}(\cdot)) \geq p^*(T)'(\tilde{x}(T) - x^*(T)).$$

The last relation implies, by letting $T \to \infty$, that $x^*(\cdot)$ is overtaking optimal.

b) $\lim_{t\to\infty} \tilde{x}(t) \neq \lim_{t\to\infty} x^*(t) = \overline{x}$.

Then, by the same argument as the one used in the proof of Theorem 4.1 one can show that

$$\lim_{T\to\infty} \int_0^T \left(f_0(\tilde{x}(t), \tilde{u}(t)) - f_0(\overline{x}, \overline{u})\right) dt = -\infty,$$

while, ($\overline{x}$ being reachable)

$$\lim_{T\to\infty} \int_0^T \left(f_0(x^*(t), u^*(t))\, dt - f_0(\overline{x}, \overline{u})\right) dt > -\infty,$$

Which again implies that $x^*(\cdot)$ is overtaking optimal. ∎

4.8 Existence of optimal trajectories

In this section the existence of weakly overtaking or overtaking optimal trajectories will be proved under the same kind of assumptions which have permitted the proof of the convergence property. The lines of reasoning are very close to those of [27] dealing with a discrete time capital accumulation model. See also [28].

The idea is to construct an *Associated Problem of Lagrange* (APL) with infinite time interval but with a bounded value of the supremum of the integral, then show that a solution to this problem exists, and finally check that the solution of the APL is either weakly overtaking or overtaking optimal.

Using the function $\mathcal{L}(\cdot,\cdot) : (x, v) \to \mathbb{R}$ introduced in Section 4.2, the control problem can be *deparametrized* and replaced by a formulation akin to the Lagrange problem in the calculus of variations.

Assumption 4.1′ will play a fundamental role in the proof of the existence theorems. As in Section 3.2, we associate with $\mathcal{L}(x, v)$ the Hamiltonian

$$\mathcal{H}(x,p) \triangleq \sup \{\mathcal{L}(x,v) + p'v \,:\, v \in \mathbb{R}^n\}. \tag{4.81}$$

Rockafellar [159], [161] has shown that the optimality conditions are that the state-costate trajectories are solutions of the subdifferential inclusions

$$\dot{x}(t) \in \partial_p \mathcal{H}(x(t), p(t))$$
$$\dot{p}(t) \in -\partial_x \mathcal{H}(x(t), p(t)).$$

The subdifferential, $\partial_x h$, of a function $x \to h(x)$ at a point x is the set of all subgradients of the function $h(\cdot)$ at that particular point; a vector p is said to be a subgradient of a convex function $h(\cdot)$ at x if $h(z) \geq h(x) + p'(z - x)$ for every z. If $\mathcal{H}$ happens to be differentiable the subdifferential inclusions reduce to the classical Hamiltonian system

$$\dot{x} = \frac{\partial \mathcal{H}}{\partial p},$$
$$\dot{p} = -\frac{\partial \mathcal{H}}{\partial x}.$$

The following two assumptions summarize the conditions for existence of weakly overtaking or overtaking trajectories respectively.

Assumption 4.5 *$\mathcal{L}(\cdot,\cdot)$ is concave, and there exists a saddle point $(\overline{x}, \overline{p})$ of $\mathcal{H}$ where $\overline{x}$ is the unique solution to*

$$\max \{f_0(x,u) \,:\, 0 = f(x,u), \quad u \in U(x)\}.$$

Assumption 4.6 *$\mathcal{L}(\cdot,\cdot)$ is concave, and there exists a unique saddle point $(\overline{x}, \overline{p})$ of $\mathcal{H}$ in the neighborhood of which $\mathcal{H}$ is strictly concave, strictly convex.*

Since $\mathcal{L}(\cdot,\cdot)$ is concave, $\mathcal{H}(\cdot,\cdot)$ defined in (4.81) is concave in x, convex in p. The saddle point condition for $\mathcal{L}$ is equivalent to (see Rockafellar [160] Theorem 37.5)

$$\mathcal{L}(x,v) \leq \mathcal{L}(\overline{x}, 0) - \overline{p}'v \text{ for every } (x,v) \in \mathbb{R}^n \times \mathbb{R}^n \tag{4.82}$$

and thus the functional $\mathcal{L}_0 : \mathbb{R}^n \times \mathbb{R}^n \to \mathbb{R}$, defined by

$$\mathcal{L}_0(x,v) \triangleq \mathcal{L}(x,v) - \mathcal{L}(\overline{x},0) + \overline{p}'v \tag{4.83}$$

satisfies

$$\mathcal{L}_0(x,v) \leq 0 \text{ for every } (x,v) \in \mathbb{R}^n \times \mathbb{R}^n. \tag{4.84}$$

The *Associated Problem of Lagrange* (APL) can now be defined as that of maximizing the improper integral,

$$\int_0^\infty \mathcal{L}_0(x(t),\dot{x}(t))\, dt \tag{4.85}$$

over all admissible trajectories emanating from x_0.

Definition 4.2 *An absolutely continuous function* $x^*(\cdot) : [0, \infty) \to \mathbb{R}^n$ *is a solution of the APL at* x_0 *if* $x^*(0) = x_0$, *the integral* (4.85) *is finite and maximal for* $x^*(\cdot)$, *i.e., for any absolutely continuous function* $x : [0, \infty) \to \mathbb{R}^n$ *with*

$$x(0) = x_0$$

the following holds

$$\liminf_{T\to\infty} \left[\int_0^T \mathcal{L}_0(x^*(t),\, \dot{x}^*(t))\, dt - \int_0^T \mathcal{L}_0(x(t),\, \dot{x}(t))\, dt\right] \geq 0.$$

To prove the existence of an optimal solution to the APL, we need the following result, due to Varaiya [177], concerning the compactness of admissible trajectories.

Theorem 4.6 *Assume that the control system* (4.1) *satisfies Assumption* 4.2 *and further, assume that the velocity set*

$$\varphi(x) \triangleq \{f(x,u) : u \in U(x)\} \tag{4.86}$$

is convex (possibly empty). Then the set of all admissible trajectories is a compact subset of the space of all continuous function of $[0, \infty)$ *with values in* $\mathbb{R}^n$, *endowed with the topology of uniform convergence on compact subsets of* $[0, \infty)$.

The existence of an optimal solution to the APL can now be established.

Theorem 4.7 *Let Assumption* 4.5 *hold and assume that* $\overline{x}$ *is reachable from* x_0 *in a finite time. Assume also that the velocity set* $\varphi(x)$, *defined as in* (4.86), *is convex (possibly empty) for all* x. *Then there exists a solution* $x^*(\cdot)$ *for the* APL *at* x_0.

Proof: Let

$$g(x_0) \triangleq \sup\Big\{\int_0^\infty \mathcal{L}_0(x(t),\dot{x}(t))\,dt \;:\; x(0)=x_0\Big\} \tag{4.87}$$

where the supremum is over all absolutely continuous functions $x(\cdot) \;:\; [0,\,\infty) \to \mathbb{R}^n$ with the given intitial value and the reachability condition on $\overline{x}$ implies that

$$g(x_0) > -\infty.$$

Thus there exists a sequence of functions $x^k(\cdot) \;:\; [0,\,\infty) \to \mathbb{R}^n$ such that

$$\lim_{n\to\infty} \int_0^\infty \mathcal{L}_0(x^k(t),\,\dot{x}^k(t))\,dt = g(x_0)$$

and we may assume that for every $k \geq 1$

$$\int_0^\infty \mathcal{L}_0(x^k(t),\dot{x}^k(t))\,dt > -\infty.$$

This last inequality implies that one has

$$\mathcal{L}_0(x^k(t),\,\dot{x}^k(t)) \neq -\infty \text{ a.e. on } [0,\,\infty)$$

so we conclude that the functions $x^k(\cdot)$ are trajectories emanating from x_0.

In view of Theorem 4.6, the set of trajectories emanating from x_0 is compact in the topology of uniform convergence on bounded intervals. One can thus extract a converging subsequence from $\{x^k(\cdot)\}_{k\in\mathbb{N}}$ whose limit $x^*(\cdot) \;:\; [0,\,\infty) \to \mathbb{R}^n$ is also a trajectory emanating from x_0. Moreover it follows that in this situation also

$$\lim_{k\to\infty} \dot{x}^k(t) = \dot{x}^*(t).$$

holds for almost every t in $[0,\infty)$.[2]

It is straightforward (see [28], Lemma 2.1) that $\mathcal{L}_0(x,v)$ is upper semicontinuous. Using Fatou's lemma one obtains $\int_0^T \mathcal{L}_0(x^*(t),\,\dot{x}^*(t))\,dt \geq g(x_0)$ for every $T>0$, for every $T>0$, from which follows that

$$\int_0^\infty \mathcal{L}_0(x^*(t),\dot{x}^*(t))\,dt = g(x_0)$$

proving that $x^*(\cdot)$ is a solution to the APL. ∎

Lemma 4.6 *Let Assumption* 4.6 *hold and suppose that $\overline{x}$ is reachable from x_0 in a finite time. Then any trajectory $x(\cdot)$ emanating from x_0 and for which*

$$\lim_{t\to\infty} x(t) \neq \overline{x}$$

is such that

$$\int_0^\infty \mathcal{L}_0(x(t),\,\dot{x}(t))\,dt = -\infty.$$

Proof: In proving this lemma we follow Rockafellar [159]. The maximum of $\mathcal{H}(\cdot,\overline{p})$ over $\mathbb{R}^n$ is $\mathcal{H}(\overline{x},\overline{p})$ and is attained only at $\overline{x}$, so it is possible to construct a continuous increasing function $\gamma : \mathbb{R}^+ \to \mathbb{R}^+$ with $\gamma(0) = 0$ such that

$$\mathcal{H}(x,\overline{p}) - \mathcal{H}(\overline{x},\overline{p}) \leq -\gamma(\|x-\overline{x}\|).$$

By virtue of (4.81) and since

$$\mathcal{H}(\overline{x},\overline{p}) = \mathcal{L}(\overline{x},0)$$

the inequality above yields

$$\mathcal{L}(x,v) + \overline{p}'v - \mathcal{L}(\overline{x},0) \leq -\gamma(\|x-\overline{x}\|) \text{ for every } v \in \mathbb{R}^n$$

and according to (4.83)

$$\mathcal{L}_0(x,v) \leq -\gamma(\|x-\overline{x}\|) \text{ for every } v \in \mathbb{R}^n. \tag{4.88}$$

This establishes the lemma. ■

Theorem 4.8 *Let Assumption* 4.6 *hold. Moreover suppose that the state x is constrained to stay in a compact subset X of $\mathbb{R}^n$. Then there exists an overtaking optimal trajectory emanating from x_0.*

Proof: By Theorem 4.7 there exists a solution $x^*(\cdot)$ to the APL at x_0 which satisfies

$$J_T(x_0,u^*(\cdot)) = \int_0^T \mathcal{L}(x^*(t),\, \dot{x}^*(t))\, dt, \text{ for any } T > 0.$$

Let $x(\cdot)$ be any trajectory emanating from x_0 and generated by $u(\cdot)$. It follows from the construction of $\mathcal{L}(\cdot,\cdot)$, that

$$J_T(x_0,u(\cdot)) \leq \int_0^T \mathcal{L}(x(t),\, \dot{x}(t))\, dt, \text{ for every } T > 0.$$

Therefore

$$J_T(x_0,u^*(\cdot)) - J_T(x_0,u(\cdot)) \geq \int_0^T \left[\mathcal{L}(x^*(t),\, \dot{x}^*(t)) - \mathcal{L}(x(t),\, \dot{x}(t))\right] dt,$$

which yields, by virtue of (4.82)

$$\begin{aligned} J_T(x_0,u^*(\cdot)) - J_T(x_0,u(\cdot)) \geq & \int_0^T \Big[\mathcal{L}_0(x^*(t),\, \dot{x}^*(t)) - \mathcal{L}_0(x(t),\, \dot{x}(t)) \\ & -\overline{p}'(\dot{x}^*(t) - \dot{x}(t))\Big]\, dt \\ = & \int_0^T \left[\mathcal{L}_0(x^*(t),\, \dot{x}^*(t)) - \mathcal{L}_0(x(t),\, \dot{x}(t))\right] dt \\ & -\overline{p}'(x^*(T) - x(T)). \end{aligned} \tag{4.89}$$

By Lemma 4.6, the solution of the APL must converge to $\overline{x}$. If $x(\cdot)$ satisfies

$$\lim_{t\to\infty} x(t) = \overline{x}$$

then there exists an $\varepsilon > 0$ such that for every sufficiently large T

$$\int_0^T \left[\mathcal{L}_0(x^*(t),\, \dot{x}^*(t)) - \mathcal{L}_0(x(t),\, \dot{x}(t))\right] dt > -\frac{\varepsilon}{2}, \tag{4.90}$$

since $x^*(\cdot)$ solves the APL, and $x(\cdot) \neq x^*(\cdot)$. Moreover, if T is large enough then

$$\overline{p}'(x^*(T) - x(T)) < \frac{\varepsilon}{2} \tag{4.91}$$

since both $x^*(t)$ and $x(t)$ tend to $\overline{x}$. By (4.89)–(4.91) it follows that for T large enough

$$J_T(x_0, u^*(\cdot)\,) - J_T(x_0, u(\cdot)\,) > -\varepsilon$$

and therefore $x^*(\cdot)$ overtakes $x(\cdot)$. If $x(t)$ does not converge to $\overline{x}$, then by Lemma 4.6

$$\int_0^\infty \mathcal{L}_0(x(t),\, \dot{x}(t))\, dt = -\infty. \tag{4.92}$$

Since both trajectories remain bounded as T tends to infinity, (4.89) and (4.92) imply that $x^*(\cdot)$ is overtaking optimal. ∎

Remark 4.8 *The existence of an overtaking trajectory emanating from x_0 follows from the weaker hypothesis that*

$$(x, v) \neq (\overline{x}, 0) \Rightarrow \mathcal{L}_0(x, v) < 0. \tag{4.93}$$

Actually this inequality would lead to a proof of Lemma 4.6 without the assumption of strict concavity of $\mathcal{L}(\cdot,\cdot)$ with respect to x. The proof would be very similar to the proof of Theorem 4.1 (see [28]). We will generalize this in the next section.

Under Assumption 4.4 alone and without further assumptions like (4.93) or the more general one given in the next section, it is not possible to show that a solution to the APL is overtaking. Only weak overtaking optimality can be guaranteed. This is because it no longer follows from

$$\int_0^\infty \mathcal{L}_0(x(t),\, \dot{x}(t))\, dt > -\infty$$

that $x(t)$ tends to $\overline{x}$ as t tends to infinity. However a weaker convergence lemma can be established

Lemma 4.7 *Let Assumption 4.4 hold, and suppose that the bounded trajectory $x(\cdot)$ is such that*

$$\liminf_{T\to\infty} \int_0^T \Big[\mathcal{L}(x(t),\,\dot{x}(t)) - \mathcal{L}(\overline{x},0)\Big]\,dt > -\infty. \tag{4.94}$$

Then

$$\lim_{t\to\infty} \frac{1}{T}\int_0^T x(t)\,dt = \overline{x}. \tag{4.95}$$

Proof: Since $x(\cdot)$ is a trajectory $\mathcal{L}(x(t),\,\dot{x}(t))$ is finite for almost all t. By (4.93) and by the concavity of $\mathcal{L}(\cdot,\cdot)$, there exists a real number A such that

$$\begin{aligned}\mathcal{L}\Big(\frac{1}{T}\int_0^T x(t)\,dt,\,\frac{1}{T}\int_0^T \dot{x}(t)\,dt\Big) - \mathcal{L}(\overline{x},0) &\geq \frac{1}{T}\int_0^T \Big[\mathcal{L}(x(t),\,\dot{x}(t)) - \mathcal{L}(\overline{x},0)\Big]\,dt \\ &\geq \frac{A}{T} \text{ for every } T > 0.\end{aligned} \tag{4.96}$$

Thus any cluster point $(\hat{x},0)$ of the set

$$\Big\{\Big(\frac{1}{T}\int_0^T x(t)\,dt,\,\frac{x(T)-x(0)}{T}\Big)\,:\,T>0\Big\}$$

maximizes $x \to \mathcal{L}(x,0)$ since, by (4.96) $\mathcal{L}(\hat{x},0) \geq \mathcal{L}(\overline{x},0)$ and, by definition $\mathcal{L}(\,\cdot\,,0)$ is maximued by $\overline{x}$. Since it is a unique maximizer we conclude that

$$\hat{x} = \overline{x}$$

for every cluster point of the set

$$\Big\{\frac{1}{T}\int_0^T x(t)\,dt\,:\,T \geq 0\Big\}.$$

Since $x(t)$ remains in a compact set X, there are cluster points and the lemma is proved.

■

Theorem 4.9 *Assume the hypotheses of Theorem 4.7, except replace Assumption 4.5 by Assumption 4.4. Then there exists a weakly overtaking optimal trajectory emanating from x_0.*

Proof: Let $x^*(\cdot)$ be the solution to the APL at x_0. Assuming that $x^*(\cdot)$ is not weakly overtaking optimal, then there exists another trajectory $x(\cdot)$ emanating from x_0 such that for certain $\varepsilon > 0$ and T_0

$$\int_0^T \Big[\mathcal{L}(x^*(t), x^*(t)) - \mathcal{L}(x(t), \dot{x}(t))\Big]\,dt \leq -\varepsilon \tag{4.97}$$

holds for every $T > T_0$.

Since $x^*(\cdot)$ solves the APL and by (4.89) there exists T_1 such that

$$\int_0^T \left[\mathcal{L}(x^*(t), \dot{x}^*(t)) - \mathcal{L}(x(t), \dot{x}(t))\right] dt > -\frac{\varepsilon}{2} - \overline{p}'(x^*(T) - x(T)) \tag{4.98}$$

holds for every $T > T_1$.

Take $T_2 = \max\{T_0, T_1\}$ and obtain from (4.97) and (4.98) that, for $T > T_2$

$$\overline{p}'(x^*(T) - x(T)) \geq \frac{\varepsilon}{2}. \tag{4.99}$$

It follows from (4.97) that the trajectory $x(\cdot)$ satisfies (4.94) since $x^*(\cdot)$ satisfies (4.94) as the solution to the APL. We thus conclude from Lemma 4.7

$$\lim_{T\to\infty} \frac{1}{T} \int_{T_2}^T \overline{p}'(x^*(t) - x(t))\, dt = 0,$$

which contradicts (4.99) and concludes the proof of the theorem. ∎

4.9 Overtaking optimality under relaxed assumptions

Leizarowitz has shown in [122], that the boundedness assumptions on $x(\cdot)$ as well as the strict concavity assumptions on $\mathcal{L}(\cdot,\cdot)$ can be removed. In this section, we briefly summarize these results. For simplicity we adhere to the calculus of variation format and consider the equivalent Lagrange problem as given in equation (4.7) and (4.8). Throughout, we make the following assumption.

Assumption 4.7 *The function $\mathcal{L}(\cdot,\cdot) : \mathbb{R}^n \times \mathbb{R}^n \to \mathbb{R}$ is upper semicontinuous and concave, and has the following coercivity condition:*

$$\mathcal{L}(x,v)/[\,\|x\| + \|v\|\,] \to -\infty \text{ as } \|x\| + \|v\| \to \infty. \tag{4.100}$$

We also assume that the set

$$\mathcal{D} = \{(x,v) \in \mathbb{R}^n \times \mathbb{R}^n : \mathcal{L}(x,v) > -\infty\} \tag{4.101}$$

is closed. (Observe that if $x(\cdot)$ is an admissible trajectory then $(x(t), \dot{x}(t)) \in \mathcal{D}$ a.e. in $[0, \infty)$.) We further assume that there exists at least one bounded trajectory $x(\cdot)$ such that $\mathcal{L}(x(t), \dot{x}(\cdot)) > -\infty$ for a.e. $t \geq 0$.

The above hypothesis allows us to prove the following.

Proposition 4.1 *If Assumption 4.7 holds then there exists $\overline{x} \in \mathbb{R}^n$ such that $(\overline{x}, 0) \in \mathcal{D}$.*

Proof: Let $x(\cdot) : [0, \infty) \to \mathbb{R}^n$ be a bounded trajectory satisfying $\mathcal{L}(x(t), \dot{x}(t)) > -\infty$ a.e. $t \geq 0$. Then the convexity of $\mathcal{D}$ implies that for all $T > 0$

$$\Big(\frac{1}{T}\int_0^T x(t)dt, \frac{1}{T}\int_0^T \dot{x}(t)dt\Big) \in \mathcal{D}.$$

It follows from the boundedness of $x(t)$ that there exists a sequence $\{T_j\}_{j=1}^{\infty}, T_j \to \infty$ such that

$$\lim_{j\to\infty} \frac{1}{T_j}\int_0^{T_j} x(t)\, dt = \overline{x}.$$

for some $\overline{x} \in \mathbb{R}^n$. Moreover, this boundedness also implies

$$\lim_{j\to\infty} \frac{1}{T_j}\int_0^{T_j} \dot{x}(t)\, dt = \lim_{j\to\infty} \frac{1}{T_j}\,[x(T_j) - x_0] = 0,$$

yielding (since $\mathcal{D}$ in closed) $(\overline{x}, 0) \in \mathcal{D}$, as desired. ∎

The above proposition makes it plausible to assume further that

Assumption 4.8 *There exists a unique point* $(\overline{x}, 0)$ *such that*

$$\mathcal{L}(\overline{x}, 0) = \max\{\mathcal{L}(x, 0) : x \in \mathbb{R}^n\}, \tag{4.102}$$

and moreover, there exists a vector $\overline{p} \in \mathbb{R}^n$ *such that*

$$\mathcal{L}(x, v) \leq \mathcal{L}(\overline{x}, 0) - \overline{p}'v \tag{4.103}$$

holds for all $(x, v) \in \mathcal{D}$.

As in the previous section this hypothesis allows us to consider the APL as posed in Definition 4.2. As we shall see shortly, a special role will be played by the Von Neumann set (defined here in the (x, v) space)

$$\mathcal{G} = \{(x, v) \in \mathcal{D} : \mathcal{L}_0(x, v) = 0\}, \tag{4.104}$$

where $\mathcal{L}_0(x, v) = \mathcal{L}(x, v) - \mathcal{L}(\overline{x}, 0) + \overline{p}'v$. Clearly, $\mathcal{G}$ is a closed convex set. Moreover from the coercivity relation (4.99), it is easy to see that

$$\mathcal{L}_0(x, v)/[\,\|x\| + \|v\|\,] \to -\infty \text{ as } [\,\|x\| + \|v\|\,] \to \infty,$$

and thus $\mathcal{G}$ is necessarily bounded. In the results to be presented below, we assume that $\mathcal{G}$ has the following property S.

Definition 4.3 *Let $\mathcal{T}$ be the family of trajectories $x(\cdot) : [0, \infty) \to \mathbb{R}^n$ such that $(x(t), \dot{x}(t)) \in \mathcal{G}$ for all $t > 0$. We say that the set $\mathcal{G}$ has property S if $x(t) \to \overline{x}$ as $t \to \infty$ for all trajectories $x(\cdot) \in \mathcal{T}$ and the convergence is uniform in the trajectories $x(\cdot) \in \mathcal{F}$ (i.e. $\forall \varepsilon > 0$, there exists $T = T(\varepsilon) > 0$ such that for each $x(\cdot) \in \mathcal{F}$ $\|x(t) - \overline{x}\| < \varepsilon$ for all $t \geq T$).*

Finally, before presenting the existence result we need the following proposition concerning the compactness of admissible trajectories.

Proposition 4.2 *Let $\mathcal{L}(\cdot,\cdot) : \mathbb{R}^n \times \mathbb{R}^n \to \overline{\mathbb{R}}$ be an integrand which satisfies the coercivity relation $\mathcal{L}(x,v)/\|v\| \to -\infty$ as $\|v\| \to \infty$ uniformly in x. Then for every real number α and β, the set of absolutely continuous functions $x(\cdot) : [0,T) \to \mathbb{R}^n$ satisfying $\max_{0 \leq t \leq T} \|x(t)\| \leq \beta$ and $\int_0^T \mathcal{L}(x(t), \dot{x}(t))\, dt \geq \alpha$ forms a compact subset of the space of all continuous functions on $[0,T]$ with values in $\mathbb{R}^n$ with respect to the usual uniform convergence topology.*

This result represents a modern version of the classical Nagumo-Tonelli result in the calculus of variation [46]. In fact the above coercivity relation may be replaced by the condition that for some function $\varphi : \mathbb{R}^+ \to \mathbb{R}^+$ increasing to infinity we have

$$\mathcal{L}(x,v)/\varphi(\|v\|) \to -\infty$$

uniformly in x.

We now give the following generalization of Theorem 4.8.

Theorem 4.10 *Assume that Assumptions 4.7 and 4.8 both hold and that the Von Neumann set $\mathcal{G}$, defined by (4.104) has property S. If $\overline{x}$ is reachable from x_0, then there exists an overtaking optimal solution emanating from x_0.*

Proof: In the first part of this proof we establish that the APL has an optimal solution. As a consequence of the fact that $\overline{x}$ is reachable from x_0, it follows that

$$g(x_0) = \sup \Big(\int_0^\infty \mathcal{L}_0(x(t), \dot{x}(t))\, dt \Big) > -\infty,$$

where the supremum is taken over all admissible trajectories emanating from x_0. Let M be a negative number chosen so that $\mathcal{F}$, the set of all trajectories emanating from x_0 such that

$$\int_0^\infty \mathcal{L}_0(x(t), \dot{x}(t))\, dt \geq M,$$

is not empty. We claim that $\mathcal{F}$ is a collection of uniformly continuous and uniformly bounded functions of $[0, \infty)$. To establish the first claim suppose it is not true, then there exists sequences $\{T_k\}_{k=1}^{\infty}$ and $\{t_k\}_{k=1}^{\infty}$ such that $t_k > T_k$, $\lim_{k\to\infty} t_k - T_k = 0$, and trajectories $\{x_{k(\cdot)}\}$ such that

$$\|x_k(t_k) - x_k(T_k)\| \geq \varepsilon \tag{4.105}$$

for all $k \geq 1$ and for some $\varepsilon > 0$. From Jensen's inequality, it follows that

$$\begin{aligned}\int_{T_k}^{t_k} \mathcal{L}_0(x_k(t), \dot{x}_k(t))\, dt &\leq (t_k - T_k)\mathcal{L}_0\Big(\frac{1}{t_k - T_k}\int_{T_k}^{t_k} x_k(s)\, ds, \frac{x_k(t_k) - x_k(T_k)}{t_k - T_k}\Big) \\ &\leq \varepsilon\mathcal{L}_0\Big(\frac{1}{t_k - T_k}\int_{T_k}^{t_k} x_k(s)\, ds, \frac{x_k(t_k) - x_k(T_k)}{t_k - T_k}\Big)\Big/\Big\|\frac{x_k(t_k) - x_k(T_k)}{t_k - T_k}\Big\|.\end{aligned}$$

This last expression tends to $-\infty$ as $k \to \infty$ by the coercivity condition (4.100). This proves the uniform continuity of $\mathcal{F}$.

Further, the coercivity condition (4.100) implies $\mathcal{L}_0(x, v) \to -\infty$ as $\|x\| \to \infty$, uniformly in v. From this and the uniform continuity of $\mathcal{F}$ proved above, it follows that $\mathcal{F}$ is also uniformly bounded. Now let $\{x_k(\cdot)\}_{k=1}^{\infty} \subset \mathcal{F}$ be a maximizing sequence (i.e., $\lim_{k\to\infty} \int_0^{\infty} \mathcal{L}_0(x_k(t), \dot{x}_k(t))\, dt = g(x_0)$). Then by the uniform boundedness and continuity of $\{x_k(\cdot)\}_{k=1}^{\infty}$ there exists a function $x^*(\cdot)$ such that some subsequence, say still $\{x_k(\cdot)\}_{k=1}^{\infty}$, converges uniformly on every finite interval $[0, T]$. Proposition 4.2 guarantees that $x^*(\cdot)$ is a trajectory, and therefore we have by Fatou's Lemma

$$g(x_0) = \lim_{k\to\infty} \int_0^{\infty} \mathcal{L}_0(x_k(t), \dot{x}_k(t))\, dt \geq \int_0^{\infty} \mathcal{L}_0(x^*(t), \dot{x}^*(t))\, dt,$$

implying $x^*(\cdot)$ solves the APL.

We now prove $x^*(\cdot)$ is an overtaking optimal trajectory. From the definition of $\mathcal{L}_0(\cdot, \cdot)$, if $x(\cdot)$ is any trajectory emanating form x_0, we have

$$\begin{aligned}&\int_0^T \mathcal{L}(x^*(t), \dot{x}^*(t)) - \int_0^T \mathcal{L}(x(t), \dot{x}(t))\, dt \\ &= \int_0^T \mathcal{L}_0(x^*(t), \dot{x}^*(t)) - \mathcal{L}_0(x(t), \dot{x}(t))\, dt + \overline{p}'(x^*(T) - x(T)).\end{aligned} \tag{4.106}$$

Now, if $\int_0^{\infty} \mathcal{L}_0(x(t), \dot{x}(t))\, dt = -\infty$, then clearly both sides of (4.106) will be greater than ε for all T sufficiently large since $x(\cdot)$ and $x^*(\cdot)$ are bounded by assumption. This implies that $x^*(\cdot)$ overtakes $x(\cdot)$. Therefore, we assume $\int_0^{\infty} \mathcal{L}_0(x(t), \dot{x}(t)) > -\infty$, and further that

$x^*(\cdot)$ is not overtaking optimal. In this case there exists times $t_k \to \infty$ and $\varepsilon > 0$ such that

$$\|x(t_k) - \overline{x}\| \geq \varepsilon \text{ for all } k \geq 1. \tag{4.107}$$

Since $\mathcal{G}$ has property S, there exists some $t_0 \geq 0$ such that for every trajectory $s(\cdot) \in \mathcal{T}$ (i.e. satisfying $(s(t), \dot{s}(t)) \in \mathcal{G}$ for almost all $t \geq 0$) we have

$$|s(t) - \overline{x}| \leq \frac{\varepsilon}{2} \text{ for } t \geq t_0. \tag{4.108}$$

Define for $t_k \geq t_0$, the functions $x_k(\cdot) : [0, \infty) \to \mathbb{R}^n$ by $x_k(t) = x(t_k - t_0 + t)$. Then the sequence $\{x_k(\cdot)\}_{k=1}^{\infty}$ is clearly uniformly bounded and uniformly continuous on $[0, \infty)$. Therefore there exists a function $s(\cdot)$ and subsequence, say still $\{x_k(\cdot)\}$, which converges uniformly to $s(\cdot)$ on each compact interval of $[0, \infty)$. Moreover, for every $T > 0$

$$\lim_{k\to\infty} \int_0^T \mathcal{L}_0(x_k(t), \dot{x}_k(t))\, dt = \lim_{k\to\infty} \int_{t_k - t_0}^{t_k - t_0 + T} \mathcal{L}_0(x(t), \dot{x}(t))\, dt = 0$$

and therefore by Proposition 4.2 we have

$$\int_0^T \mathcal{L}_0(s(t), \dot{s}(t))\, dt = 0 \text{ for all } T > 0.$$

This implies that $(s(t), \dot{s}(t)) \in \mathcal{G}$ for almost all $t > 0$ and so from (4.108) we have $\|s(t_0) - \overline{x}\| < \frac{\varepsilon}{2}$. Hence, for k sufficiently large $\|x(t_k) - \overline{x}\| = \|x_k(t_0) - \overline{s}\| < \varepsilon$ for all k sufficiently large. This contradicts (4.107) and therefore we must have that $x^*(\cdot)$ is overtaking optimal. ∎

Remark 4.9 *The above result is similar to Theorem* 4.8 *except for the fact that in the earlier result it is assumed that the admissible states are confined to lie in a compact set* X *and that the set* $\mathcal{G}$ *is the singleton* $\{(\overline{x}, 0)\}$. *The above result generalizes Theorem* 4.8 *since it allows* $\mathcal{G}$ *to be larger and permits the set of admissible states to be unbounded.*

We now present one simple example from [122] where the result is applicable but Theorem 4.8 is not.

Example 4.3 Define $\mathcal{L}(\cdot,\cdot) : \mathbb{R}^n \times \mathbb{R}^n \to \overline{R}$ by the following

$$\mathcal{L}(x, v) = \begin{cases} -x^2 & \text{if } (x, v) \in [0,1] \times [0,1] \\ \infty & \text{otherwise.} \end{cases}$$

In this case it is easy to see that

(i) $\overline{x} = 0$ and $\overline{p} = 0$, and

(ii) $\mathcal{G} = \{(0, v) : -1 \leq v \leq 1\}$.

In addition, the only trajectory $x(\cdot) : [0, \infty) \to \mathbb{R}$ satisfying $(x(t), \dot{x}(t)) \in \mathcal{G}$ is $x(t) \equiv 0$, $0 \leq t$, and so $\mathcal{G}$ clearly has property S. The remaining hypothesis of Theorem 4.10 are easily seen to be satisfied, and so the above problem has an overtaking optimal solution, $x^*(\cdot)$, for every initial value x_0. In fact, $x^*(\cdot)$ is given by

$$\dot{x}^*(t) = \begin{cases} -\operatorname{sgn} x^*(t) & \text{if } x^*(t) \neq 0 \\ 0 & \text{if } x^*(t) = 0. \end{cases}$$

The hypothesis that the set $\mathcal{G}$ has property S is crucial to the proof of Theorem 4.10. The following examples, still from [122], further illuminate its significance.

Example 4.4 In this example we consider the general problem for the scalar case (i.e., $\mathcal{L}(\cdot,\cdot) : \mathbb{R}^n \times \mathbb{R}^n \to \overline{\mathbb{R}}$ satisfying Assumption 4.7). We assume that there exists $\overline{x} \in \mathbb{R}$ and $\overline{p} \in \mathbb{R}^1$ satisfying (4.102) and (4.103). We establish the following fact:

If $\overline{x}$ can be reached from x_0 in finite time, then there exists an overtaking optimal trajectory $x^*(\cdot)$ with $x^*(0) = x_0$.

As a result of Theorem 4.10, it is sufficient to show that the set $\mathcal{G}$ satisfies property S. To see this we notice that $\mathcal{G}$ is a compact convex subset of $\mathbb{R}^2$ and we have $(\overline{x}, 0) \in \mathcal{G}$. There are two cases to consider.

Case 1. *$\mathcal{G}$ has empty interior.* In this case $\mathcal{G}$ is a line segment containing the point $(\overline{x}, 0)$. If this line has nonnegative slope or is vertical, then the only trajectory satisfying $(x(t), \dot{x}(t)) \in \mathcal{G}$ for all $t \geq 0$ is $x(t) \equiv \overline{x}$. If this line has negative slope, then all trajectories satisfying $(x(t), \dot{x}(t) \in \mathcal{G}$ for all $t > 0$ converge to $\overline{x}$ exponentially and uniformly as $t \to \infty$. To see this we notice that in this case, if $(x(t), \dot{x}(t)) \in \mathcal{G}$ there exists $\alpha < 0$

$$\dot{x}(t) = \alpha(x(t) - \overline{x})$$

which implies,

$$x(t) = (x_0 - \overline{x})e^{\alpha t} + \overline{x},$$

giving us the desired result. That is, $\mathcal{G}$ has property S.

Case 2. *$\mathcal{G}$ has nonempty interior.* In this case either $\mathcal{G}/\{(\overline{x}, 0)\}$ is contained in the open upper half plane or it is contained in the lower half plane of the (x, v) plane (recall $(\overline{x}, 0)$ is the only point of this form contained in $\mathcal{G}$). For brevity we consider only the former case. Then it is easy to see that there exists a nonnegative, continuous, convex function $\alpha(\cdot) : \mathbb{R} \to \mathbb{R}$ such that

(i) $\alpha(z) = 0$ if and only if $x = \overline{x}$,

and
(ii) for each $(x, v) \in \mathcal{G}$ we have $v \geq \alpha(x)$.

From this it follows that if $(x(t), \dot{x}(t)) \in \mathcal{G}$, then

$$\dot{x}(t) \geq \alpha(x(t))$$

which implies $x(\cdot)$ is strictly increasing, unless $x(t) \equiv \overline{x}$. If for some time $t > 0$, $x(t) > \overline{x}$ it follows that $x(t)$ increases without bound and must therefore leave the compact set $\mathcal{G}$. Therefore $x(t) \leq \overline{x}$ for all t which implies $\lim_{t \to \infty} x(t) = \overline{x}$ since $x(\cdot)$ is increasing. The other case is handled similarly. Thus we see that again $\mathcal{G}$ has property S and so the desired conclusion follows from Theorem 4.10.

The above example shows that in the scalar case, the existence of $\overline{x}$ and $\overline{p}$, along with the usual growth and upper semicontinuity assumptions, is sufficient to guarantee the existence of an overtaking optimal solution. The next example shows this is not true in higher dimensions.

Example 4.5 In this example we let $x = (x_1, x_2) \in \mathbb{R}^2$ and $v = (v_1, v_2) \in \mathbb{R}^2$, and define $\mathcal{L}(\cdot, \cdot) : \mathbb{R}^2 \times \mathbb{R}^2 \to \overline{\mathbb{R}}$ be defined by

$$\mathcal{L}(x, v) = \begin{cases} -(x_1 - v_2)^2 - (x_2 + v_1)^2 - \overline{p}'v & \text{if } (x, v) \in D_1 \times D_2 \\ -\infty & \text{otherwise} \end{cases}$$

where $D_1 = \{x \in \mathbb{R}^2 : |x| \leq 1\}$ and $D_2 = \{v \in \mathbb{R}^2 : v_1^2 \leq 1,\ v_2^2 \leq 1\}$, and $\overline{p} = (2, 0) \in \mathbb{R}^2$.

In the above situation, it is easy to see that $\overline{x} = 0$ and $\overline{p}$ as above satisfy equations (4.101) and (4.102). This implies

$$\mathcal{L}_0(x, v) = \begin{cases} -(x_1 - v_2)^2 - (x_2 + v_1)^2 & \text{if } (x, v) \in D_1 \times D_2 \\ -\infty & \text{otherwise.} \end{cases}$$

Now consider the initial condition $x_0 = \begin{pmatrix} 1 \\ 0 \end{pmatrix}$. We claim that if the above problem has an overtaking optimal trajectory emanating from x_0, then it is given by

$$x^*(t) = \begin{pmatrix} \cos(t) \\ \sin(t) \end{pmatrix}.$$

To see this suppose there exists an overtaking optimal trajectory $x(\cdot)$, different from $x^*(\cdot)$. Then $\mathcal{L}_0(x(t), \dot{x}(t))$ is not identically zero and hence for any $T > 0$

$$\begin{aligned} D(T) &= \int_0^T \mathcal{L}(x(t), \dot{x}(t)) - \mathcal{L}(x^*(t), \dot{x}^*(t))\, dt \\ &= \int_0^T \mathcal{L}_0(x(t), \dot{x}(t))\, dt + 2[\cos(T) - x_1(T)]. \end{aligned}$$

For each $k = 1, 2, \ldots$, define $T_k = (2k+1)\pi$. Then we have

$$\begin{aligned} D(T_k) &= \int_0^{T_k} \mathcal{L}_0(x(t), \dot{x}(t))\, dt - 2(1 + x_1(T_k)) \\ &< -2(1 + x_1(T_k)) \\ &\leq 0. \end{aligned}$$

This implies $\liminf_{T\to\infty} D(T) < 0$, which contradicts the overtaking optimality of $x(\cdot)$.

We now show that $x^*(\cdot)$ is not overtaking optimal. To see this we define the admissible trajectory

$$\xi(t) = \begin{cases} \begin{pmatrix} 1-t \\ 0 \end{pmatrix} & \text{for } 0 \leq t \leq 1 \\ \begin{pmatrix} 0 \\ 0 \end{pmatrix} & \text{for } 1 \leq t, \end{cases}$$

and observe that for all $T \geq 1$,

$$\int_0^T \mathcal{L}(x^*(t), \dot{x}^*(t))\, dt - \int_0^T \mathcal{L}(\xi(t), \dot{\xi}(t))\, dt - 2(\cos(T) - \xi_1(T)) = \frac{4}{3} - 2\cos(T).$$

By taking $T_k = 2k\pi$, we get for $k = 1, 2, \ldots$,

$$\int_0^{T_k} \mathcal{L}(x^*(t), \dot{x}^*(t))\, dt - \int_0^{T_k} \mathcal{L}(\xi(t), \dot{\xi}(t))\, dt = -\frac{2}{3} < 0$$

which implies

$$\liminf_{T\to\infty} \int_0^T \mathcal{L}(x^*(t), \dot{x}^*(t)) - \mathcal{L}(\xi(t), \dot{\xi}(t))\, dt < 0,$$

so that $x^*(\cdot)$ is not overtaking optimal.

In the above example it is easy to see that, except for property S, all the hypothesis of Theorem 4.10 are satisfied. This implies that property S is essential.

In [121] (also stated in [122]) a set of conditions which assure that property S holds are given. These conditions concern the asymptotic stability properties of the differential inclusion

$$\dot{x}(t) \in G(x(t)) \text{ a.e. on } [0, \infty), \tag{4.109}$$

where $G(x) = \{v : (x, v) = \mathcal{G}\} = \{v : \mathcal{L}_0(x, v) = 0\}$.
Observe that the graph of $G(\cdot)$ is precisely the set $\mathcal{G}$. To give the above mentioned result we need the following definition.

Definition 4.4 *For a differential inclusion of the type described by* (4.109) *we shall say that* $x(\cdot) : [0, \infty) \to \mathbb{R}^n$ *is an elliptic solution for* $G(\cdot)$ *if there exists a scalar* $\alpha \neq 0$ *and* $a, b, \in \mathbb{R}^n$ *satisfying* $\|a\| + \|b\| \neq 0$ *such that for all* $t > 0$,

$$x(t) = a\cos(\alpha t) + b\sin(\alpha t) \tag{4.110}$$

and

$$(x(t), \dot{x}(t)) \in \mathcal{G}. \tag{4.111}$$

We now state the following result.

Theorem 4.11 *Let $x \to G(x)$ be a set valued function having a convex, compact graph $\mathcal{G} = \{(x, v) : v \in G(x)\}$, which satisfies*

$$(x, 0) \in \mathcal{G} \text{ if and only if } x = \overline{x}, \tag{4.112}$$

for some fixed $\overline{x} \in \mathbb{R}^n$. If $G(\cdot)$ has no elliptic solutions, then $\mathcal{G}$ has property S.

Proof: See [121] Theorem 3.2. ∎

Notes and Comments

1) When the constraint on the control depends on the state x e.g. $U(x) \triangleq \{u \in \mathbb{R}^n : h(x(\cdot), u(\cdot))\} \geq 0$, it is convenient to define the optimal control in a feedback form

$$\hat{u}(x,t) \in U(x,t).$$

Along the optimal trajectory x^* the control is thus given by

$$u^*(t) = \hat{u}(x^*(t),t), \quad \forall t \geq 0.$$

Assume that $\hat{u}(\cdot,\cdot)$ is C^1 in x, $f_0(\cdot,\cdot)$ and $f_1(\cdot,\cdot)$ are C^1 in both x and u, then the Maximum Principle becomes: There exists $\mu \geq 0$ and $\lambda(\cdot)$ piecewise differentiable such that if one defines

$$H(x,u,\lambda(t),\mu) \triangleq \mu f_0(x(\cdot), u(\cdot)) + \lambda'(t) f(x(\cdot), u(\cdot))$$

then

$$H(x^*(t),u^*(t),\lambda(t),\mu) \geq H(x^*(t),u,\lambda(t),\mu) \quad \forall u \in U(x)$$

and

$$\dot{\lambda}(t) = -\left[\frac{\partial H}{\partial x} + \frac{\partial H}{\partial u}\frac{\partial \hat{u}}{\partial x}\right]\Bigg|_{\substack{x = x^*(t)\\ u = u^*(t)}}$$

(see G. Leitmann, An Introduction to Optimal Control, Mc Graw-Hill, 1966, for a derivation of this maximum principle).

Now, if the constraint qualification conditions are satisfied at $(x^*(t), u^*(t))$ then there exists a Lagrange multiplier $\nu(t) \geq 0$ such that

$$\begin{aligned} \left[\frac{\partial H}{\partial u} + \nu(t)'\frac{\partial h}{\partial u}\right]\Bigg|_{\substack{x = x^*(t)\\ u = u^*(t)}} &= 0 \\ \nu(t)'h(x^*(t),u^*(t)) &= 0. \end{aligned}$$

This last equality can be rewritten

$$\nu(t)'h\Big(x^*(t), \hat{u}(x^*(t)), t\Big) = 0$$

which shows that the differentiable function of x

$$\nu(t)'h(x,\hat{u}(x),t)$$

reaches a minimum at $x^*(t)$. Therefore,

$$0 = \left[\nu(t)'\frac{\partial h}{\partial x} + \nu(t)'\frac{\partial h}{\partial u}\frac{\partial \hat{u}}{\partial x}\right]\Bigg|_{\substack{x = x^*(t)\\ u = u^*(t)}}.$$

Now let us introduce the maximized Hamiltonian

$$\mathcal{H}(x,\lambda) \triangleq \max_{u\in U(x)} H(x,u,\lambda,1)$$

Assume that there exists a unique solution of the maximization of the RHS, yielding a function $\vec{u}(x,\lambda)$ which is C^1 in both x and λ. Thus

$$\mathcal{H}(x,\lambda) = H(x,\vec{u}(x,\lambda),\lambda,1).$$

We want to show that, along the optimal trajectory $x^*(\cdot)$ we also have

$$\dot{\lambda}(t)' = -\frac{\partial \mathcal{H}'}{\partial x}\bigg|_{\substack{x = x^*(t)\\ \lambda = \lambda(t)}}.$$

The optimal control $u^*(t)$ is given by

$$u^*(t) = \hat{u}(x^*(t),t) = \vec{u}(x^*(t),\lambda(t))$$

and therefore the complementarity condition

$$\nu(t)'h(x^*(t),u^*(t)) = 0$$

shows that, for $\lambda(t)$ kept fixed, the differentiable function of x

$$\nu(t)'h\Big(x,\vec{u}(x,\lambda(t))\Big)$$

reaches a minimum at $x^*(t)$. Therefore,

$$\left[\nu(t)'\frac{\partial h}{\partial x} + \nu(t)'\frac{\partial h}{\partial u}\frac{\partial \vec{u}}{\partial x}\right]\bigg|_{\substack{x = x^*(t)\\ u = u^*(t)\\ \lambda = \lambda(t)}} = 0$$

and hence

$$\begin{aligned}
\dot{\lambda}(t)' &= -\left[\frac{\partial H'}{\partial x} + \nu(t)'\frac{\partial h}{\partial x}\right]\bigg|_{\substack{x = x^*(t)\\ u = u^*(t)}} = 0 \\
&= -\left[\frac{\partial H'}{\partial x} - \nu(t)'\frac{\partial h}{\partial u}\frac{\partial \vec{u}}{\partial x}\right]\bigg|_{\substack{x = x^*(t)\\ u = u^*(t)\\ \lambda = \lambda(t)}} \\
&= -\left[\frac{\partial H'}{\partial x} + \frac{\partial \mathcal{H}}{\partial u}\frac{\partial \vec{u}}{\partial x}\right]\bigg|_{\substack{x = x^*(t)\\ u = u^*(t)\\ \lambda = \lambda(t)}} \\
&= -\frac{\partial \mathcal{H}'}{\partial x}\bigg|_{\substack{x = x^*(t)\\ \lambda = \lambda(t)}}.
\end{aligned}$$

Now look at the derivative

$$\frac{\partial \mathcal{H}}{\partial \lambda}\Bigg|_{\substack{x = x^*(t) \\ \lambda = \lambda(t)}} = \left[\frac{\partial H}{\partial \lambda} + \frac{\partial H}{\partial u}\frac{\partial \vec{u}}{\partial \lambda}\right]\Bigg|_{\substack{x = x^*(t) \\ u = u^*(t) \\ \lambda = \lambda(t)}}$$

A repetition of the same argument would show that

$$\left[\nu(t)'\frac{\partial h}{\partial u}\frac{\partial \vec{u}}{\partial \lambda}\right]\Bigg|_{\substack{x = x^*(t) \\ u = u^*(t) \\ \lambda = \lambda(t)}} = 0$$

and therefore

$$\left[\frac{\partial H}{\partial u}\frac{\partial \vec{u}}{\partial \lambda}\right]\Bigg|_{\substack{x = x^*(t) \\ u = u^*(t) \\ \lambda = \lambda(t)}} = 0$$

which yields

$$\frac{\partial \mathcal{H}}{\partial \lambda}\Bigg|_{\substack{x = x^*(t) \\ \lambda = \lambda^*(t)}} = \left[\frac{\partial H}{\partial \lambda}\right]\Bigg|_{\substack{x = x^*(t) \\ u = u^*(t) \\ \lambda = \lambda(t)}}$$

Finally the optimal state-costate trajectories are solutions of the Hamiltonian system

$$\begin{aligned} \dot{x} &= \frac{\partial \mathcal{H}}{\partial \lambda}, \\ \dot{\lambda} &= -\frac{\partial \mathcal{H}}{\partial x}. \end{aligned}$$

Notice that this lengthy, but simple, derivation of the Hamiltonian form for the maximum principle is based on very restrictive assumptions which make the maximum principle equilvalent to a Dynamic Programming equation.

Under convexity assumptions, the Hamiltonian system formulation is obtained under much weaker differentiability assumptions as is noticed in Section 4.8.

2) This proof has been given by Makato Yano, in a Note on the Existence of an Optimal Capital Accumulation, *Journal of Economic Theory*, Vol. 27, pp. 421–429, 1981.

Consider a sequence $\{x^n(\cdot)\}_{n\in\mathbb{N}}$ converging to $x^*(\cdot)$ uniformly on every bounded interval. Consider an arbitrary interval $[0, T]$. We will show that $\dot{x}^n(\cdot)$ converges to $\dot{x}^*(\cdot)$ in μ-measure on $[0, T]$. More precisely we will show that:

$$\mu[t \in [0, T] : \|\dot{x}^n(t) - \dot{x}^*(t)\| > \varepsilon] \to 0 \text{ as } n \to \infty.$$

Denote $x_i^n(t)$ the ith component of $x^n(t)$ and use the norm:

$$\|\dot{x}^n(t) - \dot{x}^*(t)\| = \max_{i=1,2,\ldots,n} |\dot{x}_i^n(t) - \dot{x}_i^*(t)|.$$

Now consider the sets:

$$\begin{aligned} A_{in}^+ &\triangleq \{t \in [0,T] : \dot{x}_i^n(t) - \dot{x}_i^*(t) > \varepsilon\} \\ A_{in}^- &\triangleq \{t \in [0,T] : \dot{x}_i^n(t) - \dot{x}^*(t) < -\varepsilon\} \\ A_{in}^- &\triangleq A_{in}^+ \cup A_{in}^-. \end{aligned}$$

If there is no convergence in μ-measure then for $\sigma > 0$, for some i and infinitely many n's one has:

$$\mu[A_{in}] > \sigma$$

and therefore, one can extract a sequence $\{n_\ell\}_{\ell \in \mathbb{N}}$ such that either

$$\mu[A_{in_\ell}^+] > \sigma \quad \forall \ell$$

or

$$\mu[A_{in_\ell}^-] > \sigma \quad \forall \ell.$$

For brevety we assume the first case holds. Hence there exists an interval $[a,b]$ such that $b - a > \sigma'$ for some $\sigma' > 0$ and that: $[a,b] \subset A_{in_\ell}^+$ for infinitely many ℓ's. If t is in $[a,b]$ one has: $\dot{x}_i^{n_\ell}(t) - \dot{x}_i^*(t) \geq \varepsilon > 0$. Thus $\|x_i^{n_\ell}(b) - x_i^*(b) - x_i^{n_\ell}(a) + x_i^*(a)\| \geq \sigma'\varepsilon$ for all ℓ. This contradicts the uniform convergence of $x^n(\cdot)$ to $x_i^*(\cdot)$ in $[0,T]$. Thus $\dot{x}^n(t)$ converges to $\dot{x}^*(t)$ in μ-measure, so that one can choose a subsequence which converges to $x^*(t)$ almost everywhere.

Chapter 5

The Reduction to Finite Rewards

5.1 Introduction

This chapter proposes a unifying approach to be used when one is dealing with infinite horizon optimization problems with unbounded rewards. This approach which is called the *reduction to finite rewards* encompasses the turnpike property that was the main tool used in Chapter 4. It is also more general as it permits the consideration of problems with slowly decreasing discount rates, problem of tracking periodic trajectories and also problems which are non convex w.r.t. the state variable x. An important aspect of the method developed here is that it exploits a connection which exists between continuous and discrete-time control systems. This link is particularly attractive when one considers that many early results on the turnpike property have been obtained for discrete time economic growth models (see e.g. Gale [81]). Also the method establishes a link with the Dynamic Programming approach.

5.2 The Property $\mathcal{R}$

In Chapter 4 we considered the *associated problem of Lagrange* by introducing the following integrand function $\mathcal{L}_0(\cdot,\cdot) : \mathbb{R}^n \times \mathbb{R}^n \to \mathbb{R}$:

$$\mathcal{L}_0(x,v) = \mathcal{L}(x,v) - \overline{p}'v - \mu$$

where we denote $\mu = \max_{x \in \mathbb{R}^n} \mathcal{L}(x,0)$. With every trajectory $x(\cdot)$ on $[0,\infty)$ and for every two points of time $0 < t_1 < t_2 < \infty$ there is associated the reward expression

$$C(x(\cdot)\,; t_1, t_2) = \int_{t_1}^{t_2} \mathcal{L}(x(t), \dot{x}(t))\, dt,$$

and the analogous reward which corresponds to $\mathcal{L}_0(\cdot,\cdot)$:

$$C_0(x(\cdot)\,;t_1,t_2) = \int_{t_1}^{t_2} \mathcal{L}_0(x(t),\dot{x}(t))\,dt.$$

The following relation holds between these two expressions:

$$C(x(\cdot)\,;t_1,t_2) = C_0(x(\cdot)\,;t_1,t_2) + [\overline{p}'x(t_2) - \overline{p}'x(t_1)] + \mu(t_2 - t_1). \tag{5.1}$$

Thus the reward $C(x(\cdot)\,;t_1,t_2)$ is split into three terms:

1. The reward associated with a nonpositive integrand function.
2. The difference between the evaluation of a fixed function $p(\cdot) : \mathbb{R}^n \to \mathbb{R}$ at the two end states of the interval, $x(t_1)$ and $x(t_2)$. (In the present situation $p(x) = \overline{p}'x$ for some $\overline{p}' \in \mathbb{R}^n$.)
3. A term which is independent of the choice of the trajectory and is proportional to the length of the time interval.

While studying overtaking optimality one considers expressions of the form $C(x_1(\cdot)\,;0,T) - C(x_2(\cdot)\,;0,T)$ for large values of T. The difference between this and the expression $C_0(x_1(\cdot)\,;0,T) - C_0(x_2(\cdot)\,;0,T)$, which correspond to $\mathcal{L}(\cdot,\cdot)$ and $\mathcal{L}_0(\cdot,\cdot)$ respectively, is clearly $p(x_2(T)) - p(x_1(T))$, and is due to the second term above. It follows from the turnpike property that only those trajectories $x(\cdot)$ which satisfy $\lim_{t\to\infty} x(t) = \bar{x}$ are relevant as far as optimality considerations are concerned. Hence in the limit where T tends to infinity one may replace $C(x_1(\cdot)\,;0,T) - C(x_2(\cdot)\,;0,T)$ by $C_0(x_1(\cdot)\,;0,T) - C_0(x_2(\cdot)\,;0,T)$. However, the maximization of $\int_0^\infty \mathcal{L}_0(x(t),\dot{x}(t))\,dt$ is in the ordinary sense, since by nonpositivity of $\mathcal{L}_0(\cdot,\cdot)$, the expression $C_0(x(\cdot)\,;0,\infty)$ is well defined for every $x(\cdot)$ (possibly negative infinity), while there exist trajectories $x(\cdot)$ for which $C_0(x(\cdot)\,;0,\infty) > -\infty$. We restrict attention to bounded trajectories (which is natural in the presence of the turnpike property) and define the *modified reward flow*, using the constant μ in (5.1)

$$C_T(x(\cdot)) = C(x(\cdot)\,;0,T) - \mu T. \tag{5.2}$$

It has the following property:

Property $\mathcal{R}$ *If $x(\cdot)$ is a bounded trajectory then either $T \to C_T(x(\cdot))$ is a bounded function on $[0,\infty)$, or $C_T(x(\cdot)) \to -\infty$ as $T \to \infty$. Moreover, there exist trajectories $x(\cdot)$ for which $T \to C_T(x(\cdot))$ is bounded on $[0,\infty)$.*

The above statement follows easily from (5.1), the nonpositivity of $\mathcal{L}_0(\cdot,\cdot)$, and the fact that there are trajectories $x(\cdot)$ for which $T \to C_0(x(\cdot)\,;0,T)$ is bounded. Clearly those are better than the others. Moreover, the boundedness of $T \to C_T(x(\cdot))$ provides several natural ways to compare trajectories. Thus, e.g., one may define a functional

$$\Phi(x(\cdot)) = \liminf_{T\to\infty} C_T(x(\cdot))$$

and consider the maximization of $\Phi(x(\cdot))$ over all trajectories $x(\cdot)$ which satisfy $x(0) = x_0$ for a prescribed initial value x_0.

We have the following uniqueness result for the constant μ:

Proposition 5.1 *There exists at most one constant μ for which Property $\mathcal{R}$ holds.*

Proof: Suppose that Property $\mathcal{R}$ holds for μ and let $\mu' < \mu$. Let $C_T(x(\cdot))$ be as in (5.2) and choose a trajectory $x(\cdot)$ such that $T \to C_T(x(\cdot))$ is bounded on $[0,\infty)$. Then the function $T \to C(x(\cdot), 0, T) - \mu' T$ will diverge to infinity when T tends to infinity, implying that Property $\mathcal{R}$ does not hold for μ'.

If however, $\mu' > \mu$ then it is easy to see that $C(x(\cdot); 0, T) - \mu' T$ diverges to $-\infty$ as T tends to infinity, for every trajectory $x(\cdot)$. Therefore Property $\mathcal{R}$ does not hold in this case too, and the proof is complete. ∎

The existence of such a constant is, in our opinion, important enough to deserve a name, and we introduce the following terminology:

Definition 5.1 *Consider a reward flow $T \to C(x(\cdot); 0, T)$ defined on a collection of bounded trajectories. If there exists a constant μ such that the modified reward flow $T \to C_T(x(\cdot))$ in (5.2) has the Property $\mathcal{R}$ then we say that we achieved a reduction to finite rewards.*

The approach of introducing the APL while considering the maximization of $\int_0^T \mathcal{L}(x, \dot{x})dt$ as $T \to \infty$ will be extended to more general situations. The reduction to finite rewards constitutes a first step. When this is achieved then we still have to determine a function $p(\cdot) : \mathbb{R}^n \to \mathbb{R}$ such that the following will hold for every $x(\cdot)$:

$$C(x(\cdot); t_1, t_2) = C_0(x(\cdot); t_1, t_2) + p(x(t_2)) - p(x(t_1)) + \mu(t_2 - t_1) \tag{5.3}$$

for some nonpositive reward flow $C_0(x(\cdot); t_1, t_2)$. Once a relation (5.3) is established we can concentrate on maximizing $C_0(x(\cdot); 0, \infty)$ in the ordinary sense. In the presence of turnpike properties this will lead to optimal (in the overtaking sense) trajectories for the original functional $C(x(\cdot); t_1, t_2)$.

Rather than considering the continuous-time framework we will follow a similar program for discrete-time control systems. The analogous results for such systems and their relevance to the study of continuous-time systems are described in the next section.

5.3 The connection between continuous and discrete time control systems

We start by describing the discrete-time version of the problem considered in Section 5.1. A function

$$s(\cdot,\cdot) : \mathbb{R}^n \times \mathbb{R}^n \to \mathbb{R} \cup \{-\infty\}$$

is given, with the interpretation that $s(x,y)$ is the reward for steering the system from state $x \in \mathbb{R}^n$ to state $y \in \mathbb{R}^n$ in a fixed time interval $[0, T_0]$. We call $s(\cdot,\cdot)$ *a transition function.* The control of the system is exercised by choosing a *program* $\mathbf{x} = \{x_k\}_{k=0}^{\infty}$ which is a bounded sequence in $\mathbb{R}^n$. With every program $\mathbf{x}$ there is associated a *reward flow*

$$C_N(\mathbf{x}) = \sum_{k=0}^{N-1} s(x_n, x_{n+1}) \tag{5.4}$$

and $C_N(\mathbf{x})$ is the reward gained while applying the program $\mathbf{x}$ on the $[0, NT_0]$ time interval. The typical situation is that $C_N(\mathbf{x}) \to -\infty$ as $N \to \infty$ for every program $\mathbf{x}$. Since ordinary maximization is not possible we wish to maximize $C_N(\mathbf{x})$ as $N \to \infty$ in another sense, e.g., using the overtaking optimality criterion.

As described in section 5.1 a major step toward maximization in the overtaking sense is the reduction to finite rewards, which we now redefine in the discrete-time framework:

Definition 5.2 *We say that we achieved a* **reduction to finite rewards** *for the reward flow* (5.4) *if there exists a constant* μ *with a* modified reward flow

$$\tilde{C}_N(\mathbf{x}) = C_N(\mathbf{x}) - \mu N$$

such that for a program $\mathbf{x}$ *either* $\tilde{C}_N(\mathbf{x}) \to -\infty$ *as* $N \to \infty$, *or* $\{\tilde{C}_N(\mathbf{x})\}_{N=1}^{\infty}$ *is a bounded sequence. Moreover, there are programs* $\mathbf{x}$ *for which* $\{\tilde{C}_N(\mathbf{x})\}_{N=1}^{\infty}$ *is bounded.*

Analogous to the structure of $\mathcal{L}(\cdot,\cdot)$ in Assumption 4.4, we consider in the present discrete time version a transition function $s(\cdot,\cdot)$ of the following form

$$s(x,y) = \mu + p(x) - p(y) + \theta(x,y) \tag{5.5}$$

where

(i) $p(\cdot) : \mathbb{R}^n \to \mathbb{R}$ is a continuous function.

(ii) μ is a constant.

(iii) $\theta(\cdot,\cdot) : \mathbb{R}^n \times \mathbb{R}^n \to \mathbb{R} \cup \{-\infty\}$ is nonpositive and continuous on the set $D = \{(x,y) : \theta(x,y) > -\infty\}$.

In virtue of (5.5) the modified reward flow $\tilde{C}_N = C_N - \mu N$ can be expressed as follows:

$$\tilde{C}_N(\mathbf{x}) = \sum_{k=0}^{N-1} \theta(x_k, x_{k+1}) + p(x_0) - p(x_N). \tag{5.6}$$

Clearly the right hand side of (5.6) is either bounded or tends to negative infinity, depending on whether or not $\sum_{n=0}^{\infty} \theta(x_k, x_{k+1})$ is finite. If $s(\cdot,\cdot)$ in (5.5) is such that $\sum_{n=0}^{\infty} \theta(x_k, x_{k+1}) > -\infty$ for some program $\mathbf{x}$ then there exists a reduction to finite rewards for $s(\cdot,\cdot)$.

The nonpositive transition function $\theta(\cdot,\cdot)$ defines naturally a discrete time dynamical system as follows: for every $x \in \mathbb{R}^n$ let

$$S(x) = \{y \in \mathbb{R}^n \, : \, \theta(x, y) = 0\} \tag{5.7}$$

and consider programs $\mathbf{x} = \{x_k\}_{k=0}^{\infty}$ which satisfy

$$x_{k+1} \in S(x_k), \quad k = 0, 1, 2 \ldots . \tag{5.8}$$

Turnpike properties of the infinite horizon optimization problem are related to stability properties of the dynamical system defined by (5.7) and (5.8). Namely, the asymptotic behavior of solutions of (5.8) may provide useful information about the overtaking optimization problem. For a program $\mathbf{x}$ the term $p(x_N)$ in (5.6) is bounded, hence the collection of programs $\mathbf{x}$ for which $\sum_{k=0}^{\infty} \theta(x_k, x_{k+1}) > -\infty$ is the class in which one would look for an optimal program. But then $\theta(x_k, x_{k+1}) \to 0$ as $k \to \infty$ and in certain situations one may deduce from this some convergence properties of $\mathbf{x}$.

To illustrate this idea suppose that there exists a $x \in \mathbb{R}^n$ such that

(1) $\theta(x, x) = 0$, and

(2) Every solution $\mathbf{x}$ of (5.8) satisfies $x_k \to x$ as $k \to \infty$, uniformly for initial values x_0 in bounded sets.

We claim that then the turnpike property holds, that is

Proposition 5.2 *Let $s(\cdot,\cdot)$ have the form* (5.5) *and suppose that $\theta(\cdot,\cdot)$ and x are such that* (1) *and* (2) *above hold. If $\mathbf{x}$ is a program for which $x_k \not\to x$ as $k \to \infty$ then $\tilde{C}_N(\mathbf{x}) \to -\infty$ as $N \to \infty$ (where, as above, $\tilde{C}_N(\mathbf{x}) = C_N(\mathbf{x}) - \mu N$).*

Proof: Since $\mathbf{x}$ is a program it is bounded, say $|x_k| \le B$ for some $B > 0$ and all $k \ge 0$. We assume that $\sum_{k=0}^{\infty} \theta(x_k, x_{k+1}) > -\infty$ and $x_k \not\to x$ as $k \to \infty$, say $|x_{k_j} - x| \ge \varepsilon$ for some $\varepsilon > 0$ and every $j \ge 1$. For every $j \ge 1$ we consider the finite sequence $\{y^{(j)}\}_{i=-k_j}^{0}$ defined by

$$(y^{(j)})_i = x_{i+k_j},$$

and construct a sequence $\{z_i\}_{i=-\infty}^{0}$ such that $z_i = \lim_{j\to\infty} (y^{(j)})_i$, for every $-\infty < i \le 0$. Then it is easy to see that $z_{i+1} = S(z_i)$ for every $i \le -1$, nevertheless $z_0 \ne x$. This, however, contradicts the uniform convergence to x of programs $\mathbf{x}$ with bounded starting points. ∎

Remark 5.1 *In the above discussion we* assumed *the rather special form* (5.5) *for the transition function* $s(\cdot,\cdot)$. *In Section* 5.5 *we will* prove *the existence of such a representation under a certain controllability assumption. In fact, in the framework of the subsequent sections the reduction to finite rewards will be proved to be equivalent to the possibility of representing* $s(\cdot,\cdot)$ *as in* (5.5), *with some program* $\mathbf{x}$ *satisfying* $\sum_{k=0}^{\infty} \theta(x_k, x_{k+1}) > -\infty$.

We conclude this section by describing how the discrete-time results may be applied to study continuous-time problems, which is our main concern. We will describe below three problems for which this procedure will prove to be fruitful.

Problem I: A nonlinear control system.

We consider the state equation

$$\begin{aligned} \dot{x}(t) &= f(x(t), u(t)) \\ x(0) &= x_0 \end{aligned} \tag{5.9}$$

with a reward flow

$$C_T(x(\cdot)) = \int_0^T f_0(x(t), u(t)).\,dt \tag{5.10}$$

Both f and f_0 are assumed to be continuous. The state function $x(\cdot)$ and the control function $u(\cdot)$ are subject to the constraints $x(t) \in X$ and $u(t) \in U$, where X and U are compact sets in $\mathbb{R}^n$ and $\mathbb{R}^m$ respectively. The class of admissible controls is composed of the measurable functions $u(\cdot)$ which satisfy $u(t) \in U$ for all $t \ge 0$ and whose responses $x(\cdot)$ (i.e. the corresponding solutions of (5.9)) satisfy $x(t) \in X$ for all $t \ge 0$. The goal is to "maximize" $C_T(x(\cdot))$ as $T \to \infty$.

Define the transition function $s(\cdot,\cdot) : X \times X \to \mathbb{R}$ by

$$s(y,z) = \inf\Big\{\int_0^1 f_0(x(t), u(t))\,dt\Big\} \tag{5.11}$$

where the infimum is over all controls $u(\cdot)$ such that $u(t) \in U$, $x(t) \in X$ holds for $0 \le t \le 1$, and moreover, $x(0) = y$ and $x(1) = z$. Consider the discrete-time reward flow

$$C_N(\mathbf{x}) = \sum_{k=0}^{N-1} s(x_k, x_{k+1}) \tag{5.12}$$

for a program $\mathbf{x} = \{x_k\}_{k=0}^{\infty} \subset X$. It is clear that a reduction to finite rewards is possible for (5.9), (5.10) if and only if there is such a reduction for the discrete-time system (5.11), (5.12), and then both reductions use the same constant μ. Moreover, one may try to use optimal programs $\mathbf{x}^*$ for (5.11), (5.12) to construct optimal solutions for (5.9), (5.10) as follows: for every $k > 0$ let $u^*(\cdot)$ be a control defined on $[k, k+1)$ with the response $x^*(\cdot)$ such that

$$x^*(k) = y,\ x^*(k+1) = z, \quad \text{and } s(y,z) = \int_k^{k+1} f_0(x^*(t), u^*(t))\, dt.$$

If indeed $u^*(\cdot)$ may be defined in this way for every $k \ge 0$, then it is an admissible control on $[0, \infty)$, and in certain situations it will turn out to be optimal.

Problem II: A tracking problem.

A linear, controllable, time invariant system

$$\begin{aligned} \dot{x}(t) &= Ax(t) + Bu(t), \\ x(0) &= x_0 \end{aligned} \tag{5.13}$$

is given, where $x(t) \in \mathbb{R}^n$, $u(t) \in \mathbb{R}^m$, and the admissible controls are the measurable functions $u(\cdot)$ on $[0, \infty)$ such that $\int_0^T \|u(t)\|^2\, dt < \infty$ for every $0 < T < \infty$. A periodic trajectory $\Gamma(\cdot)$ is prescribed and the purpose is to generate a trajectory $x(\cdot)$ of (5.13) which will be close to $\Gamma(\cdot)$. If a quadratic cost criterion is adopted then the cost incurred while applying the control $u(\cdot)$ on $[t_1, t_2]$ is

$$C_{t_1,t_2}(u(\cdot)) = \int_{t_1}^{t_2} \left\{ [x(t) - \Gamma(t)]' Q [x(t) - \Gamma(t)] + u' R u \right\} dt \tag{5.14}$$

where $u(\cdot)$ and $x(\cdot)$ satisfy (5.13), Q is positive semidefinite and R is positive definite. In this problem one tries to *minimize* the cost over the $[0, T]$ time interval for large T, in contrast to the situation elsewhere in this book, where the goal is to *maximize* over the infinite horizon. Typically, no matter how one chooses $u(\cdot)$, the cost $C_{0,T}(u(\cdot))$ will diverge to infinity as $T \to \infty$.

The associated discrete-time control problem is defined as follows. Let τ be the period of $\Gamma(\cdot)$ and, using the cost expression (5.14), we define $s(x, y)$ as the minimal value of steering the system from state x to state y within the time interval $[0, \tau]$. The goal is then to find optimal sequences $\mathbf{x}^*$ for the reward flow $C_N(\mathbf{x}) = \sum_{k=0}^{N-1} s(x_k, x_{k+1})$ as $N \to \infty$, and to use them, employing the procedure described in Problem I above, to construct optimal solutions for the continuous-time tracking problem. This problem was studied by Artstein and Leizarowitz [6].

Problem III: Problem of Lagrange with slowly decreasing discounting factor.

Given a continuous integrand function

$$(x, v) \to \mathcal{L}(x, v)$$

we consider the reward flow

$$C_T(x(\cdot)) = \int_0^T \varphi(t)\mathcal{L}(x(t), \dot{x}(t))\, dt$$

where $\varphi(\cdot)$ is a monotone decreasing funciton $\varphi : [0, \infty) \to (0, \infty)$ such that $\lim_{t\to\infty} \varphi(t) = 0$. We assume that $x(\cdot)$ and $\dot{x}(\cdot)$ are measurable and take values in compact sets. (See, e.g., Rockafellar [155] for the case where $\varphi(\cdot)$ is exponentially decreasing.)

The function $\varphi(\cdot)$ may decrease very slowly so that still $C_T(x(\cdot)) \to -\infty$ as $T \to \infty$, for every choice of an admissible trajectory $x(\cdot)$. We then look for a trajectory $x^*(\cdot)$ which will be optimal in the overtaking sense. This was considered by Leizarowitz [124].

To bring this problem under the discrete-time framework we define $s(y, z)$ as the maximal value of $\int_0^1 \mathcal{L}(x, \dot{x})dt$ over all the admissible trajectories $x(\cdot)$ which satisfy $x(0) = y$ and $x(1) = z$ (assuming that the maximal value is attained). Then we consider a discrete-time version of the problem where the reward flow is

$$C_N(\mathbf{x}) = \sum_{k=0}^{N-1} \alpha_k s(x_k, x_{k+1})$$

and $\alpha_k \to 0$, monotonically decreasing as $k \to \infty$.

5.4 Existence of a reduction to finite rewards

In this section, we will establish the existence of a reduction to finite rewards for discrete-time problems of the type which is described in Section 5.3. We will assume that the domain of $s(\cdot, \cdot)$ is "square shaped", so that the following holds:

Assumption I *The transition function $s(\cdot,\cdot)$ is upper-semicontinuous and is defined and bounded on $X \times X \subset \mathbb{R}^n \times \mathbb{R}^n$, where X is a compact subset of $\mathbb{R}^n$.*

Remark 5.2 *Since X is bounded, the programs are just sequences $\mathbf{x} = \{x_k\}_{k=0}^{\infty} \subset X$. By upper-semicontinuity we mean that*

$$s(\lim_{k\to\infty}(x_k, y_k)) \geq \limsup_{k\to\infty} s(x_k, y_k)$$

whenever $\{(x_k, y_k)\}_{k=0}^{\infty}$ is a convergent sequence.

The assumption that $s(x,y)$ is finite for $(x,y) \in X \times X$ is a controllability type assumption, which means that every state $x \in X$ can be steered to every state $y \in X$ in a fixed time interval.

Remark 5.3 *We may replace the compactness assumption that every program $\mathbf{x}$ is contained in X for some compact $X \subset \mathbb{R}^n$ by the condition*

$$\lim s(x,y) = -\infty \text{ as } |x| + |y| \to \infty.$$

One can then establish the existence of a compact set $X \subset \mathbb{R}^n$ with the property that for every sequence $\mathbf{x}$ which is not contained in X there is a program in X which overtakes $\mathbf{x}$. Thus from the optimality point of view we can confine ourselves to considering only sequences in X. (For the proof of this claim see [9], Theorem 8.1.)

Theorem 5.1 *There exist constants μ and $M > 0$ such that*

(1) *For every program $\{x_i\}_{i=1}^{\infty}$ and every integer $N \geq 0$ the inequality $\sum_{i=0}^{N} [s(x_i, x_{i+1}) - \mu] \leq M$ holds.*

(2) *For every program $\{x_i\}_{i=0}^{\infty}$ the sequence $\left\{\sum_{i=0}^{N} [s(x_i, x_{i+1}) - \mu]\right\}_{N=0}^{\infty}$ is either bounded or it diverges to $-\infty$. Moreover, the constant M can be chosen so that for every initial value x_0 there is a program $\{x_i^*\}_{i=0}^{\infty}$ with $x_0^* = x_0$ which satisfies*

$$\left|\sum_{i=0}^{N} [s(x_i^*, x_{i+1}^*) - \mu]\right| \leq M \text{ for all } N \geq 0.$$

To prove Theorem 5.1 we need the following three lemmas. Denote by $\lambda(N)$ the maximal average reward over all periodic programs of period N:

$$\lambda(N) = \max\left\{\frac{1}{N}\sum_{i=0}^{N-1} s(x_i, x_{i+1}) \,:\, x_0 = x_N\right\}. \tag{5.15}$$

Let μ be defined as the supremal growth-rate of reward flows over all programs, namely

$$\mu = \sup\left\{\limsup_{N\to\infty}\frac{1}{N}\sum_{i=0}^{N-1} s(x_i, x_{i+1}) \,:\, \{x_i\}_{i=0}^{\infty} \subset X\right\}. \tag{5.16}$$

Remark 5.4 *The quantity μ is a natural candidate to satisfy Theorem* 5.1. *In fact, if a quantity μ does satisfy Theorem* 5.1, *then it satisfies* (5.16) *too.*

The next lemma describes a close relation between the sequence $\{\lambda(N)\}$ in (5.15) and the constant μ defined in (5.16).

Lemma 5.1 *The following relation holds*

$$\mu = \sup_{N\geq 1} \lambda(N).$$

Proof: On the one hand, it is easy to see that $\mu \geq \lambda(N)$ for every N, and therefore $\mu \geq \sup_{N\geq 1} \lambda(N)$. On the other hand, given an $\varepsilon > 0$ there is a finite program with N arbitrarily large such that

$$\frac{1}{N}\sum_{i=0}^{N-1} s(x_i, x_{i+1}) > \mu - \varepsilon$$

and therefore

$$\frac{1}{N+1}\left[\sum_{i=0}^{N-1} s(x_i, x_{i+1}) + s(x_N, x_0)\right] > \mu - 2\varepsilon,$$

demonstrating that

$$\lambda(N+1) > \mu - 2\varepsilon,$$

so that

$$\sup_{N\geq 1} \lambda(N) \geq \mu.$$

■

In the sequel we will use freely the property $\mu = \sup_{N\geq 1} \lambda(N)$.

Lemma 5.2 *The sequence $\lambda(N)$ converges to μ.*

Proof: Since $s(x,y)$ is bounded so is the sequence $\{\lambda(N)\}_{N=1}^{\infty}$. Let $\alpha = \limsup_{N\to\infty} \lambda(N)$. We want to prove that $\liminf_{N\to\infty} \lambda(N) \geq \alpha$. Given $\varepsilon > 0$, there is an integer N such that $\lambda(N) > \alpha - \varepsilon$. Since every program which is periodic of period N is also periodic of period kN for every integer k, and as its average reward over a time interval of length kN is the same as that over a period of length N it follows from (5.15) that

$$\lambda(kN) \geq \lambda(N) \text{ for every } k \geq 1 \text{ and } N \geq 1. \tag{5.17}$$

Further denote

$$a = \max\{s(y,z) : (y,z) \in X \times X\}, \quad b = \inf\{s(y,z) : (y,z) \in X \times X\}. \tag{5.18}$$

Given any finite program $\{x_i\}_{i=0}^N$, replacing x_0 by $x_0' \in X$ may decrease the reward by at most $(a-b)$, since

$$s(x_0', x_1) - s(x_0, x_1) \geq -(a-b).$$

Similarly the replacement of x_N by any $x_N' \in X$ may decrease the reward by at most $(a-b)$. Let N', N_1, N_2 be integers such that $N' = N_1 + N_2$. Let $\{x_i\}_{i=0}^{N_1}$ be a program for which $\lambda(N_1)$ is attained and $\{y_i\}_{i=0}^{N_2}$ be one for which $\lambda(N_2)$ is attained. We replace y_0 and y_{N_2} by x_0 (which is equal to x_{N_1}), identify y_0 with x_{N_1} and form a finite program of length $N_1 + N_2 = N'$

$$z_i = \begin{cases} x_i & 0 \leq i \leq N_1 \\ y_{i-N_1} & N_1 < i < N_1 + N_2 \\ x_0 & i = N_1 + N_2. \end{cases}$$

Using this program, for which $z_0 = z_{N_1+N_2}$, we get $N'\lambda(N') \geq N_1\lambda(N_1) + N_2\lambda(N_2) - 2(a-b)$. In particular take $N' = kN + r$ where $0 \leq r < N$, then $N'\lambda(N') \geq (kN)\lambda(kN) + r\lambda(r) - 2(a-b)$. Dividing by N' we obtain

$$\lambda(N') \geq \frac{kN}{kN+r}\lambda(kN) + \frac{r\lambda(r) - 2(a-b)}{kN+r}.$$

For k large enough the second term in the last inequality is larger than $-\varepsilon$, and so, by (5.17) and the way we chose N we get

$$\lambda(N') \geq (\alpha + \varepsilon)\frac{kN}{kN+r} - \varepsilon > \alpha - 2\varepsilon$$

for all large k and every $1 \leq r \leq N$. Thus we conclude that $\liminf_{N\to\infty} \lambda(N) \geq \alpha$ since the preceding inequality is true for every $\varepsilon > 0$.

The inequality $\alpha \leq \mu$ follows from Lemma 5.1, while the inequality $\alpha \geq \mu$ follows from (5.17), hence $\alpha = \mu$ and we obtain $\lim_{N\to\infty} \lambda(N) = \mu$, concluding the proof. ∎

Lemma 5.2 gives a growth estimate to the reward of periodic programs. The next lemma gives a result which is finer than a mere growth-rate.

Lemma 5.3 *The following inequality holds:*

$$\liminf_{N\to\infty} N[\lambda(N) - \mu] > -\infty.$$

Proof: Denote by $r(N)$ the maximal average reward over all finite sequences of length N (without any restrictions on the end values x_0 and x_N):

$$r(N) = \max\left\{\frac{1}{N}\sum_{i=0}^{N-1} s(x_i, x_{i+1}) \; : \; \{x_i\}_{i=0}^N \subset X\right\}. \tag{5.19}$$

We claim that $r(N) \geq \mu$ for every $N \geq 1$. Suppose to the contrary that $r(N) < \mu$ for some N. Then for every integer k

$$\frac{1}{kN} \sum_{i=0}^{kN-1} s(x_i, x_{i+1}) \leq \frac{1}{k} (kr(N)) = r(N),$$

and this being true for every program $\{x_i\}$ implies $r(kN) \leq r(N)$. Also from the definitions of r and λ in (5.19) and (5.15) we obtain $r(kN) \geq \lambda(kN)$, so we get $\mu > r(N) \geq r(kN) \geq \lambda(kN)$ for all $k \geq 1$. Letting $k \to \infty$ we get a contradiction since $\lim_{k\to\infty} \lambda(kN) = \mu$ by Lemma 5.2.

Recall the definition of a and b in (5.18). Let $N \geq 1$ be an integer and let $\{x_i\}_{i=0}^{N}$ be a finite program for which $r(N)$ in (5.19) is attained. If we replace in this program x_N by x_0 we get a periodic program of period N whose reward in the interval $[0, N]$ is not smaller than that of the former by more than $a - b$, so we get for every $N \geq 1$

$$N\lambda(N) - Nr(N) \geq -(a - b).$$

Combining this with $r(N) \geq \mu$ leads to

$$\liminf_{N\to\infty} N[\lambda(N) - \mu] \geq \liminf_{N\to\infty} N[\lambda(N) - r(N)] \geq -(a - b). \qquad \blacksquare$$

Before proving Theorem 5.1 let us introduce an abbreviating notation. For $N \geq 1$ we denote

$$m_N(\mathbf{x}) = \sum_{i=0}^{N-1} [s(x_i, x_{i+1}) - \mu]$$

and call it *the modified reward* of $\mathbf{x}$. Note that the modified reward of a finite periodic program is nonpositive, since from $x_0 = x_N$ and $\mu = \sup_{N\geq 1} \lambda(N)$ it follows that $\frac{1}{N} \sum_{i=0}^{N-1} s(x_i, x_{i+1}) \leq \mu$, implying $m_N(x) \leq 0$.

Proof of Theorem 5.1: We will prove first the claim in (1), namely that for every program $\mathbf{x}$ and every $N \geq 1$ the inequality $m_N(x) < a - b$ holds (recall (5.18)). Given $\mathbf{x}$ and N we replace x_N by x_0 and thus decrease the modified reward on the interval $[0, N]$ by no more than $a - b$. After the replacement, we get a periodic program, whose modified reward is nonpositive, so we conclude that $m_N(\mathbf{x}) < a - b$ as claimed. Expressing this formally we have

$$\begin{aligned} m_N(\mathbf{x}) = & \left\{ \sum_{i=0}^{N-2} \left[s(x_i, x_{i+1}) - \mu\right] + \left[s(x_{N-1}, x_0) - \mu\right] \right\} \\ & + \left[s(x_{N-1}, x_N) - s(x_{N-1}, x_0)\right] \end{aligned} \tag{5.20}$$

where the first term is nonpositive, while the second is bounded by $a - b$.

We prove now the assertion made in (2). We only have to prove that for every initial value x_0 there exists a program $\mathbf{x}$ for which $\{m_N(\mathbf{x})\}_{N=1}^{\infty}$ is bounded, and that the bound is independent of the initial value. (It follows from (1) that if $\{m_N(\mathbf{x})\}_{N=1}^{\infty}$ is not bounded then it diverges to negative infinity.) Using Lemma 5.3 we choose $\gamma > 0$ so that for all $N \geq 1$ we have

$$N[\lambda(N) - \mu] > -\gamma. \tag{5.21}$$

Let $\{x_0^N, \ldots, x_N^N\}$ be chosen so that $x_0^N = x_N^N$ and $\lambda(N) = \frac{1}{N} \sum_{i=0}^{N-1} s(x_i^N, x_{i+1}^N)$. By (5.21) we get $\sum_{i=0}^{N-1} [s(x_i^N, x_{i+1}^N) - \mu] > -\gamma$ for all $N \geq 1$. We claim that if $\ell \geq N \geq 1$ then

$$\sum_{i=0}^{N-1} \left[s(x_i^\ell, x_{i+1}^\ell) - \mu\right] > -\gamma + b - a \tag{5.22}$$

namely, the modified reward of the finite programs $\{x_0^\ell, \ldots, x_\ell^\ell\}$ computed on the initial interval of length N has a bound, uniform for all ℓ, N satifying $1 \leq N \leq \ell$. Let us assume, to get a contradiction, that for some $\ell > N$ the opposite of (5.22) holds. We compute

$$\sum_{i=0}^{\ell-1} \left[s(x_i^\ell, x_{i+1}^\ell) - \mu\right] = \sum_{i=0}^{N-1} \left[s(x_i^\ell, x_{i+1}^\ell) - \mu\right] + \sum_{i=N}^{\ell-1} \left[s(x_i^\ell, x_{i+1}^\ell) - \mu\right].$$

As discussed in (5.20) the second term is bounded by $a - b$, while by assumption the first one does not exceed $-\gamma + b - a$, so we get

$$\sum_{i=0}^{\ell-1} [s(x_i^\ell, x_{i+1}^\ell) - \mu] \leq -\gamma + b - a + (a - b) = -\gamma$$

contradicting (5.21).

Now let $\ell_k \to \infty$ be an increasing sequence of integers such that $x_i^{\ell_k} \to x_i$ as $k \to \infty$ for every $i \geq 0$. Then

$$m_N(\mathbf{x}) = \sum_{i=0}^{N-1} \left[s(x_i, x_{i+1}) - \mu\right] \geq \limsup_{k \to \infty} \sum_{i=0}^{N-1} \left[s(x_i^{\ell_k}, x_{i+1}^{\ell_k}) - \mu\right] \geq -\gamma + b - a$$

for every $N \geq 1$. Thus conditions (1) and (2) are satisfied with $M = |\gamma| + a - b$. ∎

Remark 5.5 *The question of computing μ is, in general, a difficult one. In a special case when the reward of steering x to y equals that of steering y to x the computation is simple. Suppose that $s(x, y) = s(y, x)$ for all $(x, y) \in X \times X$. We claim that if $b = \max\{s(x, y) : (x, y) \in X \times X\}$ then $\mu = b$. To show this observe that for every program $\mathbf{x}$ inequality $\frac{1}{N} \sum_{i=0}^{N-1} s(x_i, x_{i+1}) \leq b$ holds. Therefore, $\lambda(N) \leq b$ for all $N > 1$. By the property $s(x, y) = s(y, x)$ we get that $\lambda(2) = b$, so that $\sup_{N \geq 1} \lambda(N) = \lambda(2) = b$, implying $\mu = b$.*

The following is an adaptation of Definition 1.2 to our framework. In (i) and (ii) below we follow Gale [81], von Weizsäcker [180] and Koopman [111].

Definition 5.3 *We define the following partial order among programs:* $x \succ y$ *if* $x_0 = y_0$ *and given an* $\varepsilon > 0$ *there is an* N_0 *such that for all* $N \geq N_0$

$$\sum_{i=0}^{N} s(x_i, x_{i+1}) > \sum_{i=0}^{N} s(y_i, y_{i+1}) - \varepsilon.$$

(i) *An* **overtaking optimal program** *is a program* $\mathbf{x}$ *such that* $\mathbf{x} \succ \mathbf{y}$ *for all* $\mathbf{y}$ *with* $y_0 = x_0$.

(ii) *We say that a program* $\mathbf{x}$ *is a* **weakly overtaking optimal program** *if for every* $\mathbf{y}$ *satisfying* $y_0 = x_0$ *and for every* $\varepsilon > 0$ *there is a sequence of integers* $N_k \to \infty$ *such that*

$$\sum_{i=0}^{N_k} s(x_i, x_{i+1}) > \sum_{i=0}^{N_k} s(y_i, y_{i+1}) - \varepsilon \tag{5.23}$$

for all k.

(iii) *For a fixed* $\delta > 0$ *a program* $\mathbf{x}$ *is called* δ-**weakly overtaking optimal program** *if there exists a sequence of integers* $N_k \to \infty$ *such that for every* $\mathbf{y}$ *with* $y_0 = x_0$ *the inequality*

$$\sum_{i=0}^{N_k} s(x_i, x_{i+1}) > \sum_{i=0}^{N_k} s(y_i, y_{i+1}) - \delta$$

holds for all large enough k.

Remark 5.6 *The distinction between weakly overtaking optimal programs and* δ*-weakly optimal programs is the following: If* $\mathbf{x}$ *is a weakly overtaking optimal program then its reward is larger, up to an arbitrary* $\varepsilon > 0$*, than any other program for infinitely many intervals* $[0, N_k]$*. These intervals depend on the compared program. If* $\mathbf{x}$ *is a* δ*-weakly optimal program then all the rewards are compared up to the fixed value* δ *and the sequence of intervals is independent of the program which is compared to* $\mathbf{x}$*. Therefore, a weakly overtaking optimal program is not necessarily a* δ*-weakly optimal program for every* $\delta > 0$.

Here is an immediate consequence of the reduction to finite rewards:

Proposition 5.3 *Under Assumption* I*, for every* $\delta > 0$ *there is a* δ*-weakly optimal program* $\mathbf{x}_\delta$ *with a prescribed initial state* x_0.

Proof: Define the functional $\Phi(\mathbf{x}) = \limsup_{N\to\infty} m_N(\mathbf{x})$. By Theorem 5.1, $\Phi(\cdot)$ is bounded from above by $M > 0$. Therefore we can choose $\mathbf{x}_\delta$ with $(\mathbf{x}_\delta)_0 = x_0$ such that $\Phi(\mathbf{x}_\delta) > \Phi(\mathbf{x}) - \delta$ for every $\mathbf{x}$ with the prescribed initial state. Let $N_k \to \infty$ be such that $m_{N_k}(\mathbf{x}_\delta) \to \Phi(\mathbf{x}_\delta)$ as $k \to \infty$. Then N_k is the sequence needed in (iii) and x_δ is a δ-weakly optimal program. ∎

The following example exhibits the possibility of existence of a weakly overtaking optimal program along with nonexistence of an overtaking optimal program.

Example 5.1 Consider the function $s : [-1,1] \times [-1,1] \to \mathbb{R}$, given by $s(x,y) = y - x - (x+y)^2$. Since for every periodic program the reward is nonpositive, we deduce that $\mu \leq 0$. But computing the reward for the program $x_i = 0$ for all $i \geq 0$, leads to $\mu \geq 0$, hence $\mu = 0$. For a program $\mathbf{x}$ we have $m_N(\mathbf{x}) = x_N - x_0 - \sum_{i=0}^{N-1} (x_i + x_{i+1})^2$. The minimization of $\sum_{i=0}^{N-1} (x_i + x_{i+1})^2$ under the constraints $x_0 = A$ and $x_N = B$ is obtained by

$$x_k = (-1)^k (A + k\Delta), \quad 0 \leq k \leq N,$$

where $\Delta = \frac{B-A}{N}$ if N is even, and $\Delta = -\frac{A+B}{N}$ if N is odd. We claim: if $\mathbf{x}$ is weakly overtaking optimal then $\mathbf{x} = \{x_0, -x_0, x_0, -x_0, \ldots\}$. For suppose that $\mathbf{x}$ is weakly overtaking optimal and there is an $N \geq 1$ such that $x_N \neq (-1)^N x_0$. Assume that N is even and let $\delta = |(x_N - x_0)/N|$ (For odd N we employ a similar argument). We choose an even integer N_1 so large that $2/N_1 < \delta$. As mentioned above the minimum of $\sum_{i=0}^{N_1-1} (x_i + x_{i+1})^2$ with prescribed end states values x_0 and x_{N_1} is obtained by

$$z_k = (-1)^k \Big[x_0 + \frac{k}{N_1}(x_{N_1} - x_0)\Big].$$

But $\{x_0, x_1, \ldots, x_{N_1}\} \neq \{z_0, z_1, \ldots, z_{N_1}\}$ since

$$|z_N - z_0| = N \Big|\frac{x_{N_1} - x_0}{N_1}\Big| \leq N \frac{2}{N_1} < N\delta = |x_N - x_0|.$$

We therefore can replace the segment $\{x_0, \ldots, x_{N_1}\}$ by $\{z_0, \ldots, z_{N_1}\}$ (recall that $x_{N_1} = z_{N_1}$) and thus strictly increase the reward for all large times. This proves the claim.

We assert now that for $x_0 > 0$ the program $\{x_0, -x_0, x_0, -x_0, \ldots\}$ is weakly overtaking optimal. Otherwise, by (5.23) there is a program $\mathbf{y}$, and numbers $\varepsilon > 0$ and N_0 such that

$$m_{2N}(\mathbf{y}) \geq -2x_0 + \varepsilon \tag{5.24}$$

and

$$m_{2N+1}(\mathbf{y}) \geq \varepsilon \tag{5.25}$$

both for all $N \geq N_0$. Substituting the explicit expression for $m_N(\mathbf{y})$ we get from (5.24)

$$-\sum_{i=0}^{2N-1} (y_i + y_{i+1})^2 - x_0 + y_{2N} \geq -2x_0 + \varepsilon,$$

and from this $y_{2N} \geq \varepsilon - x_0$. Similarly, substitution of $m_{2N+1}(\mathbf{y})$ in (5.25) leads to $y_{2N+1} \geq \varepsilon + x_0$. Adding the last two inequalities we obtain

$$y_{2N} + y_{2N+1} \geq 2\varepsilon$$

which contradicts the finiteness of $\sum_{i=0}^{\infty} (y_i + y_{i+1})^2$. We thus conclude that

$$\mathbf{x} = \{x_0, -x_0, x_0, -x_0, \ldots, \}$$

is weakly overtaking optimal.

If there exists an overtaking optimal program, it should be this same program $\mathbf{x}$. But for the program $\mathbf{y} = \{x_0, -\frac{1}{2}x_0, 0, 0, 0, \ldots\}$ we do not have $\mathbf{x} \succ \mathbf{y}$, since $m_{2N}(\mathbf{x}) = -2x_0 < \frac{1}{2}x_0^2 - x_0 = m_{2N}(\mathbf{y})$ for $N \geq 1$.

The next example will demonstrate nonexistence of weakly overtaking optimal programs.

Example 5.2 Let $s : [-1,1] \times [-1,1] \to \mathbb{R}$, $s(x,y) = 2x^2 - 2y^2 - (x+y)^2$. Reasoning as in Example 5.1, the only candidate to be a weakly overtaking optimal program is $\mathbf{x} = \{x_0, -x_0, x_0, -x_0, \ldots\}$. If we compare it with $\mathbf{y} = \{x_0, 0, 0, 0, \ldots\}$ we see that $m_N(\mathbf{y}) = -x_0^2$ for every N, while $m_N(\mathbf{x}) = 0$, so $\mathbf{x}$ is not weakly overtaking optimal.

5.5 A representation formula and turnpike properties of optimal controls

We will now confine ourselves to continuous transition functions $s(x,y)$. (We will formulate the analogous result for $s(x,y)$ which are upper-semicontinuous where the proof requires minor modifications.) In (5.5) we considered transition functions of a special form. Here we intend to represent every continuous function on a compact square-shaped domain in this special form. As discussed above, this has a close relation to the reduction to finite rewards.

Theorem 5.2 *Let $X \subset \mathbb{R}^n$ be compact and $s(\cdot,\cdot) : X \times X \to \mathbb{R}$ be a continuous function. Then $s(\cdot,\cdot)$ can be represented as follows:*

$$s(x,y) = \mu + p(x) - p(y) + \theta(x,y) \qquad \textit{where}$$

(a) *μ is a constant,*

(b) *$p : X \to \mathbb{R}$ and $\theta : X \times X \to \mathbb{R}$ are continuous functions, and*

(c) *θ is nonpositive and the set $S(x) = \{y \in X : \theta(x,y) = 0\}$ is nonempty for every $x \in X$.*

Proof: Let μ be as guaranteed in Theorem 5.1 and defined in (5.16). Define $p : X \to \mathbb{R}$ by

$$p(x) = \sup_{x_0=x} \{\limsup_{N\to\infty} m_N(\mathbf{x})\}, \tag{5.26}$$

where the supremum is over all programs $\mathbf{x}$ which satisfy $x_0 = x$. Given any pair $(x,y) \in X \times X$ we claim that $p(x) \geq [s(x,y) - \mu] + p(y)$, since if we confine ourselves to programs $\mathbf{x}$ such that $x_0 = x$, $x_1 = y$ and compute the supremum in the right hand side of (5.26) over these programs only, then we get $[s(x,y) - \mu] + p(y)$. Of course $p(x)$ is not smaller than this value. If we define $\theta(\cdot,\cdot) : X \times X \to \mathbb{R}$ by

$$\theta(x,y) = s(x,y) - \mu + p(y) - p(x) \tag{5.27}$$

then we get (5.5) with θ nonpositive. The uniform continuity of $s(\cdot,\cdot)$ on $X \times X$ implies the continuity of $p(\cdot)$, and thus, by (5.27) the function $\theta(\cdot,\cdot)$ is continuous too.

It only remains to prove that $S(x)$ is nonempty for every $x \in X$. Suppose to the contrary that for some $x \in X$

$$\max\{\theta(x,y) : y \in X = -\delta < 0\}. \tag{5.28}$$

There is a program $\mathbf{x}$ such that $x_0 = x$ and $\limsup_{N\to\infty} m_N(\mathbf{x}) > p(x) - \frac{1}{2}\delta$. We compute

$$\begin{aligned} p(x) - \frac{1}{2}\delta &< \limsup_{N\to\infty} m_N(\mathbf{x}) \\ &= [\theta(x,x_1) + p(x) - p(x_1)] + \limsup_{N\to\infty} \sum_{i=1}^{N} [s(x_i,x_{i+1}) - \mu] \\ &\leq [-\delta + p(x) - p(x_1)] + p(x_1), \end{aligned}$$

implying $p(x) - \frac{1}{2}\delta < p(x) - \delta$, a contradiction; hence (5.28) is false. ∎

For $s(x,y)$ which is merely upper-semicontinuous, we have the following analogue of Theorem 5.2:

Theorem 5.2′ *Let $s(\cdot,\cdot) : X \times X \to \mathbb{R}$ be bounded and upper-semicontinuous. Then $s(\cdot,\cdot)$ admits a representation* (5.5) *with* (c) *replaced by the following property:*
(c′) *The function $\theta(\cdot,\cdot) : X \times X \to \mathrm{R}$ is nonpositive and for every $x \in X$, $\sup_{y\in X}\{\theta(x,y) : y \in X\} = 0$.*

The proof is essentially the same as that for Theorem 5.2 with only minor modifications. We leave out the details.

We will employ the following terminology, which was introduced by Gale [81]:

Definition 5.4 *A program* $\mathbf{x}$ *is called* **a good program** *if* $\{m_N(\mathbf{x})\}_{N=1}^{\infty}$ *forms a bounded sequence.*

We apply Theorems 5.2 and 5.2′ to show how one can construct good programs. Generally, for upper-semicontinuous functions, this can be achieved by choosing $\mathbf{x}$ so that $\theta(x_i, x_{i+1}) > -1/2^i$. If $s(x, y)$ is continuous this can be done using dynamic programming, by constructing an $\mathbf{x}$ which satisfies $x_{i+1} \in S(x_i)$, that is as a solution to (5.8). For a program so generated we get

$$\left|\sum_{i=0}^{N-1} [s(x_i, x_{i+1}) - \mu]\right| = \left|p(x_0) - p(x_N)\right| \leq M,$$

proving that $\mathbf{x}$ is a good program.

If it happens that the transition function $s(x, y)$ has the form which is described below in Proposition 5.4, (which is more restrictive than that in (5.5)), then we can establish the existence of overtaking optimal programs.

Proposition 5.4 *Let* $s(\cdot,\cdot) : X \times X \to \mathbb{R}$ *be a continuous function which admits a representation* $s(x, y) = \mu + p(x) - p(y) + \theta(x, y)$, *where*

(i) μ *is a constant,* $p(\cdot) : X \to \mathbb{R}$ *and* $\theta(\cdot,\cdot) : X \times X \to \mathbb{R}$ *are continuous functions.*

(ii) $\theta(x, y)$ *is nonpositive and there is an element* $\overline{x} \in X$ *such that* $(x, y) = (\overline{x}, \overline{x})$ *if and only if* $\theta(x, y) = 0$.

Then for every initial state there is an overtaking optimal program.

Proof: If $\mathbf{x}$ is a good program then there is a bound on

$$\left\{p(x_0) - p(x_N) + \sum_{i=0}^{N-1} \theta(x_i, x_{i+1})\right\}_{N=1}^{\infty},$$

consequently $\theta(x_i, x_{i+1}) \to 0$, and condition (ii) implies that $x_i \to \overline{x}$. So for every good program $\mathbf{x}$ we have

$$\sum_{i=0}^{\infty} [s(x_i, x_{i+1}) - \mu] = p(x_0) - p(\overline{x}) + \sum_{i=0}^{\infty} \theta(x_i, x_{i+1}).$$

The functional $\mathbf{x} \to \sum_{i=0}^{\infty} \theta(x_i, x_{i+1})$ is upper-semicontinuous on the set of programs endowed with the topology of pointwise convergence. (If $\mathbf{x}^k \to \mathbf{x}$ and $\sum_{i=0}^{\infty} \theta(x_i, x_{i+1}) = \alpha$ then given $\varepsilon > 0$, $\sum_{i=0}^{N} \theta(x_i, x_{i+1}) < \alpha + \varepsilon$ for some N, so $\sum_{i=0}^{\infty} \theta(x_i^k, x_{i+1}^k) < \alpha + 2\varepsilon$ for sufficiently large k.) Let $\mathbf{z}$ be a program which maximizes this functional. Clearly $\mathbf{z}$ is a good program so it maximizes the functional $\mathbf{x} \to \sum_{i=0}^{\infty} [s(x_i, x_{i+1}) - \mu]$ too and the overtaking optimality of $\mathbf{z}$ follows. ∎

As a corollary we get the following result, concerning concave functions $s(x, y)$:

Proposition 5.5 *Let $s(x, y)$ be strictly concave in (x, y) such that the function $x \to s(x, x)$ has a maximum at $\bar{x} \in X$. Then for every initial value there is a unique overtaking optimal program.*

Proof: It is easy to see that there is a $\eta \in \mathbb{R}^n$ such that $s(x, y) \geq s(\bar{x}, \bar{x}) + \eta'(x - y)$. Define $\mu = s(\bar{x}, \bar{x})$ and $\theta(x, y) = s(x, y) - \mu - \eta'(x - y)$. Then θ is nonpositive and strictly concave. Since $\theta(\bar{x}, \bar{x}) = 0$ it follows that whenever $(x, y) \neq (\bar{x}, \bar{x})$, then $\theta(x, y) > 0$.

Example 5.3 Let $s(x, y) = x^2 - (1 + x^2 + y^2)^{\frac{1}{2}}$, $0 \leq x \leq 1$ and $0 \leq y \leq 1$. We claim that there is an overtaking optimal program for every initial value. Denote $r^2 = x^2 + y^2$ and let $\rho = (1 + r^2)^{\frac{1}{2}}$. Then we can write

$$s(x, y) = \frac{1}{2}[x^2 - y^2 - \rho(2 - \rho) - 1]$$

which may be given the form required in Proposition 5.5, by defining

$$\mu = 1 - \sqrt{3},$$

$$p(x) = \frac{1}{2}x$$

and

$$\theta(r) = -\frac{1}{2}[\rho(2 - \rho) + 3 - 2\sqrt{3}].$$

We have $\theta(r) < 0$ for $1 \leq r < \sqrt{2}$ and $\theta(\sqrt{2}) = 0$, so the claim follows from Proposition 5.3.

5.6 Systems with unbounded rewards and with discounting factors

It is customary in mathematical economics to introduce a discounting factor into the rewards expressions. Usually this factor is chosen to be an exponentially decreasing function, which reflects the result of a constant interest rate. The introduction of an exponentially decreasing discounting factor induces on the interval $[0, \infty)$ a finite measure, and in many cases makes the reward functional upper-semicontinuous in a convenient topology for the trajectories.

We discuss here a discounting factor which tends to zero, but may do so very slowly. Therefore compactness arguments which work for the exponential decaying situation would not work. We consider the following problem. Let $s(\cdot,\cdot) : X \times X \to \mathbb{R}$ be a continuous function and $\{\alpha_i\}_{i=0}^{\infty}$ be a sequence of positive numbers converging monotonically to zero. For a program $\mathbf{x}$ we study the reward flow

$$\Big\{\sum_{i=0}^{N} \alpha_i\, s(x_i, x_{i+1})\Big\}_{N=0}^{\infty}.$$

Analogous to Definition 5.3 we define here an overtaking optimal program as a program $\{z_k\}_{k=0}^{\infty}$ such that for every $\varepsilon > 0$ and every other program $\mathbf{x}$ with $x_0 = z_0$ we have

$$\sum_{k=0}^{N} \alpha_k\, s(z_k, z_{k+1}) \geq \sum_{k=0}^{N} \alpha_k\, s(x_k, x_{k+1}) - \varepsilon \tag{5.29}$$

for all N large enough.

Using the representation formula (5.4) we can establish the existence of an overtaking optimal program for the present problem. We emphasize that $\{\alpha_i\}$ may decrease to zero very slowly so that still $\sum_{i=0}^{\infty} \alpha_i\, s(x_i, x_{i+1}) = \infty$ for every program.

Theorem 5.3 *Let $s(\cdot,\cdot) : X \times X \to \mathbb{R}$ be continuous and $\{\alpha_i\}_{i=1}^{\infty}$ be a decreasing sequence with $\alpha_i \to 0$ as $i \to \infty$. Then for every initial value x_0 there is an overtaking optimal program $\mathbf{z}$ with $z_0 = x_0$.*

Proof: Using (5.5) we compute for $\mathbf{x}$

$$\begin{aligned}\sum_{i=0}^{N-1} \alpha_i[s(x_i, x_{i+1}) - \mu] &= \sum_{i=0}^{N-1} \alpha_i\, \theta(x_i, x_{i+1}) + \alpha_0\, p(x_0) - \alpha_{N-1}\, p(x_N)\\ &\quad + \sum_{i=1}^{N-1} (\alpha_i - \alpha_{i+1}) p(x_i).\end{aligned} \tag{5.30}$$

The series $\sum_{i=0}^{\infty} \alpha_i \, \theta_i(x_i, x_{i+1})$ is either convergent or diverges to negative infinity, by non-positivity, and the series $\sum_{i=1}^{\infty} (\alpha_i - \alpha_{i-1}) \, p(x_i)$ converges, since $\sum_{i=1}^{\infty} (\alpha_i - \alpha_{i-1})$ converges absolutely. We denote

$$\varphi(\mathbf{x}) = \sum_{i=0}^{\infty} \alpha_i[s(x_i, x_{i+1}) - \mu]$$

$$\varphi_1(\mathbf{x}) = \sum_{i=1}^{\infty} (\alpha_i - \alpha_{i-1})p(x_i)$$

$$\varphi_2(\mathbf{x}) = \sum_{i=0}^{\infty} \alpha_i \, \theta(x_i, x_{i+1})$$

and have the relation $\varphi(\mathbf{x}) = \varphi_1(\mathbf{x}) + \varphi_2(\mathbf{x}) + \alpha_0 p(x_0)$. We consider the set of programs endowed with the topology of pointwise convergence, and claim that in this topology both φ_1 and φ_2 are upper-semicontinuous.

(i) Upper-semicontinuity of φ_2: This follows from the nonpositivity and continuity of $\theta(\cdot, \cdot)$.

(ii) Upper-semicontinuity of φ_1: Let $\mathbf{x}^k \to \mathbf{x}$ and let $\varepsilon > 0$. Then there is an N such that $\left|\sum_{i=N}^{\infty} (\alpha_i - \alpha_{i-1})p(y_i)\right| < \varepsilon/3$ for every program $\mathbf{y}$. For this N

$$\begin{aligned} \varphi_1(\mathbf{x}) - \varphi_1(\mathbf{x}^k) \; = \; & \sum_{i=N}^{\infty} (\alpha_i - \alpha_{i-1})p(x_i) - \sum_{i=N}^{\infty} (\alpha_i - \alpha_{i-1})p(x_i^k) \\ & + \sum_{i=1}^{N-1} (\alpha_i - \alpha_{i-1})[p(x_i) - p(x_i^k)]. \end{aligned}$$

If k is large enough the third term is not less than $-\varepsilon/3$ so that $\varphi_1(\mathbf{x}) > \varphi_1(\mathbf{x}^k) - \varepsilon$ for all large k.

Thus we conclude that also φ itself is upper semicountinuous in this topology and by the compactness of X and the boundedness above of φ there is a program $\mathbf{z}$ such that $\varphi(\mathbf{z}) \geq \varphi(\mathbf{x})$ for all $\mathbf{x}$.

To prove (5.29), given any program $\mathbf{x}$ and an $\varepsilon > 0$ there is N_ε such that for all $N > N_\varepsilon$

$$\sum_{i=0}^{N} \alpha_i[s(z_i, z_{i+1}) - \mu] > \sum_{i=0}^{N} \alpha_i[s(x_i, x_{i+1}) - \mu] - \varepsilon.$$

Now (5.29) follows by adding $\mu \sum_{i=0}^{N} \alpha_i$ to both sides of the inequality. ∎

We will employ now Theorem 5.3 to study the continuous-time control systems which were described in Problem III in Section 5.3. Slightly more generally we consider a system

$$\begin{cases} \dot{x}(t) = f(x(t), u(t)) \\ x(0) = x_0 \\ c_T(u(\cdot)) = \int_0^T \varphi(t) f_0(x(t), u(t))\, dt \end{cases} \tag{5.31}$$

where $x(t) \in X$, $u(t) \in U$ for $0 \le t \le T$, while X and U are compact subsets of $\mathbb{R}^n$ and $\mathbb{R}^m$ respectively, and $t \to \varphi(t)$ is a positive scalar function which decreases monotonically to zero. The admissible controls are all the measurable functions $u(\cdot)$ for which the constraints $u(t) \in U$ and $x(t) \in X$ are satisfied (where $x(\cdot)$ and $u(\cdot)$ are related as in (5.31)). Let $\tau > 0$ be the sampling time in the discretization of the time. We assume the following:

(1) The functions $f(\cdot,\cdot)$ and $f_0(\cdot,\cdot)$ are continuous on $X \times U$.

(2) A controllability assumption: For every $(x, y) \in X \times X$ there is an admissible control $u(\cdot)$ with a corresponding trajectory $x(\cdot)$ which satisfies $x(0) = x, x(\tau) = y$.

(3) The function $s(\cdot,\cdot)$ which is defined on $X \times X$ by

$$s(x,y) = \sup\Big\{\int_0^\tau f_0(x(t), u(t))dt \; : \; \dot{x}(t) = f(x,u),\; x(t) \in X,$$
$$u(t) \in U,\; x(0) = x, x(\tau) = y\Big\}$$

is assumed to be continuous on $X \times X$.

(4) The function $w_T(\cdot,\cdot)$, where

$$w_T(x,y) = \max\Big\{\int_T^{T+\tau} \varphi(t) f_0(x(t), u(t)\, dt \; : \; \dot{x}(t) = f(x,u),\; x(t) \in X,$$
$$u(t) \in U,\; x(T) = x, x(T+\tau) = y\Big\},$$

is well defined (namely, the maximum is attained by a certain admissible control), and is continuous on $X \times X$. This is assumed for every $T > 0$.

Remark 5.7 *There can be given explicit assumptions concerning $f(\cdot,\cdot)$ and $f_0(\cdot,\cdot)$ which guarantee the validity of* (3) *and* (4). *However, they seem to be too restrictive and we prefer the implicit assumptions* (3) *and* (4). *Assumption* (3) *guarantees a constant μ with a reduction to finite rewards.*

Remark 5.8 *For the system which is described in Problem* III, *which is a special case of* (5.31), *the controllability assumption* (2) *holds provided that zero is an interior point of* U, *and* τ, *the sampling time, is large enough. Moreover, in this situation* (3) *and* (4) *hold if it is assumed that* $v \to \mathcal{L}(x,v)$ *is concave for every fixed* x.

Theorem 5.4 *Let the control system* (5.31) *satisfy assumptions* (1)–(4). *Then for every initial state* $x_0 \in X$ *there exists an overtaking optimal solution* $x^*(\cdot)$ *satisfying* $x^*(0) = x_0$.

To prove Theorem 5.4 we shall need the following Lemma:

Lemma 5.4 *Given an* $\varepsilon > 0$ *then there is a time* $T_0 > 0$ *such that*

$$\int_{T_1}^{T_2} \varphi(t)\big[f_0(x(t),u(t)) - \mu\big]\, dt < \varepsilon \text{ for every } T_2 > T_1 \geq T_0 \tag{5.32}$$

for every control $u(\cdot)$ *with a corresponding response* $x(\cdot)$ *which satisfies* $u(t) \in U$ *and* $x(t) \in X$ *for all* $t \geq 0$.

Proof: It is enough to prove the claim for T_1, T_2 integers. Denote by $\psi(\cdot)$ the function $\psi(t) = \varphi(k)$ for $k \leq t < k+1$. For all integers $M > N$ and functions $x(t) \in X$, $u(t) \in U$ the following equality holds

$$\int_N^M \varphi(t)\big[f_0(x,u) - \mu\big]\, dt = \int_N^M \big[\varphi(t) - \psi(t)\big]\big[f_0(x,u) - \mu\big]\, dt + \int_N^M \psi(t)\big[f_0(x,u) - \mu\big]\, dt.$$

Let A be a bound on $|f_0(x,u) - \mu|$, and notice that $\int_N^M |\varphi(t) - \psi(t)|\, dt \leq \varphi(N)$ so the first term in the last equation is bounded by $A\varphi(N)$ and is smaller than $\frac{\varepsilon}{2}$ for large enough N. For the second term we estimate

$$\begin{aligned}\int_N^M \psi(t)\big[f_0(x,u) - \mu\big]\, dt &= \sum_{k=N}^{M-1} \varphi(k) \int_k^{k+1} \big[f_0(x,u) - \mu\big]\, dt \\ &\leq \sum_{k=N}^{M-1} \varphi(k)\big[s(x(k), x(k+1)) - \mu\big].\end{aligned}$$

By (5.30) the last expression can be estimated by

$$\varphi(N)p(x_N) - \varphi(M-1)p(x_M) + \sum_{i=N+1}^{M-1} \big[\varphi(i) - \varphi(i-1)\big]p(x(i))$$

and by boundedness of $|p(x)|$ and the property $\varphi(t) \to 0$ this estimate is less than ε in absolute value if N is large enough, thus the claim is proved. ∎

Proof of Theorem 5.4: From (5.32) it follows that for every $x(t) \in X$, $u(t) \in U$, the expression $\int_0^\infty \varphi(t)\big[f_0(x(t),u(t)) - \mu\big]\,dt$ either converges, or diverges to negative infinity. Moreover, for good programs this integral is finite (because the convergence of $\sum_{i=0}^{\infty} \theta(x_i, x_{i+1})$ clearly implies that of $\sum_{i=0}^{\infty} \alpha_i\,\theta(x_i, x_{i+1})$ if $\alpha_i \to 0$). Let

$$m = \sup \int_0^\infty \varphi(t)[f_0(x,u) - \mu]\,dt,$$

where $u(\cdot)$ and $x(\cdot)$ are related by the differential equation $\dot{x} = f(x,u)$ and initial condition $x(0) = x_0 \in X$, and where the supremum is taken over all admissible controls. Then there are admissible pairs (x_k, u_k) such that

$$\int_0^\infty \varphi(t)\big[f_0(x_k(t), u_k(t)) - \mu\big]\,dt = m_k \to m.$$

We assume that the sampling time is $\tau = 1$. It can be assumed that $\lim_{k\to\infty} x_k(i) = x(i)$ for every integer $i \geq 0$, and by assumption (4), define $x(t)$ in $[i, i+1]$ as a solution to the maximization of $\int_i^{i+1} \varphi(t) f_0(z(t), u(t))dt$ over the pairs $(u(\cdot)\,, z(\cdot)\,)$ satisfying $z(i) = x(i)$, $z(i+1) = x(i+1)$. Now by assumption (4) (the continuity of $w_N(x,y)$) and Lemma 5.4 it follows that, given an $\varepsilon > 0$,

$$\int_0^T \varphi(t)\big[f_0(x(t), u(t)) - \mu\big] \geq m - \varepsilon$$

for all large T, so we conclude

$$\int_0^\infty \varphi(t)\big[f_0(x(t), u(t)) - \mu\big]\,dt = m$$

and $x(\cdot)$ is an overtaking optimal solution. ∎

Remark 5.9 *This result can be interpreted as follows: Given an initial value $x_0 \in X$ there exists a constant m so that for every admissible pair $(x(\cdot)\,, u(\cdot)\,)$ with $x(0) = x_0$ the following limit exists and satisfies*

$$\lim_{T\to\infty} \Big\{ \int_0^T \varphi(t) f(x(t), u(t))\,dt - m - \mu \int_0^T \varphi(t)\,dt \Big\} \leq 0,$$

while there exists an admissible pair $(x^*(\cdot)\,, u^*(\cdot)\,)$ with $x^*(0) = x_0$ for which the equality holds. The function $T \to m + \mu \int_0^T \varphi(t)dt$ expresses the maximal reward-growth as $T \to \infty$.

5.7 Infinite horizon tracking of periodic signals

We consider the tracking problem described as Problem II in Section 5.3. Thus the plant is (5.13) and the cost expression is (5.14) and we consider here an infinite horizon minimization problem. Since all the cost flows $\{C_{0,T}(u)\}_{T\geq 0}$ diverge to infinity as T grows to infinity, our goal is to minimize in the overtaking sense. The overtaking notion determines a partial order on the admissible controls. We will prove the existence of a unique global minimizer with respect to this partial order. The following three theorems present the results on this problem.

Theorem 5.5 *For each initial state x_0 there exists a unique overtaking optimal control $u^*(\cdot) = u^*(x_0, \cdot)$. It possesses the following stronger version of overtaking optimality: for any other control $u(\cdot)$ there is a time T_0 and an $\varepsilon > 0$, depending on $u(\cdot)$, such that*

$$C_{0,T}(u^*(\cdot)) + \varepsilon < C_{0,T}(u(\cdot)) \tag{5.33}$$

for all $T > T_0$.

The stronger version of overtaking means that if a different policy is chosen, the eventual loss cannot be recouped even asymptotically. The next result is concerned with the feedback representation.

Theorem 5.6 *The overtaking optimal control $u^*(\cdot)$ can be expressed in a closed form as follows*

$$u^*(t,x) = -R^{-1}B'(Kx - g(t)) \tag{5.34}$$

where K is the symmetric positive definite solution of the Riccati equation

$$KA + A'K - KBR^{-1}B'K + Q = 0 \tag{5.35}$$

$g(\cdot)$ is periodic of period τ, (the period of $\Gamma(\cdot)$), and is given on $[0,\tau]$ by

$$g(t) = [I - e^{-\tau G}]^{-1} \int_0^\tau e^{-sG} Q\Gamma(t+s)\, ds \tag{5.36}$$

and $G = KBR^{-1}B' - A'$.

The next result is concerned with the asymptotic behavior of the optimal solution $t \to x^*(t)$ as $t \to \infty$.

Theorem 5.7 *The response $x^*(\cdot)$ to the overtaking optimal control $u^*(\cdot)$ converges as t tends to infinity to a periodic trajectory $\gamma^*(\cdot)$ of period τ, this regardless of the initial state x_0. The trajectory $\gamma^*(\cdot)$ is given on $[0,\tau]$ as the unique solution of the following variational problem:*

minimize

$$\int_0^\tau \left\{[x(t)-\Gamma(t)]'Q[x(t)-\Gamma(t)]+u(t)'Ru(t)\right\}\,dt$$

subject to

$$x(0)=x(\tau) \tag{5.37}$$

where $x(\cdot)$ is the response to $u(\cdot)$.

The engineering literature offers some approaches to our problem. Anderson and Moore ([5] Section 1.3) note that the optimal solution to the finite time tracking on $[0,T]$ converges as $T\to\infty$. They suggest this limit as a solution, which has useful engineering properties, although it cannot be compared to other controls via the cost functional. It follows easily from the results of this section that this limit can be identified as the overtaking optimal solution (see Artstein and Leizarowitz [6] Theorem 3).

A different approach (see e.g., Gilbert [86]) is to seek for an optimal solution among the periodic controls. The passage from the initial condition x_0 to the optimal periodic trajectory is regarded as transient. (Obviously this transient becomes significant if the system is subject to persistent disturbances.) Maffezzoni [129] solves the optimal periodic control problem for the tracking problem and offers a feedback stabilizer, one of many that exist. Theorem 5.7 is a justification to the optimal periodic control approach, as it is easy to show that the solution to (5.37) is the optimal periodic trajectory. In addition our approach provides a criterion for handling the transients.

Consider the function $s(\cdot,\cdot)$ defined on $\mathbb{R}^n\times\mathbb{R}^n$ by

$$s(y,z)=\min\int_0^\tau \left\{[x(t)-\Gamma(t)]'Q[x(t)-\Gamma(t)]+u(t)'Ru(t)\right\}\,dt$$

subject to

$$x(0)=y,\, x(\tau)=z. \tag{5.38}$$

Here τ is the period of $\Gamma(\cdot)$ and $x(\cdot)$ is the response to $u(\cdot)$. It is easy to verify that s is well defined, strictly convex and tends to infinity quadratically as $|y|+|z|\to\infty$. Let D be the diagonal of $\mathbb{R}^n\times\mathbb{R}^n$, that is $D=\{(x,x)\,:\,x\in\mathbb{R}^n\}$. Then s has a unique minimum on D which we denote by $(\overline{x},\overline{x})$.

Proposition 5.6 *The function $s(\cdot,\cdot)$ can be expressed as*

$$s(x,y)=s(\overline{x},\overline{x})+p'(x-y)+\theta(x,y) \tag{5.39}$$

with $p \in \mathbb{R}^n$ and $\theta(\cdot,\cdot)$ being a strictly convex function which grows quadratically to infinity when $\|x\| + \|y\| \to \infty$. Moreover, $\theta(x,y) > 0$ whenever $(x,y) \neq (\overline{x},\overline{x})$, and $\theta(\overline{x},\overline{x}) = 0$.

Proof: The strict convexity of $s(\cdot,\cdot)$ implies the existence of a supporting vector (p_1, p_2) to the epigraph of s at $(\overline{x},\overline{x})$, namely

$$s(x,y) = s(\overline{x},\overline{x}) + (p_1,p_2)'((x,y) - (\overline{x},\overline{x})) \tag{5.40}$$

with equality only at $(\overline{x},\overline{x})$. We claim that $p_1 = -p_2$. Indeed, it is easy to check that $s(x,x)$ converges to $s(\overline{x},\overline{x})$ quadratically as $x \to \overline{x}$, while, if $p_1 \neq -p_2$, the right hand side of (5.40) with $y = x$ would converge to $s(\overline{x},\overline{x})$ only linearly. The function $\theta(x,y)$ can now be defined as $s(x,y)$ minus the right hand side of (5.40). Its properties follow directly from those of $s(\cdot,\cdot)$ and (5.39). This completes the proof. ∎

Note that the form of $s(\cdot,\cdot)$ in (5.39) is as considered in Proposition 5.4.

Consider now the functional

$$\Phi(\sigma) = \sum_{i=0}^{\infty} \theta(y_i, y_{i+1}) \tag{5.41}$$

defined for sequences $\sigma = \{y_i\}_{i=0}^{\infty}$ with $y_i \in \mathbb{R}^n$.

Lemma 5.5 *Let x_0 be a given initial condition. Among the sequences $\sigma = \{x_0, y_1, y_2, \ldots,\}$ the functional $\Phi(\cdot)$ has a unique finite minimum at $\sigma^* = \{x_0, x_1, x_2, \ldots\}$ which satisfies $x_i \to \overline{x}$ as $i \to \infty$.*

Proof: There are sequences σ with $\Phi(\sigma) < \infty$, e.g. $\{x_0, \overline{x}, \overline{x}, \overline{x}, \ldots\}$. The continuity and nonnegativity of $\theta(\cdot,\cdot)$ imply that $\Phi(\cdot)$ is lower-semicontinuous with respect to coordinatewise convergence of sequences. Hence there exists a minimum, which is unique by the strict convexity of $\theta(\cdot,\cdot)$. The last statement follows since $y_i \not\to \overline{x}$ implies $\Phi(\sigma) = \infty$. ∎

Proposition 5.7 *For every initial condition x_0 the minimizer σ^* of $\sigma \to \Phi(\sigma)$ is the unique overtaking optimal solution for the cost flow $C_N(\sigma) = \sum_{i=0}^{N-1} s(y_i, y_{i+1})$. It possesses the following stronger property: for every other sequence $\sigma = \{x_0, y_1, y_2, \ldots\}$ there are $\varepsilon > 0$ and N_0 such that $C_N(\sigma^*) < C_N(\sigma) - \varepsilon$ if $N > N_0$.*

Proof: From (5.39) we know that

$$C_N(\sigma) - C_N(\sigma^*) = -p'y_N + p'x_N + \sum_{i=0}^{N-1} \theta(y_i, y_{i+1}) - \sum_{i=0}^{N-1} \theta(x_i, x_{i+1}). \tag{5.42}$$

The quadratic growth of $\theta(\cdot,\cdot)$ (by Proposition 5.6) implies that $-p'y_N + \theta(y_{N-1}, y_N)$ is bounded from below. Hence, if y_k does not converge to $\bar{x}$, then the quantity in (5.41) diverges to infinity and the result holds. If $y_k \to \bar{x}$ as $k \to \infty$ then $p'(x_k - y_k)$ tends to zero. Since $\Phi(\sigma^*) < \Phi(\sigma)$ it then follows that σ^* overtakes σ, and clearly the stronger overtaking property holds. ∎

Proof of Theorem 5.5: Let x_0 be an initial condition. Let $\sigma^* = \{x_0, x_1, \dots,\}$ be the overtaking solution of the discrete-time problem guaranteed by Proposition 5.7. We claim that the derived overtaking optimal solution is the unique control $u^*(\cdot)$ whose response $x(\cdot)$ satisfies $x(k\tau) = x_k$ for all k, and such that the cost of steering x_k to x_{k+1} over $[k\tau, (k+1)\tau]$ is $s(x_k, x_{k+1})$. The controllability and standard properties of finite time tracking imply that such a control u^* exists. Proposition 5.7 implies that u^* overtakes any other control, say u, on the discrete sequences $\{k\tau\}_{k=0}^{\infty}$. Let $y(\cdot)$ be the response to $u(\cdot)$, given $y(0) = x_0$, and let $y_k = y(k\tau)$. If $\{y_k\}$ does not converge to $\bar{x}$, or if it converges but $C_{k\tau,(k+1)\tau}(u(\cdot))$ does not converge to $s(\bar{x}, \bar{x})$, then by our previous results $C_{k\tau}(u^*(\cdot)) - C_{k\tau}(u(\cdot)) \to -\infty$ as $k \to \infty$, and the desired overtaking property clearly holds for all t. Otherwise, $y_k \to \bar{x}$ and $u(\cdot)$ steers y_k to y_{k+1} near optimally. Then the uniqueness of the optimal steering implies that the cost $C_{k\tau,k\tau+t}(u(\cdot))$ converges uniformly for $0 \le t \le \tau$ to the cost $C_{k\tau,k\tau+t}(u^*(\cdot))$. The stronger version of overtaking on the discrete set $\{k\tau\}_{k=0}^{\infty}$ implies this property for all t. This completes the proof. ∎

We next consider the limit trajectory and establish Theorem 5.7.

Proof of Theorem 5.7: We notice that $\bar{x}$ was constructed such that $\gamma^*(0) = \gamma^*(\tau) = \bar{x}$. Hence by Lemma 5.5 and the preceding paragraph, any overtaking optimal trajectory $x(\cdot)$ satisfies $x(k\tau) \to \gamma^*(0)$ as $k \to \infty$. It is also clear that since steering $x(k\tau)$ to $x((k+1)\tau)$ is done optimally along the overtaking solution, then $x(t)$ converges to $\gamma^*(t)$ for all t, by the obvious continuous dependence of the optimal solution on the boundary values.

The computation of $\gamma^*(\cdot)$ is rather a simple exercise which is described as follows. A necessary condition for optimality is the existence of an $\eta(\cdot)$ which, together with $x(\cdot)$, solves the system

$$\frac{d}{dt}\begin{bmatrix} x \\ \eta \end{bmatrix} = \begin{bmatrix} A & BR^{-1}B' \\ Q & -A' \end{bmatrix}\begin{bmatrix} x \\ \eta \end{bmatrix} + \begin{bmatrix} 0 \\ -Q\Gamma(t) \end{bmatrix}. \tag{5.43}$$

Denote by W the $(2n \times 2n)$ coefficient matrix in (5.43). The variation of parameter formula implies that $\gamma^*(\cdot)$ with its adjoint $\eta^*(\cdot)$ have the representation

$$\begin{bmatrix} \gamma^*(t) \\ \eta^*(t) \end{bmatrix} = e^{tW}\begin{bmatrix} \bar{x} \\ \eta^*(0) \end{bmatrix} + \int_0^t e^{(t-s)W}\begin{bmatrix} 0 \\ -Q\Gamma(s) \end{bmatrix} ds. \tag{5.44}$$

The initial conditions $\overline{x} = \gamma^*(0)$ and $\eta^*(0)$ can be derived from the periodicity, by using $t = \tau$. The outcome is

$$\begin{bmatrix} \overline{x} \\ \eta^*(0) \end{bmatrix} = \left[I - e^{-\tau W}\right]^{-1} \int_0^\tau e^{-sW} \begin{bmatrix} 0 \\ Q\Gamma(s) \end{bmatrix} ds. \tag{5.45}$$

■

We derive now the optimal feedback promised in Theorem 5.6. The method modifies standard techniques.

Proof of Theorem 5.6: Let K be the positive definite solution of (5.35). Since (5.43) is a necessary condition we get, by uniqueness, and in a standard way, that the optimal control has the form $u^*(t) = R^{-1}B'\eta_0(t)$, where $\eta_0(\cdot)$ and the overtaking optimal trajectory $x_0(\cdot)$ solve (5.43) on $[0, \infty)$. Define

$$g(t) = \eta_0(t) + Kx_0(t). \tag{5.46}$$

Then it is straightforward that

$$\dot{g}(t) = \left[KBR^{-1}B' - A'\right] g(t) - Q\Gamma(t). \tag{5.47}$$

Denote by G the coefficient matrix in (5.47), that is $G = KBR^{-1}B' - A'$. It is well known that $-G$ is stable. Expressing $g(\cdot)$ in the variation of constants formula yields

$$g(t) = e^{tG}\left[g(0) - \int_0^t e^{-sG}Q\Gamma(s)\, ds\right]. \tag{5.48}$$

The stability of $-G$ implies that the quantity in parenthesis in (5.48) converges. It must converge to zero, otherwise $g(t)$ diverges as $t \to \infty$, which we know is not the case by (5.46) and the fact that $x_0(t)$ and $\eta_0(t)$ converge to $\gamma^*(t)$ and $\eta^*(t)$ respectively (as $t \to \infty$). Hence

$$g(0) = \int_0^\infty e^{-sG}Q\Gamma(s)\, ds \tag{5.49}$$

and with (5.48) we get

$$g(t) = \int_t^\infty e^{(t-s)G}Q\Gamma(s)\, ds. \tag{5.50}$$

From this it is clear that $g(\cdot)$ is indeed periodic of period τ. The representation (5.36) can be obtained by simple manipulation, and the feedback (5.34) by substitution. Inserting (5.34) into the system yields the following equation for the overtaking trajectory:

$$\dot{x} = \left[A - BR^{-1}B'K\right] x + BR^{-1}B'g(t). \tag{5.51}$$

It implies a strengthening of Theorem 5.7, that the convergence of $x(t) - \gamma^*(t)$ to zero is exponential, due to the fact that also $\gamma^*(\cdot)$ satisfies (5.51) and $-G$ is stable.

5.8 Optimal trajectories and turnpike properties of infinite horizon autonomous nonconvex systems

We consider the maximization of $\int_0^T \mathcal{L}(x, \dot{x})\, dt$ as $T \to \infty$, which is a special case of Problem I in Section 5.3. The situation where $\mathcal{L}(\cdot,\cdot)$ is jointly concave was discussed in Chapter 4, and was studied by Brock and Haurie [28] and Leizarowitz [119] with the overtaking criterion. In [119] a certain differential inclusion $\dot{x} \in G(x)$ was associated with the integrand $\mathcal{L}(\cdot,\cdot)$ and optimality properties of the maximization problem were related to dynamical properties of this differential inclusion. The concavity of $\mathcal{L}(\cdot,\cdot)$ implied the simplification that $x \to G(x)$ had a convex graph and its dynamical properties be determined by a certain *linear differential equation.*

Here we discuss a situation where rather than joint convexity of $\mathcal{L}(\cdot,\cdot)$ we merely require $v \to \mathcal{L}(x,v)$ to be concave for every fixed x. Consideration of the overtaking optimality notion leads again to examining a related differential inclusion, which, in certain cases, reduces to an ordinary differential equation. Dynamical properties of this differential inclusion (or equation) are related to optimality properties of the maximization problem as follows: if all the w-limit sets are the same, independent of the solution, then there is a weak overtaking optimal trajectory. If the common w-limit set is a singleton then there exists an overtaking optimal trajectory and the system has the turnpike property.

In the next section we will demonstrate how various optimal quantities can be explicitly computed in certain situations. Thus optimal trajectories can be computed as solutions of the above mentioned differential inclusion (equation). In the scalar case we will compute explicit expressions for the maximal reward growth-rate and for the differential inclusion which governs the behavior of the optimal trajectories. We will also apply our results to problems in $\mathbb{R}^n$ where the integrand is in a separated form $\mathcal{L}(x,v) = \ell(x) + g(v)$. In both cases the maximal reward growth-rate μ is equal to $\max_x \mathcal{L}(x,0)$.

We assume that there is a compact set $X \subset \mathbb{R}^n$, the closure of an open and convex set in $\mathbb{R}^n$, such that for every $v \in \mathbb{R}^n$

$$\mathcal{L}(x,v) > -\infty \text{ if and only if } x \in X.$$

Throughout this and the next section, we assume the following:

Assumption A

(i) *The function $(x,v) \to \mathcal{L}(x,v)$ defined on $X \times \mathbb{R}^n$ is continuous, and for every fixed $x \in X$ the function $v \to \mathcal{L}(x,v)$ is concave.*

(ii) $\mathcal{L}$ *satisfies the coercivity condition*

$$\mathcal{L}(x,v)/\|v\| \to -\infty \text{ as } \|v\| \to \infty \text{ for every } x \in X.$$

(iii) *There is a constant* $c > 0$ *such that*

$$|\mathcal{L}(x,v) - \mathcal{L}(y,v)| \leq c\|x-y\|(1+\|v\|)$$

for every $x, y \in X$ *and every* $v \in \mathbb{R}^n$.

We call a function for which Assumption A holds *an integrand.*

Definition 5.5 *A function* $x(\cdot) : [0,\infty) \to \mathbb{R}^n$ *is called* **a trajectory** *of* $\mathcal{L}$ *if it is absolutely continuous (a.c.) and if* $\mathcal{L}(x(t),\dot{x}(t)) > -\infty$ *for almost every* $t \in [0,\infty)$, *namely* $x(t) \in X$ *for every* $t \geq 0$.

The problem that we consider here is the maximization of $\int_0^T \mathcal{L}(x(t),\dot{x}(t))dt$ as $T \to \infty$. With every trajectory $x(\cdot)$ of $\mathcal{L}$ we associate the reward-flow $c(\cdot)$ defined by

$$c(t) = \int_0^t \mathcal{L}(x(s),\dot{x}(s))\,ds.$$

We are interested in the existence and characterization of overtaking, and weakly overtaking optimal trajectories for a given integrand.

The following is a heuristic discussion which is intended to motivate the consideration of equation (5.53) below.

For the finite-horizon problem of maximizing $\int_0^T \mathcal{L}(x,\dot{x})\,dt$ the value function $(x,t) \to \varphi(x,t)$ satisfies the Bellman equation

$$\begin{cases} \dfrac{\partial\varphi}{\partial t} + \max\limits_v \left\{\mathcal{L}(x,v) + v'\nabla\varphi(x,t)\right\} = 0, \\ \varphi(T,x) = 0 \end{cases} \tag{5.52}$$

(see, e.g., Theorem 4.1, p. 83, of Fleming and Rishel [78]). The function $(x_0,t_0) \to \varphi(x_0,t_0)$ expresses the maximal value of $\int_{t_0}^T \mathcal{L}(x,\dot{x})\,dt$ where $x(\cdot)$ is subject to $x(t_0) = x_0$. In the infinite-horizon case we expect the reward integral to grow at some maximal rate $T \to \mu T$, but still expect the expressions

$$\left[\int_0^T \mathcal{L}(x(t),\dot{x}(t))\,dt - \mu T\right]$$

to remain bounded as T varies in $[0, \infty)$, for the better trajectories $x(\cdot)$. Thus replacing $\mathcal{L}(x,v)$ in (5.52) by $\mathcal{L}(x,v) - \mu$ would yield an equation for the excess reward, namely the reward left after subtracting the linear part μT from the reward integral $\int_0^T \mathcal{L}(x,\dot{x})\,dt$. Moreover, since the problem is defined on an infinite time interval, φ is time independent and the term $\partial\varphi/\partial t$ in (5.52) should disappear, and we are led to consideration of the equation

$$\max_v \left\{\mathcal{L}(x,v) + v'\nabla p(x)\right\} = \mu, \quad x \in X, \ v \in \mathbb{R}^n. \tag{5.53}$$

This is the Bellman equation for our infinite-horizon problem which is an equation both for the function $p(\cdot)$ and for the scalar μ. This completes the heuristic discussion and we now return to a rigorous treatment of the problem.

We consider a Lipschitz continuous function $p(\cdot)$ in X and denote by $Dp(x_0, v_0)$ the directional derivative of $p(\cdot)$ at the point x_0 in the direction v_0, namely

$$Dp(x_0, v_0) = \lim_{h \to 0+} \frac{p(x_0 + hv_0) - p(x_0)}{h}.$$

For almost every x_0 is X the directional derivative is defined for all v_0, since, as $p(\cdot)$ is Lipschitz continuous, $\nabla p(x_0)$ is defined for almost every x_0 in X. For a given $X \subset \mathbb{R}^n$ a pair (p, μ) can be given the control interpretation described above if it is a *good solution* of (5.53), which is defined as follows:

Definition 5.6 *Let $x \to p(x)$ be Lipschitz continuous in X. The pair $(p(\cdot), \mu)$ is a* **good solution** *of* (5.53) *if:*

(i) *Equality holds in* (5.53) *for almost every x in X.*

(ii) *For every $x_0 \in X$ there is an a.c. trajectory $x(\cdot)$ such that $x(0) = x_0$, $x(t) \in X$ for all $t \geq 0$ and $\mathcal{L}(x(t), \dot{x}(t)) + Dp(x(t), \dot{x}(t)) = \mu$ for almost every t in $[0, \infty)$.*

We define the set-valued function $x \to G(x)$ by

$$G(x) = \{v \in \mathbb{R}^n \, : \, \mathcal{L}(x,v) + Dp(x,v) = \mu\} \tag{5.54}$$

whenever $Dp(x,v)$ is defined. For a prescribed initial value $x_0 \in X$ we consider the differential inclusion

$$\dot{x}(t) \in G(x(t)) \text{ for almost every } t > 0 \tag{5.55}$$

with the initial condition $x(0) = x_0$. An a.c. solution $x(\cdot)$ of (5.55) which satisfies $x(t) \in X$ for all $t \geq 0$ is called a *viable solution* of (5.55). Using this terminology condition (ii) in Definition 5.6 can be phrased as: for every $x_0 \in X$ there is a viable solution $x(\cdot)$ of (5.55) satisfying $x(0) = x_0$.

If $G(x)$ is a singleton, as in the case where $v \to \mathcal{L}(x,v)$ is strictly concave and $p(\cdot)$ is continuously differentiable (so that $Dp(x,v) = \nabla p(x)'v$ in (5.54)), then (5.55) is an ordinary differential equation. The requirement in (ii) of Definition 5.6 is that all the solutions of this equation which start in X at $t = 0$ will stay in X for all times $t > 0$.

Assume that $(p(\cdot), \mu)$ is a good solution of (5.53) and for an $x_0 \in X$ let $x(\cdot)$ be a viable solution of (5.54) satisfying $x(0) = x_0$. The function $t \to p(x(t))$ is a.c., being the composition of a Lipschitz continuous $p(\cdot)$ and an a.c. $x(\cdot)$, and its derivative is a.e. equal to $Dp(x(t), \dot{x}(t))$. It thus follows that a.e. in $[0, \infty)$ $\mathcal{L}(x(t), \dot{x}(t)) = \mu - Dp(x(t), \dot{x}(t))$ and

$$\int_0^T \mathcal{L}(x(t), \dot{x}(t))\, dt = \mu T + p(x_0) - p(x(T)). \tag{5.56}$$

Since $p(\cdot)$ is bounded on X (5.56) implies

$$\lim_{T\to\infty} \frac{1}{T} \int_0^T \mathcal{L}(x(t), \dot{x}(t))\, dt = \mu. \tag{5.57}$$

On the other hand, now let $x(\cdot)$ be any trajectory of $\mathcal{L}$ (recall Definition 5.5), so that $x(\cdot)$ is a.c. and $x(t) \in X$ for all $t \geq 0$. It is the assertion of Proposition 5.11 below that for almost every $t > 0$ the following holds

$$\mathcal{L}(x(t), \dot{x}(t)) \leq \mu - Dp(x(t), \dot{x}(t))$$

implying that

$$\limsup_{T\to\infty} \frac{1}{T} \int_0^T \mathcal{L}(x(t), \dot{x}(t))\, dt \leq \mu. \tag{5.58}$$

We thus conclude from (5.57) and (5.58) the following:

Theorem 5.8 *If $(p_i(\cdot), \mu_i)$, $i = 1, 2$ are good solutions of (5.53) then $\mu_1 = \mu_2$, namely uniqueness holds for the scalar component of the good solution. This unique value μ is the maximal reward-growth, or the maximal long-run average reward.*

We show next that the set of good solutions is not vacuous and that there indeed exist good solutions for the infinite-horizon Bellman equation (5.53). The method of proof is by defining a finite-horizon problem with the same maximal reward as that of an infinite-horizon problem which is closely related to the original one, and applying known results for finite-horizon control problems.

Theorem 5.9 *Let $\mathcal{L}(\cdot,\cdot)$ be an integrand with a constraint set $X \subset \mathbb{R}^n$. Then the Bellman equation (5.53) has a good solution $(p(\cdot), \mu)$.*

To prove Theorem 5.9 we will need the following results.

Proposition 5.8 *Under Assumption A there exist constants* μ *and* $M > 0$ *such that*

$$\int_0^T [\mathcal{L}(x(t), \dot{x}(t)) - \mu]\, dt \leq M \tag{5.59}$$

for all $T > 0$ *and all trajectories* $x(\cdot)$, *and there is a trajectory* $y(\cdot)$ *of* $\mathcal{L}$ *so that*

$$\left| \int_0^T [\mathcal{L}(y(t), \dot{y}(t)) - \mu]\, dt \right| \leq M \tag{5.60}$$

for all $T > 0$.

Proof : We first prove that (5.59) and (5.60) hold for every integer $T > 0$ and then the assertion will follow. To this end we define the function $m(\cdot, \cdot) : X \times X \to \mathbb{R}$

$$m(y, z) = \sup\left\{ \int_0^1 \mathcal{L}(x, \dot{x})\, dt \ : \ x(0) = y, x(1) = z \right\}, \tag{5.61}$$

where the supremum is over all the trajectories $x(\cdot)$ which satisfy $x(0) = y, x(1) = z$ and $x(t) \in X$ for $0 \leq t \leq 1$. The fact that X is the closure of an open and convex set implies that $m(y, z) < \infty$ for every $y, z \in X$, since

$$\int_0^1 \mathcal{L}(x(t), \dot{x}(t))\, dt < \infty \text{ for } x(t) = x + t(y - x),$$

by the continuity of $\mathcal{L}(\cdot, \cdot)$. We claim that the conditions in Assumption A guarantee that $\mathcal{L}(\cdot, \cdot)$ is bounded above on $X \times \mathbb{R}^n$. Indeed, it follows easily from (i) and (ii) of Assumption A that

$$\mathcal{L}(x, v)/\|v\| \to -\infty \text{ as } \|v\| \to \infty,$$

uniformly for $x \in X$. This, together with the continuity of $\mathcal{L}(\cdot, \cdot)$ implies that $\mathcal{L}(\cdot, \cdot)$ is bounded above on $X \times \mathbb{R}^n$. It thus follows that $m(y, z) < \infty$ for every $y, z \in X$.

The conditions in Assumption A guarantee that the supremum in (5.61) is attained by some trajectory (see Theorem 4.1, p. 68, of Fleming and Rishel [78]). Moreover, $m(\cdot, \cdot)$ is continuous on $X \times X$ (see Proposition 5.9 below).

It follows from Theorem 5.1 that there exist constants μ and $M_0 > 0$ such that

$$\begin{cases} \sum_{i=0}^{N-1} m(x_i, x_{i+1}) - \mu N \leq M_0 \\ \qquad \text{for all } N \geq 1 \text{ and every sequence } \{x_i\}_{i=0}^{\infty} \subset X, \\ \qquad \text{and there exists a sequence } \{x_i^*\}_{i=0}^{\infty} \subset X \text{ such that} \\ \left| \sum_{i=0}^{N-1} m(x_i^*, x_{i+1}^*) - \mu N \right| \leq M_0 \text{ for all } N \geq 1. \end{cases} \tag{5.62}$$

Therefore (5.59) and (5.60) hold if we consider only integers $T > 0$.

Since $\mathcal{L}(\cdot,\cdot)$ is bounded above on $X \times \mathbb{R}^n$ it follows that there is a constant α such that

$$\int_0^\delta \mathcal{L}(x(t),\dot{x}(t))\,dt < \alpha \tag{5.63}$$

for every $0 < \delta \leq 1$ and every trajectory $x(\cdot)$. It follows from (5.62) and (5.63) that (5.59) and (5.60) hold for every $T > 0$ with

$$M = M_0 + |\alpha| + |\mu|.$$

Let $x(\cdot)$ be a trajectory such that

$$x(k) = x_k^* \text{ for all } k \geq 0, \text{ and}$$

$$\int_k^{k+1} \mathcal{L}(x(t),\dot{x}(t))\,dt = m(x_k^*, x_{k+1}^*) \text{ for every } k \geq 0.$$

Then $x(\cdot)$ satisfies (5.60), which concludes the proof of the proposition. ∎

Once we have established the validity of (5.59) and (5.60) we can go on and define the function $p(\cdot) : X \to \mathbb{R}$

$$p(x_0) = \sup\Big\{\limsup_{T\to\infty} \int_0^T \big[\mathcal{L}(x(t),\dot{x}(t)) - \mu\big]\,dt \; : \; x(\cdot) \text{ is a trajectory for } \mathcal{L}, \text{ and } x(0) = x_0\Big\}. \tag{5.64}$$

It follows from Proposition 5.8 that $p(\cdot)$ is well defined. In the sequel we will need the following results from [119].

Proposition 5.9 *The function $m(\cdot,\cdot)$ is Lipschitz continuous on $X \times X$.*

Proposition 5.10 *The function $p(\cdot)$ is Lipschitz continuous on X.*

Proposition 5.11 *Let $x(\cdot)$ be any trajectory of $\mathcal{L}$ in X. Then the following holds:*

$$\mathcal{L}(x(t),\dot{x}(t)) + Dp(x(t),\dot{x}(t)) - \mu \leq 0 \tag{5.65}$$

for almost every $t \geq 0$.

These assertions are proved in [119], Propositions 4.3, 4.4 and 4.5 respectively.

Proof of Theorem 5.9 : Let $T > 0$ be fixed and consider the finite horizon maximization problem on the finite time interval $[t_0, T]$, for some $0 \le t_0 < T$

$$\max_{x(\cdot)} \left\{ \int_{t_0}^{T} \left[\mathcal{L}(x(t), \dot{x}(t)) - \mu\right] dt + p(x(T)) \; : \; x(t_0) = y \right\}, \tag{5.66}$$

where $p(\cdot)$ is the function which is defined in (5.64). It follows from the definition of $p(\cdot)$ that the maximal value in (5.66) is $p(y)$, no matter what the values of t_0 and T are.

We denote the value function of this problem by $W(t_0, y)$. It follows from standard results about the value function (see Theorem 4.1, p. 81, of Fleming and Rishel [78]) that at all points where the value function is differentiable it satisfies

$$\frac{\partial W}{\partial t_0} + \max_{v}\left\{\mathcal{L}(y, v) + \frac{\partial W}{\partial y}' v - \mu\right\} = 0.$$

(Here we use the fact that there exist optimal solutions to the problem (5.66), as implied by Assumption A.) But as remarked above we have $W(t_0, y) = p(y)$ for all $0 \le t_0 < T$, thus we obtain the equality

$$\max_{v}\left\{\mathcal{L}(y, v) + \nabla p(v)' v - \mu\right\} = 0$$

whenever $p(\cdot)$ is differentiable. By Proposition 5.10 $p(\cdot)$ is Lipschitz continuous. Hence, by Rademacher's Theorem (see p. 216 of Federer [74]), it is differentiable a.e. in X. It thus follows that (5.53) holds a.e. in X which proves the validity of (i) in Definition 5.6.

To complete the proof we have to construct a viable solution $x(\cdot)$ of (5.55) for every initial value $x_0 \in X$. Clearly, it is enough to construct a solution $x(\cdot)$ of (5.55) which satisfies $x(0) = x_0$ and $x(t) \in X$ for all $0 \le t \le 1$. Let $x(\cdot)$ be a trajectoy which maximizes

$$\int_0^1 \left[\mathcal{L}(x(t), \dot{x}(t)) - \mu\right] dt + p(x(1)), \tag{5.67}$$
$$x(0) = x_0, \quad x(t) \in X \text{ for } 0 \le t \le 1.$$

The existence of such an $x(\cdot)$ follows from Assumption A. The function $t \to p(x(t))$ is absolutely continuous, and at every point in $(0, 1)$ such that both $x(\cdot)$ and $p(x(\cdot))$ are differentiable $(d/dt)p(x(t)) = Dp(x(t), \dot{x}(t))$. Thus the directional derivative $Dp(x(t), \dot{x}(t))$ exists for almost every t in $[0, 1]$. We claim that

$$\mathcal{L}(x(t), \dot{x}(t)) + Dp(x(t), \dot{x}(t)) - \mu \ge 0 \qquad \text{a.e. in } [0, 1]. \tag{5.68}$$

Let $\mathcal{D}$ be the set

$$\mathcal{D} = \left\{t \in [0, 1] \; : \; \mathcal{L}(x(t), \dot{x}(t)) + Dp(x(t), \dot{x}(t)) - \mu < 0\right\},$$

and we have to prove that $\mathcal{D}$ has Lebesgue measure zero. Otherwise there would be a $\beta > 0$ such that

$$\mathcal{E} = \left\{t \in [0,1] \, : \, \mathcal{L}(x(t), \dot{x}(t)) + Dp(x(t), \dot{x}(t)) - \mu \leq -\beta\right\}$$

has positive Lebesgue measure, hence $\mathcal{E}$ contains a Lebesgue point $0 < \tau < 1$ of the function $t \to \mathcal{L}(x(t), \dot{x}(t)) + Dp(x(t), \dot{x}(t))$. For such a Lebesgue point τ we have

$$\lim_{\delta \to 0+} \frac{1}{\delta} \int_{\tau}^{\tau+\delta} \left[\mathcal{L}(x(t), \dot{x}(t)) + Dp(x(t), \dot{x}(t)) - \mu\right] dt \leq -\beta$$

implying that for a sufficiently small $\delta > 0$ the following holds:

$$\int_{\tau}^{\tau+\delta} \left[\mathcal{L}(x(t)\dot{x}(t)) - \mu\right] dt + p(x(\tau+\delta)) \leq p(x(\tau)) - \frac{1}{2}\delta\beta. \tag{5.69}$$

But since $x(\cdot)$ maximizes (5.67) it follows that it must also maximizes the expression

$$\int_{\tau}^{\tau+\delta} \left[\mathcal{L}(z(t), \dot{z}(t)) - \mu\right] dt + p(z(\tau+\delta))$$

subject to $z(\tau) = x(\tau)$, with the maximal value being $p(x(\tau))$. This, however, contradicts (5.69), which concludes the proof of (5.68). It follows from (5.65) and (5.68) that $x(\cdot)$ satisfies (5.54) a.e. in $[0,1]$. Since $x(0) = x_0$ and $x(t) \in X$ for all $0 \leq t \leq 1$, it follows that (ii) of Definition 5.6 holds, which completes the proof of the theorem. ∎

By Proposition 5.11 every trajectory of $\mathcal{L}$ satisfies (5.65) on $[0, \infty)$. Our claim is that trajectories for which equality in (5.65) holds for almost every $t > 0$ are of special importance from the optimality point of view. We recall the definition of the set-valued function $x \to G(x)$ in (5.54) and consider the differential inclusion

$$\begin{cases} \dot{x}(t) \in G(x(t)) & \text{for almost every } t > 0, \\ x(0) = x_0. \end{cases} \tag{5.70}$$

For every initial value there exists, by Theorem 5.9, a trajectory $x(\cdot)$ which satisfies (5.70).

Let $x(\cdot)$ be a trajectory of $\mathcal{L}$ and assume that

$$\int_{0}^{\infty} \left[\mathcal{L}(x(t), \dot{x}(t)) + Dp(x(t), \dot{x}(t)) - \mu\right] dt > -\infty. \tag{5.71}$$

Let $[a_j, b_j]$ be a sequence of time intervals, $b_j - a_j = \ell$ for some fixed positive number ℓ. Assume that $a_j \to \infty$ as $j \to \infty$ and let $x_j(\cdot)$ be defined on $[0, \ell]$ by

$$x_j(t) = x(a_j + t).$$

The following is the content of Proposition 5.1 in [119].

Proposition 5.12 *Let $x(\cdot)$ be a trajectory of $\mathcal{L}$ and let $[a_j, b_j]$ and $x_j(\cdot)$ be as above. Suppose further that* (5.71) *holds. Then there is a function $y : [0,\ell] \to X$ such that $\dot{y}(t) \in G(y(t))$ for almost every $t \in [0,\ell]$ and there is an increasing sequence of natural numbers $\{j_k\}_{k=1}^{\infty}$ such that*

$$\lim_{k\to\infty} \max_{0\le t\le \ell} |x_{j_k}(t) - y(t)| = 0.$$

We next relate optimality properties of trajectories with dynamical properties of the relation (5.70). For a solution $x(\cdot)$ of (5.70) we denote, as usual, by $w(x(\cdot))$ the set of all points $x \in X$ such that $x(t_j) \to x$ for some sequence of times $t_j \to \infty$. Using Proposition 5.12 it is not hard to deduce the following result (see [119], Theorem 5.2).

Theorem 5.10 *Suppose that there is a set $S \in X$ such that $w(x(\cdot)) = S$ for every viable solution $x(\cdot)$ of* (5.70). *Then every viable solution of the differential inclusion* (5.70) *is a weakly overtaking optimal trajectory for the infinite-horizon maximization problem for $\mathcal{L}$ in X.*

In order to conclude that a solution of (5.70) is overtaking optimal we impose a stronger assumption on the asymptotic dynamics of (5.70).

Theorem 5.11 *Suppose that there is a point $\overline{x} \in X$ so that*

$$\lim_{t\to\infty} x(t) = \overline{x} \tag{5.72}$$

for every viable solution $x(\cdot)$ of (5.70). *Moreover, the limit in* (5.72) *has a uniform rate in the sense that for every $\varepsilon > 0$ there is a $t_0 > 0$ so that $|x(t) - \overline{x}| < \varepsilon$ for all $t > t_0$ and all the viable solutions of* (5.70). *Then there is an overtaking optimal trajectory for every initial value x_0. Such a trajectory is given by a viable solution of* (5.70) *with the initial state x_0.*

Proof: Let $x(\cdot)$ be a trajectory for $\mathcal{L}$ such that $x(0) = x_0$. Let $x^*(\cdot)$ be a viable solution of (5.70) satisfying $x^*(0) = x_0$. Comparing the corresponding reward flows $c(\cdot)$ and $c^*(\cdot)$ of $x(\cdot)$ and $x^*(\cdot)$ respectively in light of the overtaking criterion it is enough to consider the case

$$\liminf_{T\to\infty} \int_0^T [\mathcal{L}(x,\dot{x}) - \mu]\, dt > -\infty.$$

It then follows from Proposition 5.12 and the uniform rate of convergence of solutions of (5.70) to $\bar{x}$ that $x(\cdot)$ also tends to $\bar{x}$ as $t \to \infty$.

We have by (5.65)

$$\int_0^T \Big[\mathcal{L}(x,\dot{x}) - \mu\Big]\, dt \le p(x_0) - p(x(T)) \tag{5.73}$$

while $x^*(\cdot)$ satisfies

$$\int_0^T \left[\mathcal{L}(x^*,\dot{x}^*) - \mu\right] dt = p(x_0) - p(x^*(T)). \tag{5.74}$$

Using the convergence of $x(T)$ and $x^*(T)$ to $\overline{x}$, the continuity of $p(\cdot)$, (5.73) and (5.74) we conclude that for every $\varepsilon > 0$ the relation

$$\int_0^T \mathcal{L}(x^*,\dot{x}^*)\,dt > \int_0^T \mathcal{L}(x,\dot{x})\,dt - \varepsilon$$

holds for every large enough T, proving the overtaking optimality of $x^*(\cdot)$. ∎

Remark 5.10 *If $\mathcal{L}(x,\cdot)$ is strictly concave in v for every fixed x, and $p(\cdot)$ is continuously differentiable then the differential inclusion* (5.70) *is in fact a differential equation. If it is sufficiently regular, then it has a unique viable solution through every initial value in X. Asymptotic dynamic properties of this equation are then related to optimality properties of the control problem as expressed in Theorems* 5.10 *and* 5.11.

5.9 Two special cases: Scalar systems, and integrands in a separated form

In this section we consider the scalar case $n = 1$. Let $\mathcal{L}(\cdot,\cdot)$ be defined on $[a,b] \times \mathbb{R}$, where $[a,b] \subset \mathbb{R}$ is a finite interval in which $x(\cdot)$ takes values. In this case we can establish simple and explicit expressions for μ and $p(\cdot)$. Specifically

$$\mu = \max_{a \le x \le b} \mathcal{L}(x,0), \tag{5.75}$$

while $p(\cdot)$ is determined by the condition that for every $x \in [a,b]$ the value of $(-p'(x))$ is equal to a slope of a tangent line to the graph of $\mathcal{L}(x,\cdot)$ which goes through the point $(0,\mu)$. Once explicit expressions for μ and $p'(x)$ are known, the set-valued function $x \to G(x)$ can be explicitly computed and used to determine the optimal trajectories.

We are looking for a good solution of the Bellman equation

$$\max_v \{\mathcal{L}(x,v) + p'(x)v\} = \mu. \tag{5.76}$$

A solution (p,μ) of (5.76) is a good solution if the following holds:

(i) $p(\cdot)$ is continuously differentiable in (a,b) and continuous in $[a,b]$.

(ii) Denote by $g(x)$ the set of points $v \in \mathbb{R}$ for which the maximum in (5.76) is attained. Then there is a $\delta > 0$ such that

$a \leq x \leq a + \delta$ and $v \in g(x)$ implies $v > 0$,

and

$b - \delta \leq x \leq b$ and $v \in g(x)$ implies $v < 0$.

(This condition guarantees the viability of solutions of $\dot{x} \in g(x)$.)

In what follows we assume that the maximum in (5.75) is attained at a unique point $a < c < b$. For every $x \in [a, b]$, $x \neq c$, there are two tangent lines for the graph of the function $v \to \mathcal{L}(x, v)$ that go through the point $(0, \mu)$. We denote their slopes by $\ell_+(x)$ and $\ell_-(x)$. We define the set-valued function ℓ on $[a, b]$ as follows

$$\ell(x) = \begin{cases} \ell_+(x) & \text{if } a \leq x < c, \\ \text{the set of slopes of tangent lines} & \\ \text{to } \mathcal{L}(c, \cdot) \text{ at } v = 0 & \text{if } x = c, \\ \ell_-(x) & \text{if } c < x \leq b. \end{cases}$$

It is easy to see that $\ell(\cdot)$ is continuous in $[a, c)$ and in $(c, b]$. Let $g(x)$ be the points $v \in \mathbb{R}$ where the following maximum is attained

$$\max_v \{\mathcal{L}(x, v) - \ell(x)v\}. \tag{5.77}$$

It follows from the uniform divergence to $-\infty$ of $\mathcal{L}(x, v)/|v|$ as $|v| \to \infty$ that the sets $g(x)$ are uniformly bounded for all $x \in [a, b]$.

Proposition 5.13 *Assume that the maximum in* (5.75) *is attained at a unique point* $a < c < b$ *and let the functions* $x \to \ell(x)$ *and* $x \to g(x)$ *be defined as above. Then the set-valued function* $x \to g(x)$ *is upper-semicontinuous in* $[a, b]$.

Proof: Since the sets $g(x)$ are uniformly bounded, the assertion will follow once we have shown that $g(\cdot)$ has a closed graph. Since $x \to \ell(x)$ is continuous at every $x \neq c$, the graph is closed at every such point by the continuity of $\mathcal{L}(\cdot, \cdot)$ and the fact that the maximum in (5.77) has a constant value μ. We have to prove that the graph is closed at $x = c$. We consider a sequence $\{x_j\}_{j=1}^{\infty}$ such that $x_j \to c$ as $j \to \infty$, and let $v_j \in g(x_j)$ be such that $v_j \to v$ as $j \to \infty$. We assume, to get a contradiction, that $v \notin g(c)$. Since $0 \in g(c)$ we must have $v \neq 0$, and we can assume, without loss of generality, that $v > 0$. It follows from the concavity of $\mathcal{L}(c, \cdot)$ and the fact $\mu = \mathcal{L}(c, 0)$ that

$$\mathcal{L}\Big(c, \frac{v}{2}\Big) > \frac{1}{2}\, [\mu + \mathcal{L}(c, v)] \tag{5.78}$$

since equality in (5.78) would imply that $v \in g(c)$. We also have for every $j \geq 1$

$$\mathcal{L}\Big(x_j, \frac{v_j}{2}\Big) \leq \mu + \ell_j \cdot \frac{v_j}{2} = \frac{1}{2}[\mu + \mathcal{L}(x_j, v_j)]$$

with the last equality following from $v_j \in g(x_j)$, (and where ℓ_j is the slope of the relevant tangent line to $\mathcal{L}(x_j, \cdot)$). Letting $j \to \infty$ in the inequality $\mathcal{L}(x_j, v_j/2) \le \frac{1}{2}[\mu + \mathcal{L}(x_j, v_j)]$ and using the continuity of $\mathcal{L}(\cdot,\cdot)$ leads into a contradiction to (5.78) and completes the proof of the theorem. ∎

Let μ be as in (5.75) and we define a continuous function $p(\cdot)$ on $[a,b]$ by requiring that

$$\begin{cases} p'(x) = -\ell(x) & \text{for } x \neq c, \\ p(a) = 0. \end{cases} \tag{5.79}$$

Theorem 5.12 *Assume that the maximum in* (5.75) *is attained at a unique point* $a < c < b$. *Then* $(p(\cdot)\,, \mu)$ *which are defined in* (5.79) *and* (5.75) *provide a good solution, and* μ *is the maximal long-run average reward growth-rate. Any solution of* $\dot{x}(t) \in g(x(t))$, $x(0) = x_0$, *is an overtaking optimal trajectory. Such a solution exists for every* $x_0 \in [a,b]$.

Proof: In order to show that $(p(\cdot)\,, \mu)$ is a good solution it is enough to demonstrate the existence of viable solutions to $\dot{x}(t) \in g(x(t))$, $x(0) = x_0$, for every $x_0 \in [a,b]$. It follows from Proposition 5.13 and from standard results about differential inclusions that for every $x_0 \in (a,b)$ there exists a solution of $\dot{x}(t) \in g(x(t))$, $x(0) = x_0$, in some time interval $[0,\tau]$. The fact that

$$v \in g(a) \text{ implies } v > 0$$

and

$$v \in g(b) \text{ implies } v < 0$$

guarantees that there indeed exists a viable solution in $[a,b]$ for every initial value $x_0 \in [a,b]$.

The overtaking optimality of the solutions of $\dot{x}(t) \in g(x(t))$ will follow from Theorem 5.11 once we have proved that $x(t) \to c$ as $t \to \infty$ in a uniform rate for all the solutions. But this follows from the fact that for every $\delta > 0$ there is an $\eta > 0$ such that

$$x \ge c + \delta \text{ and } v \in g(x) \text{ implies } v \le -\eta,$$

and also

$$x \le c - \delta \text{ and } v \in g(x) \text{ implies } v \ge \eta.$$

∎

We now consider the optimization problem in $\mathbb{R}^n$ for integrands which are in a separated form, and demonstrate how explicit optimal quantities can be computed for a wide class of such systems.

Theorem 5.13 *Let the integrand* $\mathcal{L}(\cdot,\cdot)$ *have the form*

$$\mathcal{L}(x,v) = \ell(x) + g(v),$$

where $\ell(\cdot)$ *is continuous on* $X \subset \mathbb{R}^n$, $g(\cdot)$ *is strictly concave on* $\mathbb{R}^n$, *and* $g(v)/|v| \to \infty$ *as* $|v| \to \infty$. *Then the maximal reward growth-rate is*

$$\mu = g(0) + \max_{x\in X} \ell(x)\Big(= \max_{x\in X} \mathcal{L}(x,0)\Big). \tag{5.80}$$

Let $\mathcal{H}(\cdot) : \mathbb{R}^n \to \mathbb{R}$ *be defined by*

$$\mathcal{H}(\eta) = \max_{v\in\mathbb{R}^n} \big\{g(v) + v'\eta\big\}.$$

Then the differential inclusion (5.55) *has the form*

$$\dot{x} \in \nabla\mathcal{H}(\nabla p(x)) \tag{5.81}$$

for almost every $x \in X$, *and if* $g(\cdot)$ *is continuously differentiable in* X *then it reduces to the differential equation* (5.81).

Proof: The Bellman equation is

$$\max_{v} \big\{g(v) + v'\nabla p(x)\big\} = \mu - \ell(x) \tag{5.82}$$

and since the left-hand side is not smaller than $g(0)$ we have

$$\mu \geq g(0) + \max_{x\in X} \ell(x). \tag{5.83}$$

Let $(p(\cdot)\,,\mu)$ be a good solution of (5.82) and let $x(\cdot)$ be a trajectory such that

$$\ell(x(t)) + g(\dot{x}(t)) + Dp(x(t),\dot{x}(t)) = \mu \tag{5.84}$$

for almost every $t > 0$.

(The existence of such a trajectory is guaranteed by the fact that $(p(\cdot)\,,\mu)$ is a good solution.) Integrating both sides of (5.84) over $[0,T]$ and using the concavity of $g(\cdot)$ we get

$$\mu \leq \frac{1}{T}\int_0^T \ell(x(t))\,dt + g\Big(\frac{x(T)-x(0)}{T}\Big) + \frac{1}{T}\big[p(x(T)) - p(x(0))\big]$$

from which it follows that $\mu \leq g(0) + \max_{x\in X} \ell(x)$, by letting $T \to \infty$. This inequality and (5.83) imply (5.80).

It remains to prove (5.81). This, however, follows from the fact that $\nabla\mathcal{H}(\eta)$ is equal the value v_0 which maximizes the expression $[g(v) + \eta \cdot v]$, thus $\nabla\mathcal{H}(\nabla p(x))$ is the only point in the singleton set $G(x)$ at every point where $p(\cdot)$ is differentiable. ∎

Chapter 6

Asymptotic Stability with a Discounted Criterion; Global and Local Analysis

6.1 Introduction

In this chapter, the global asymptotic stability (GAS) property of optimally controlled systems with an infinite time horizon will be further explored by considering the case where the criterion has the following form:

$$J_T(x_0, u(\cdot)) = \int_0^T e^{-\rho t} f_0(x(t), u(t))\, dt. \tag{6.1}$$

The term $e^{-\rho t}$ is a discount factor and ρ is the discount rate. If ρ is positive and $f_0(x(\cdot), u(\cdot))$ is bounded then the integral converges when $T \to \infty$. In the case of a positive discount rate the relative weighting given to a distant future makes it negligible, hence one may suppose that high positive values of ρ can have a destabilizing effect on the optimal trajectories. Reciprocally, a negative discount rate gives the most importance to what happens in the distant future and this should amplify the stabilizing forces. With $\rho \leq 0$ the integral (6.1) is generally not convergent as $T \to \infty$.

Here we treat the case of a positive discount rate which is more important for economic applications. The GAS properties established in Chapter 4 can be extended to this case provided that more stringent assumptions be made on the Hamiltonian. The general result obtained previously was that strict concavity-strict convexity of the Hamiltonian assures GAS for an autonomous control system having a unique extremal steady state $(\overline{x}, \overline{p})$. This steady state $\overline{x}$ and the associated costate $\overline{p}$ defined a saddle point for the Hamiltonian.

When the criterion is discounted the extremal steady state does not correspond anymore to a saddle point of the Hamiltonian. In order to assure stability a *curvature condition* has to be imposed on the Hamiltonian in the neighborhood of a stationary point $(\overline{x}, \overline{p})$. These conditions have been obtained independently by Cass and Shell [37], Brock and Sheinkman [30], and Rockafellar [156] with various degrees of generality.

Another approach developed independently by Magill [131], [130] consists in analyzing the local stability property of the stationary point $(\overline{x}, \overline{p})$ with the help of an approximate linear-quadratic system describing neighboring extremals and the second variation. This approach establishes an interesting link between the optimal regulator problem and nonlinear convex systems.

Finally, a discounted version of the finite horizon turnpike property presented in Theorem 4.2 has been obtained by Feinstein and Oren [76] who showed that the so called "mirage variables" satisfy the classical turnpike theorem.

6.2 Modified Hamiltonian systems

Consider a system described by

$$\dot{x}(t) = f(x(t), u(t)) \tag{6.2}$$

$$x(0) = x_0 \tag{6.3}$$

$$h(x(t), u(t)) \geq 0 \tag{6.4}$$

$$J_T(x_0, u(\cdot)) = \int_0^T e^{-\rho t} f_0(x(t), u(t))\, dt \tag{6.5}$$

where $f(\cdot,\cdot), h(\cdot,\cdot), f_0(\cdot,\cdot)$ are C^1 in both x and u and satisfy the usual assumptions of regularity. The necessary conditions for a regular extremal trajectory can be formally written (see Note 1 Chapter 2)

$$\dot{x} = \frac{\partial}{\partial \lambda} \hat{H}(x, u, t, \lambda, \mu) \tag{6.6}$$

$$\dot{\lambda} = -\frac{\partial}{\partial x} \hat{H}(x, u, t, \lambda, \mu) \tag{6.7}$$

$$0 = \frac{\partial}{\partial u} \hat{H}(x, u, t, \lambda, \mu) \tag{6.8}$$

$$0 = \mu' h(x(\cdot), u(\cdot)), \quad \mu \geq 0 \tag{6.9}$$

where

$$\hat{H}(x, u, t, \lambda, \mu) \triangleq e^{-\rho t} f_0(x, u) + \lambda' f(x, u) + \mu' h(x, u) \tag{6.10}$$

Due to the presence of the discount factor $e^{-\rho t}$ in the expression (6.10) of the Hamiltonian, the system (6.6)–(6.9) is time-dependent. However, a simple change of variables permits the elimination of the explicit time dependence in the necessary conditions. It suffices to define

$$q(t) = e^{\rho t}\lambda(t) \quad \nu(t) = e^{\rho t}\mu(t) \tag{6.11}$$

and to introduce the *modified Hamiltonian*

$$H(x,u,q,\nu) \triangleq f_0(x,u) + q'f(x,u) + \nu'h(x,u) \tag{6.12}$$

Thus the conditions (6.6)–(6.9) reduce to

$$\dot{x} = \frac{\partial}{\partial q} H \tag{6.13}$$

$$\dot{q} = -\frac{\partial}{\partial x} H + \rho q \tag{6.14}$$

$$0 = \frac{\partial}{\partial u} H \tag{6.15}$$

$$0 = \nu' h(x(\cdot), u(\cdot)) \tag{6.16}$$

$$\nu \geq 0.$$

In economic terms, the costate variable $\lambda(t)$ and the Lagrange multiplier $\mu(t)$ indicate the marginal value of the state $x(t)$ and of the control constraint $h(x(t), u(t)) \geq 0$ at time t. Due to the discount rate ρ, these marginal values are evaluated at their *present value.* The transformation (6.11) corresponds to the consideration of the *current value* instead of the *present value* in the evaluation of state and constraint sensitivities. Thus the Hamiltonian defined in (6.12) is also called the *current-value Hamiltonian.*

Let $\mathcal{H}(x,q)$ be the function obtained (as in Note 1 of Chapter 4) after elimination of u and ν in $H(x,u,q,\nu)$ by using (6.15) and (6.16). The conditions (6.13)–(6.16) reduce then to

$$\dot{x} = \frac{\partial \mathcal{H}}{\partial q} \tag{6.17}$$

$$\dot{q} = -\frac{\partial \mathcal{H}}{\partial x} + \rho q \tag{6.18}$$

The system (6.17)–(6.18) will be called a *modified Hamiltonian System,* to account for the presence of the ρq term in (6.18).

6.3 Cass-Shell conditions for GAS of modified Hamiltonian systems

The theory presented briefly in this section is fully developed in [37] and in [130].

Consider the modified Hamiltonian system (6.17)–(6.18) with the following additional boundary conditions

$$x(0) = x_0 \tag{6.19}$$

$$\lim_{t \to \infty} e^{-\rho t} q(t)' x(t) = 0 \tag{6.20}$$

The following hypothesis is assumed.

Assumption 6.1 *There exists a unique stationary point* $(\bar{x}, \bar{q})$ *such that*

(i)

$$\begin{cases} 0 = \dfrac{\partial \mathcal{H}}{\partial q} |_{\bar{x},\bar{q}} = \nabla_q \mathcal{H}(\bar{x}, \bar{q}) \\ 0 = -\dfrac{\partial \mathcal{H}}{\partial x} |_{\bar{x},\bar{q}} + \rho \bar{q} = -\nabla_x \mathcal{H}(\bar{x}, \bar{q}) + \rho \bar{q} \end{cases}$$

(ii) $\mathcal{H}(x, q)$ *is concave in* x*, convex in* q *and such that the following holds:*
For every $\varepsilon > 0$ *there exists a* $\delta > 0$ *such that* $\|x - \bar{x}\| > \varepsilon$ *implies that for every* $q \in \mathbb{R}^n$

$$(q - \bar{q})' \nabla_q \mathcal{H}(x, q) - [\nabla_x \mathcal{H}(x, q) - \rho \bar{q}]' (x - \bar{x}) > \delta - \rho (q - \bar{q})'(x - \bar{x}). \tag{6.21}$$

Theorem 6.1 *Under Assumption* 6.1, *a uniformly bounded solution* $(x(\cdot), q(\cdot))$ *of the modified Hamiltonian system* (6.17)–(6.18) *with boundary conditions* (6.19)–(6.20) *is such that* $\lim_{t \to \infty} x(t) = \bar{x}$.

Proof: One considers the function

$$V(x, q) = -(q - \bar{q})'(x - \bar{x}) \tag{6.22}$$

The trajectory derivative of V yields

$$\begin{aligned} \dot{V} &= -(q - \bar{q})' \nabla_q \mathcal{H}(q, x) + [\nabla_x \mathcal{H}(q, x) - \rho q]'(x - \bar{x}) \\ &= -(q - \bar{q})' \nabla_q \mathcal{H}(q, x) + [\nabla_x \mathcal{H}(q, x) - \rho \bar{q}]'(x - \bar{x}) - \rho (q - \bar{q})'(x - \bar{x}) \end{aligned} \tag{6.23}$$

Thus, by the concavity-convexity of $\mathcal{H}(x, q) - \rho \bar{q} x$ one has, according to (6.22) and (6.23),

$$e^{-\rho t}(\dot{V} - \rho V) = \frac{d}{dt}(V e^{-\rho t}) \leq 0. \tag{6.24}$$

The boundedness of $x(\cdot)$ and $q(\cdot)$ with the asymptotic boundary condition (6.20) yield

$$\lim_{t\to\infty}\left(V(x(t),q(t))e^{-\rho t}\right)=0$$

and hence

$$V(x,q)\geq 0 \text{ and } \limsup_{t\to\infty}\left(V(x(t),q(t))\right)\geq 0.$$

Now, by condition (6.21) one has

$$\left.\begin{array}{l} V(\overline{x},q)=0 \\ \dfrac{d}{dt}V(x,q)<0 \text{ if } x\neq\overline{x} \end{array}\right\} \text{ for any } q,$$

and this is a sufficient condition for assuring that

$$\lim_{t\to\infty}x(t)=\overline{x}.$$

■

Remark 6.1 *The concavity in x and convexity in q of $\mathcal{H}(x,q)$ is assured if $f_0(\cdot,\cdot)$ and $f(\cdot,\cdot)$ are concave functions of x and u, and if q is positive.*

Remark 6.2 *The condition of stability* (6.21) *is a "steepness" requirement of the Hamiltonian as discussed in Cass and Shell* [37].

6.4 Brock-Sheinkman conditions for GAS of modified Hamiltonian systems

In this section the Lyapunov function approach, first described in Section 4.6 for autonomous systems, is extended to the case where the criterion is discounted.

Assumption 6.2 *There exists a unique stationary point $(\overline{x},\overline{q})$ of the modified Hamiltonian system* (6.17)–(6.18) *which is such that*

(i) *The matrix*

$$Q(x,q)\triangleq\begin{bmatrix}\mathcal{H}_{xx} & -\frac{\rho}{2}I \\ -\frac{\rho}{2}I & -\mathcal{H}_{qq}\end{bmatrix} \tag{6.25}$$

exists everywhere.

(ii) *$Q(\overline{x},\overline{q})$ is negative definite.*

(iii) *If* $(x,q) \neq (\overline{x},\overline{q})$ *is such that*

$$(q-\overline{q})'\nabla_q \mathcal{H}(x,q)' + (x-\overline{x})'(\rho q - \nabla_x \mathcal{H}(x,q)') = 0$$

then necessarily the following also holds:

$$(x-\overline{x}, q-\overline{q})' Q(x,q)(x-\overline{x}, q-\overline{q}) < 0.$$

Theorem 6.2 *Under Assumption* 6.2, *all solutions of* (6.17)–(6.18) *that are bounded for* $t \geq 0$ *converge to* $(\overline{x},\overline{q})$ *as* $t \to \infty$.

Proof: The proof is very similar to the proof of Theorem 4.4. Put

$$V(x,q) \triangleq -(q-\overline{q})'(x-\overline{x})$$

and make the change of variables

$$\begin{aligned} z &\triangleq (x-\overline{x}, q-\overline{q}) \\ F_1(z) &\triangleq \nabla_q \mathcal{H}(z) \\ F_2(z) &\triangleq -\nabla_x \mathcal{H}(z) + \rho q \\ F(z) &= (F_1(z), F_2(z)). \end{aligned}$$

One has to prove that

$$\dot{V}(z) = \nabla V(z) F(z) < 0 \text{ for all } z \neq 0.$$

For doing this one will use the same arguments as in the proof of Theorem 4.4. In particular the developments between (4.76) and (4.78) are exactly the same. One still has

$$\nabla^2 V(\lambda z) = -2A = -\begin{bmatrix} 0 & I \\ I & 0 \end{bmatrix}$$

and

$$\begin{aligned} \nabla F(\lambda z) &= \begin{bmatrix} \mathcal{H}_{qx} & \mathcal{H}_{qq} \\ -\mathcal{H}_{xx} & -\mathcal{H}_{xq} + \rho \end{bmatrix}\Bigg|_{\lambda z} \\ \nabla V(\lambda z) &= -[x'p']\begin{bmatrix} 0 & I \\ I & 0 \end{bmatrix}. \end{aligned}$$

Bringing these expressions into (4.79) and (4.80) one obtains:

$$\begin{aligned} g'(\lambda) &= -2z'AF(\lambda z) - \lambda z' \begin{bmatrix} -\mathcal{H}_{xx} & -\mathcal{H}_{xq} & +\rho \\ \mathcal{H}_{qx} & \mathcal{H}_{qq} & \end{bmatrix}\Bigg|_{\lambda z} z \\ &= -\Big[(x-\overline{x})'F_2(\lambda z) + (q-\overline{q})'F_1(\lambda z)\Big] \\ &\quad -\lambda\Big[(x-\overline{x})', (q-\overline{q})'\Big]\begin{bmatrix} -H_{xx} & \frac{\rho}{2}I \\ \frac{\rho}{2}I & H_{qq} \end{bmatrix}\Bigg|_{\lambda z}\begin{bmatrix} x-\overline{x} \\ q-\overline{q} \end{bmatrix} \\ &= -\Big\{(x-\overline{x})'F_2(\lambda z) + (q-\overline{q})F_1(\lambda z) - \lambda z' Q(\lambda z) z\Big\} \\ g''(\lambda) &= 2z'Q(\lambda z)z - 2\lambda z' A \frac{d}{d\lambda}\Big[\nabla F(\lambda z)\Big] z. \end{aligned}$$

The remainder of the proof is identical to the proof of Theorem 4.4. ∎

Remark 6.3 *The condition on the matrix $Q(x,q)$ gives more information about the steepness condition implicit in the Cass-Shell assumption. This will be still more explicit after we have introduced the concepts of α-concavity and β-convexity of the Hamiltonian.*

Definition 6.1 *A function $h(\cdot)\ :\ C \rightarrow \mathbb{R}$, $C \subseteq \mathbb{R}^n$, is α-convex, $\alpha \in \mathbb{R}$, if the function*

$$h(x) - \frac{1}{2}\alpha\|x\|^2$$

is convex on C. A function $g(\cdot)\ :\ C \rightarrow \mathbb{R}$, is β-concave, $\beta \in \mathbb{R}$, if its negative, $-g(\cdot)$, is β-convex.

Remark 6.4 *If in the above C is open and $h(\cdot)$ is C^2, then α-convexity is equivalent to*

$$\forall\, x_0 \in C,\ \forall\, w \in \mathbb{R}^n,\ w'\nabla^2 h(x_0)w \geq \alpha w'w. \tag{6.26}$$

If the Hamiltonian $\mathcal{H}(x,q)$ is β-concave in x and α-convex in q then by (6.26) the quadratic form

$$[x'q']Q(x,q)\begin{bmatrix} x \\ q \end{bmatrix} = x'\mathcal{H}_{xx}x - q'\mathcal{H}_{qq}q - \rho q'x$$

is dominated by the quadratic form

$$-\beta x'x - \alpha q'q - \rho q'x.$$

This last quadratic from is negative definite if

$$4\alpha\beta > \rho^2 \quad \text{and} \quad \alpha > 0,\ \beta > 0 \tag{6.27}$$

since this is the condition for the determinant

$$\begin{vmatrix} -\beta & +\rho/2 \\ -\rho/2 & -\alpha \end{vmatrix}$$

to be positive.

Hence the following has been proved.

Corollary 6.1 *If there exists a unique stationary point $(\overline{x},\overline{q})$, if the Hamiltonian is C^2, β-concave in x and, α-convex in q and if the condition (6.27) is satisfied then all solutions of (6.17)–(6.18) that are bounded for $t \geq 0$ converge to $(\overline{x},\overline{q})$ as $t \rightarrow \infty$.*

Remark 6.5 *The condition (6.27) was first given by Rockafellar [156] for GAS of modified Hamiltonian systems having the β-concavity and α-convexity property. Rockafellar used a more general formulation which does not necessitate the assumption that $\mathcal{H}$ is C^2.*

One sees immediately the intuitive meaning of condition (6.27). If the criterion is discounted with a positive discount rate ρ, then one has to assume "a lot of concavity in x and a lot of convexity in q" as compared with ρ for the Hamiltonian $\mathcal{H}(x,q)$ in order to assert asymptotic stability of the stationary point $(\overline{x},\overline{q})$.

6.5 Another useful condition for GAS

In this section another interesting result obtained by Brock and Sheinkman [29] is presented. This result is partly based on a Dynamicindexdynamic programming Programming-Hamilton-Jacobi-Bellman approach and thus it will only hold for strongly optimal trajectories. Define

$$\tilde{W}(x_0, t_0) = \sup \int_{t_0}^{\infty} e^{-\rho(t-t_0)} f_0(x(t), u(t))\, dt$$

subject to

$$\dot{x}(t) = f(x(t), u(t)), \quad x(t_0) = x_0.$$

It is easy to show (see Arrow and Kurz [7]) that $\tilde{W}(x_0, t_0) = W(x_0)$ is independent of t_0, the optimal control is of time stationary feedback form

$$u^*(t) = \hat{u}(x^*(t))$$

and is obtained through the solution of the Hamilton-Jacobi-Bellman equation

$$\rho W(x) - \max_u \left\{ (x, u) + \nabla W(x) f(x, u) \right\} = 0,$$

provided sufficient smoothness assumptions hold on the value function W.

Assumption 6.3 *We assume that the function* $W(\cdot) : \mathbb{R}^n \to \mathbb{R}$,

$$W(x_0) \triangleq \max_{(x,u)} \int_0^{\infty} e^{-\rho t} f_0(x(t), u(t))\, dt, \tag{6.28}$$

where $x(\cdot)$ *and* $u(\cdot)$ *satisfy* (6.2), (6.3), (6.4) *is well defined,* C^2, *and concave.*

Define, along an optimal trajectory $x^*(\cdot)$,

$$q^*(t) \triangleq \nabla W(x^*(t)). \tag{6.29}$$

It is well known (see [7] that $q^*(t)$ defined by (6.29) corresponds to the current-value costate variable in the necessary conditions given by the maximum principle. Furthermore differentiating (6.29) one gets

$$\dot{q}^*(t) = \nabla^2 W(x^*(t)) \dot{x}^*(t) \tag{6.30}$$

and therefore, under Assumption 6.3 the condition

$$\dot{q}^*(t)' \dot{x}^*(t) \leq 0, \tag{6.31}$$

holds along optimal trajectories.

Along an optimal trajectory the modified Hamiltonian system (6.13)–(6.14) is satisfied by the pair $(x^*(\cdot), q^*(\cdot))$. Furthermore, in the solution of the necessary optimality condition yielding the maximum principle, the control variable at t is obtained as a function of x and q:

$$u^*(t) = \gamma(x^*(t), q^*(t)) \tag{6.32}$$

and in the dynamicindexdynamic programming programming approach this last relation (6.32) becomes a feedback law:

$$u = \gamma(x, \nabla W(x)) \tag{6.33}$$

hence the optimal trajectory emanating from x_0 is a solution to the dynamical system

$$\begin{aligned} \dot{x} &= f(x, \gamma(x, \nabla W(x))) \\ x(0) &= x_0. \end{aligned} \tag{6.34}$$

A rest-point of (6.34) is a state $\overline{x}$ such that

$$f(\overline{x}, \gamma(\overline{x}, \nabla W(\overline{x}))) = 0. \tag{6.35}$$

Now the following stability result can be proved.

Theorem 6.3 *Consider a control problem for which Assumption* 6.3 *holds true. Assume that along a solution* $(x(\cdot), q(\cdot))$ *to the associated modified Hamiltonian system the matrix*

$$\left\{\mathcal{H}_{qq}^{-1}\mathcal{H}_{qx} + (\mathcal{H}_{qq}^{-1}\mathcal{H}_{qx})' + \dot{\widehat{\mathcal{H}_{qq}^{-1}}}\right\} \tag{6.36}$$

is negative definite.

Then an optimal trajectory $x^*(\cdot)$ *converges to a rest-point of the system*

$$\dot{x} = f(x, \gamma(x, \nabla W(x))) \tag{6.37}$$

Proof: This is a standard application of Lyapunov theory. Define the function

$$V(x,q) = \dot{x}'\mathcal{H}_{qq}^{-1}\dot{x} = \mathcal{H}_q'\mathcal{H}_{qq}^{-1}\mathcal{H}_q.$$

Differentiating along a trajectory yields

$$\begin{aligned} \dot{V}(x,q) &= (\mathcal{H}_{qq}\dot{q} + \mathcal{H}_{qx}\dot{x})'\mathcal{H}_{qq}^{-1}\dot{x}' + \mathcal{H}_{qq}^{-1}(\mathcal{H}_{qq}\dot{q} + \mathcal{H}_{qx}\dot{x}) + \dot{x}'\dot{\widehat{\mathcal{H}_{qq}^{-1}}}\dot{x} \\ &= 2\dot{q}'\dot{x} + \dot{x}'\left\{\mathcal{H}_{qx}'\mathcal{H}_{qq}^{-1} + \mathcal{H}_{qq}^{-1}\mathcal{H}_{qx} + \dot{\widehat{\mathcal{H}_{qq}^{-1}}}\right\}\dot{x}. \end{aligned}$$

By (6.31), a consequence of Assumption 6.3 is that along an optimal trajectory the term $\dot{q}'\dot{x}$ is nonpositive if $\dot{x} \neq 0$. The condition of negative definiteness of the matrix (6.36) assures that the trajectory derivative $\dot{V}(x,q)$ is negative if $\dot{x} \neq 0$. As $\mathcal{H}$ is convex in q, $\mathcal{H}_{qq}^{-1}$ is positive semidefinite and $V(x,q)$ is positive or null. This proves the convergence of $x^*(\cdot)$ toward a state $\overline{x}$ which is a rest point of the system (6.37). ∎

Remark 6.6 *The term* $\dot{\widehat{\mathcal{H}_{qq}^{-1}}}$ *in the matrix* (6.36) *is difficult to handle. In fact, the stability condition of Theorem* 6.3 *is useful only when it is possible to eliminate this disturbing term. This is the case in particular when the criterion takes the form*

$$J_T(x(\cdot)\,,u(\cdot)\,) = \int_0^T e^{-\rho t}\Big\{f_0(x(t)) + \frac{1}{2}u(t)'Ru(t)\Big\}\,dt \tag{6.38}$$

with R *being a negative definite symmetric matrix, while the state equation and the control constraints take the form*

$$\dot{x}(t) = f(x(t)) + u(t) \tag{6.39}$$

$$u(t) \leq h(x(t)). \tag{6.40}$$

In that case the current value Hamiltonian (6.12) *takes the form*

$$\begin{aligned} H(x(t),u(t),q(t),\nu(t)) &= f_0(x(t)) + \frac{1}{2}u(t)'Ru(t) + q(t)'\big(f(x(t)) + u(t)\big) \\ &\quad +\nu(t)'\big(h(x(t)) - u(t)\big) \end{aligned} \tag{6.41}$$

The first order conditions

$$\nabla_u H(x,u,q,\nu) = 0, \quad \nu'(h(x) - u) = 0$$

define u *as a linear function* q *and hence* $\mathcal{H}_{qq}$ *will be a constant matrix and* $\dot{\widehat{\mathcal{H}_{qq}^{-1}}}$ *is equal to zero.*

6.6 Neighboring extremals, the second variation and analysis of local asymptotic stability of a stationary point, using the optimal linear quadratic regulator problem

This section is devoted to a short presentation of the approach used by Magill [131], [130] for analysing the stability of dynamic economic equilibria. This method is appealing to control theorists as it is based on the approximation of the nonlinear control system, in the neighborhood of a stationary point $(\overline{x},\overline{q})$, by a linear quadratic system corresponding to the second variation. The stability properties of this linear system will produce *local stability properties* of the stationary point $(\overline{x},\overline{q})$ of the nonlinear system. The second variation is a classical tool for obtaining local second order conditions characterizing optimal trajectories in control problem (see e.g. [31]).

Consider the control problem (6.2), (6.3) and (6.4), where for the sake of simplifying the developments one does not consider any constraint on the control

$$\dot{x}(t) = f(x(t), u(t)) \tag{6.2}$$

$$x(0) = x_0 \tag{6.3}$$

$$J_T(x_0, u(\cdot))0 = \int_0^T e^{-\rho t} f_0(x(t), u(t))\, dt. \tag{6.4}$$

Let $(\overline{x}, \overline{u})$ be a steady-state verifying

$$f(\overline{x}, \overline{u}) = 0. \tag{6.42}$$

Assuming $f_0(\cdot, \cdot)$ and $f(\cdot, \cdot)$ are C^2 in x and u, an extremal trajectory satisfies

$$\dot{x} = \nabla_q H(x, q, u) \tag{6.43}$$

$$\dot{q} = -\nabla_x H(x, q, u) + \rho q \tag{6.44}$$

$$0 = \nabla_u H(x, q, u) \tag{6.45}$$

Let $(\overline{x}, \overline{u}, \overline{q})$ be a stationary solution to (6.43)–(6.45). Thus the neighboring extremal linear-quadratic approximation is obtained by considering the problem

maximize

$$\frac{1}{2}\int_0^\infty e^{-\rho t}\left\{ \left[\delta x(t)', \delta u(t)\right]' \begin{bmatrix} P & S \\ S' & R \end{bmatrix} \begin{bmatrix} \delta x(t) \\ \delta u(t) \end{bmatrix} \right\} dt \tag{6.46}$$

for $\delta x(\cdot)$ and $\delta u(\cdot)$ which satisfy

$$\delta \dot{x}(t) = A\delta x(t) + B\delta u(t) \tag{6.47}$$

$$\delta x(0) = \delta x_0, \tag{6.48}$$

where the matrices P, R, S, A, B are defined as follows:

$$\begin{bmatrix} P & S \\ S' & R \end{bmatrix} = \begin{bmatrix} H_{xx} & H_{xu} \\ H_{ux} & H_{uu} \end{bmatrix}_{\mid \substack{x=\overline{x} \\ u=\overline{u}}}$$

$$A = \frac{\partial f}{\partial x}_{\mid \substack{x=\overline{x} \\ u=\overline{u}}} \qquad B = \frac{\partial f}{\partial u}_{\mid \substack{x=\overline{x} \\ u=\overline{u}}}.$$

The linear-quadratic control problem (6.46)–(6.48) corresponds to the so-called *accessory problem* familiar in the calculus of variations for the analysis of conjugate points (see [116]) or for obtaining second order optimality conditions (see [31]). The steady-state trajectory

$$\delta x_0 = \delta x(t) \equiv 0, \quad \delta u(t) \equiv 0, \quad t \geq 0$$

corresponds to a stationary point of the system (6.43)–(6.45). We shall explore the conditions which guarantee the asymptotic stability of this rest point.

If one replaces $\delta x(t)$ and $\delta u(t)$ by

$$\xi(t) = e^{-\frac{\rho}{2}t}\,\delta x(t), \quad \eta(t) = e^{-\frac{\rho}{2}t}\,\delta u(t) \tag{6.49}$$

respectively, the linear-quadratic control system is reduced to a time-invariant form

$$J_T = \frac{1}{2}\int_0^T \left\{[\xi(t)', \eta(t)']\begin{bmatrix} P & S \\ S' & R \end{bmatrix}\begin{bmatrix} \xi(t) \\ \eta(t) \end{bmatrix}\right\} dt \tag{6.50}$$

where $\xi(\cdot)$ and $\eta(\cdot)$ satisfy the equations

$$\dot{\xi}(t) = \left(A - \frac{\rho}{2}I\right)\xi(t) + B\eta(t) \tag{6.51}$$

$$\xi(0) = \delta x_0 \tag{6.52}$$

We are interested in the maximization of J_∞. For this we will consider the asymptotic behavior of the solution of max J_T, s.t. (6.51), (6.52), as T tends to infinity and show that under appropriate assumptions this gives the desired solution.

The necessary conditions for optimality of a linear-quadratic system with cross-product terms in the criterion can be easily obtained as a variant of the conditions (1.11)–(1.13) (see [31]).

The optimal control for T finite is thus defined as a time-varying feedback law

$$\eta(t) = -R^{-1}(B'K(t) + S')\xi(t)$$

where $K(t)$ is the solution of the matrix Riccati equation

$$\begin{aligned}\dot{K}(t) = {} & -K(t)\left(A - \frac{\rho}{2}I\right) - \left(A' - \frac{\rho}{2}I\right)K(t) \\ & +(K(t)B + S)R^{-1}(S' + B'K(t)) - P \end{aligned} \tag{6.53}$$

$$K(T) = 0. \tag{6.54}$$

Furthermore one has for any ξ_0

$$\frac{1}{2}\xi_0' K(t)\xi_0 = \max\left\{\frac{1}{2}\int_t^T \left\{\left[\xi(t)', \eta(t)'\right]\begin{bmatrix} P & S \\ S' & R \end{bmatrix}\begin{bmatrix} \xi(t) \\ \eta(t) \end{bmatrix}\right\}\right\} dt \tag{6.55}$$

where $\xi(\cdot)$ and $\eta(\cdot)$ satisfy (6.51) and $\xi(0) = \xi_0$. To be more precise let's call $\tilde{K}(t,T)$ the solution of (6.53), (6.54) at t. The following lemma gives a sufficient condition for the existence of a limit for $\tilde{K}(t,T)$ when $T \to \infty$.

Lemma 6.1 *If in the system (6.50)–(6.51), the matrix* $\begin{bmatrix} P & S \\ S' & R \end{bmatrix}$ *is negative definite and if for each bounded trajectory* $\xi_0(\cdot)$ *generated by a control* $\eta_0(\cdot)$ *for some constants* $a > 0, b > 0$ *and* $\delta > 0$ *one has*

$$\|\xi_0(t)\| < ae^{-\delta t}, \quad \|\eta_0(t)\| < be^{-\delta t} \qquad \forall t \geq 0 \tag{6.56}$$

then

$$\lim_{T\to\infty} \tilde{K}(t,T) = \hat{K} \tag{6.57}$$

where $\hat{K}$ is negative definite.

Proof: Let λ^* be the minimal eigenvalue of $\begin{bmatrix} P & S \\ S' & R \end{bmatrix}$, then by (6.55) and (6.56)

$$\frac{1}{2}\xi_0'\tilde{K}(t,T)\xi_0 \geq J_T(\xi_0,\eta_0) \geq \frac{\lambda^*(a^2+b^2)}{\delta}(e^{-\delta t} - e^{-\delta T}). \tag{6.58}$$

When $T \to \infty$, since $\lambda^* < 0$ and $\delta > 0$, one has

$$\lim_{T\to\infty} \xi_0'\tilde{K}(t,T)\xi_0 > -\infty. \tag{6.59}$$

Furthermore the negative definiteness of $\begin{bmatrix} P & S \\ S' & R \end{bmatrix}$ implies that

$$\xi_0'\tilde{K}(t,T+h)\xi_0 \leq \xi_0'\tilde{K}(t,T)\xi_0, \qquad \forall\, h > 0 \tag{6.60}$$

(i.e., $\tilde{K}(t,\cdot)$ is a decreasing function). This implies $\lim_{T\to\infty} \tilde{K}(t,T) = \hat{K}(t)$ exists for each $t \geq 0$. Further, an easy change of variable shows that $\hat{K}(t)$ is independent of t (i.e., $\hat{K}(t) = \hat{K}$) and the desired result follows. ∎

Remark 6.7 *The condition* (6.56) *is an asymptotic controllability condition for the linear system* (6.51).

Corollary 6.2 *Under the assumptions of Lemma* 6.1 *the optimal control law satisfies, when $T \to \infty$*

$$\eta(t) = -R^{-1}(B'\hat{K} + S')\xi(t) \tag{6.61}$$

where $\hat{K}$ is the negative definite solution of the algebraic matrix Riccati equation

$$0 = -\hat{K}\left(A - \frac{\rho}{2}I\right) - \left(A' - \frac{\rho}{2}I\right)\hat{K} + (\hat{K}B + S)R^{-1}(S' + B'\hat{K}) - P. \tag{6.62}$$

Furthermore the optimal value of the criterion tends to $\frac{1}{2}\xi_0'\hat{K}\xi_0$ when $T \to \infty$.

Proof: Obvious. ∎

The solution of the accessory problem has been thus obtained. We are now interested in obtaining conditions which guarantee that $\delta x(\cdot)$ is converging to zero. (Recall that $\xi(t) = e^{-\frac{\rho}{2}t}\delta x(t)$).

For doing this we write the Hamiltonian system associated with the optimal control problem (6.46)–(6.48). Let's define the Hamiltonian

$$G(\delta x, \delta u, \delta q) = \frac{1}{2}[\delta x', \delta u']\begin{bmatrix} P & S \\ S' & R \end{bmatrix}\begin{bmatrix} \delta x \\ \delta u \end{bmatrix} + \delta q'(A\delta x + B\delta u). \tag{6.63}$$

One must have

$$\nabla_{\delta u} G(\delta x, \delta u, \delta q) = \delta x' S + \delta u' R + \delta q' B = 0, \tag{6.64}$$

thus

$$\delta u = -R^{-1}(S'\delta x + B'\delta q), \tag{6.65}$$

the maximized Hamiltonian is

$$\mathcal{G}(\delta x, \delta q) = [\delta x', \delta q']\begin{bmatrix} P - SR^{-1}S' & A' - SR^{-1}B' \\ A - BR^{-1}S' & -BR^{-1}B' \end{bmatrix}\begin{bmatrix} \delta x \\ \delta q \end{bmatrix}, \tag{6.66}$$

and the modified Hamiltonian system satisfied by $\delta x, \delta q$ is

$$\delta \dot{x} = \nabla_{\delta q}\mathcal{G}(\delta x, \delta q) \tag{6.67}$$
$$\delta \dot{q} = -\nabla_{\delta x}\mathcal{G}(\delta x, \delta q) + \rho\delta q. \tag{6.68}$$

We know by Theorem 6.2 that, if ρ is positive, a sufficient condition for global asymptotic stability of the solutions to (6.67)–(6.68) is that the matrix

$$Q_\rho = \begin{bmatrix} P - SR^{-1}S' & -\frac{\rho}{2}I \\ -\frac{\rho}{2}I & BR^{-1}B' \end{bmatrix} \tag{6.69}$$

be negative definite.

Remark 6.8 *The matrix Q_ρ given by (6.69) is the curvature matrix given in (6.25). To see this we define the maximized Hamiltonian*

$$\mathcal{H}(\overline{x}, \overline{q}) = H(x, q, \hat{u}(x, q)) = \max_{u \in \mathbb{R}^m} H(x, q, u),$$

where we assume $\hat{u}(\cdot,\cdot)$ is C^2. From this it follows that

$$\mathcal{H}_x(\overline{x}, \overline{q}) = (H_x + H_u\hat{u}_x)|_{(\overline{x},\overline{q},\overline{u})}$$

and

$$\mathcal{H}_{xx}(\overline{x}, \overline{q}) = (H_{xx} + 2H_{ux}\hat{u}_x + \hat{u}_x' H_{uu}\hat{u}_x + H_u\hat{u}_{xx})|_{(\overline{x},\overline{q},\overline{u})}.$$

Further, we notice that $H_u(\overline{x}, \overline{q}, \overline{u}) = 0$ implies

$$(H_{ux} + H_{uu}\hat{u}_x)|_{\overline{x},\overline{q},\overline{u})} = 0,$$

so that

$$\hat{u}_x|_{(\overline{x},\overline{q})} = -H_{uu}^{-1}H_{ux}|_{(\overline{x},\overline{q},\overline{u})}.$$

Therefore, combining the above results gives

$$\begin{aligned}\mathcal{H}_{xx}(\overline{x},\overline{q}) &= (H_{xx} - H_{xx}H_{uu}^{-1}H_{ux})|_{(\overline{x},\overline{q},\overline{u})} \\ &= P - SR^{-1}S'\end{aligned}$$

where the last statement follows from the definitions of the matrices P, R and S given previously.

The next theorem gives another proof of this result based on the asymptotic solution obtained in Lemma 6.1 and Corollary 6.2.

Theorem 6.4 *If the discount rate ρ is positive and if the matrix Q_ρ defined in* (6.69) *is negative definite, then the trajectories of the modified Hamiltonian system* (6.67)–(6.68) *are asymptotically stable.*

Proof: First we show that if Q_ρ is negative definite then the conditions of Lemma 6.1 are satisfied.

If Q_ρ is negative definite then the matrix $BR^{-1}B'$ is negative definite and hence B must be of rank n. Therefore the linear system

$$\delta\dot{x}(t) = A\delta x(t) + B\delta u(t)$$

is controllable and, as a consequence, the asymptotic controllability of the system (6.51) is ensured.

One has now to check that the matrix

$$\begin{bmatrix} P & S \\ S' & R \end{bmatrix} \tag{6.70}$$

is negative definite. If Q_ρ is negative definite then $BR^{-1}B'$ is negative definite and R must be negative definite.

Now the matrix (6.70) is negative definite if and only if the matrix

$$\begin{bmatrix} P - SR^{-1}S' & O \\ O & R \end{bmatrix} \tag{6.71}$$

is negative definite and this is true if Q_ρ is negative definite. Hence one can apply Lemma 6.1 and Corollary 6.2 for characterizing the optimal trajectories.

Consider the Lyapunov function

$$W(\delta x) = \frac{1}{2}\delta x'\hat{K}\delta x. \tag{6.72}$$

As we know by Corollary 6.2, $\hat{K}$ is negative definite, thus

$$\delta x \neq 0 \Rightarrow W(\delta x) < 0. \tag{6.73}$$

Compute the trajectory derivative

$$\nabla W(\delta x(t))\delta\dot{x}(t) = \delta x(t)'\hat{K}(A\delta x(t) - BR^{-1}(B'\hat{K} + S')\delta x) \tag{6.74}$$

(Notice that we have the expression (6.61) with the change of variables $\delta x(t) = e^{\frac{\rho}{2}t}\xi(t)$, $\delta u(t) = e^{\frac{\rho}{2}t}\eta(t)$.)

Now, as $\hat{K}$ is the solution of (6.62) one has

$$\hat{K}A + A'\hat{K} = \rho\hat{K} + \hat{K}BR^{-1}B'\hat{K} + SR^{-1}S' + \hat{K}BR^{-1}S' + SR^{-1}B'\hat{K} - P. \tag{6.75}$$

Bringing this expression into (6.74) one obtains

$$\nabla W(\delta x(t))\delta\dot{x}(t) = \delta x(t)'\left(\frac{\rho}{2}\hat{K} - \frac{1}{2}\hat{K}BR^{-1}B'\hat{K} + \frac{1}{2}(SR^{-1}S' - P)\right)\delta x(t). \tag{6.76}$$

It is easy to check that, according to the Hamilton-Jacobi equation, one has

$$\delta q(t) = \nabla W(\delta x(t)) = \hat{K}\delta x(t) \tag{6.77}$$

and therefore the expression (6.76) can be written

$$\nabla W(\delta x(t))\delta\dot{x}(t) = \frac{1}{2}\begin{bmatrix} \delta x(t)' \\ \delta q(t)' \end{bmatrix}\begin{bmatrix} SR^{-1}S' - P & \frac{\rho}{2}I \\ \frac{\rho}{2}I & BR^{-1}B' \end{bmatrix}\begin{bmatrix} \delta x(t) \\ \delta q(t) \end{bmatrix} \tag{6.78}$$

which is clearly positive if Q_ρ is negative definite. Hence $\delta x(\cdot)$ tends to 0 and by (6.77) $\delta q(\cdot)$ tends also to 0. ∎

Remark 6.9 *The negative definiteness of the matrix Q_ρ implies that R is of rank n and hence one has to have at least as many control variables as state variables. This shows how stringent these conditions are.*

6.7 The turnpike property for finite horizon optimal control problems with discounting

In this section, we present a generalization of Theorem 4.2 for problems with discounting. The result has been put forth by Feinstein and Oren [76] who called it the "Funnel turnpike Theorem". Specifically, we consider the autonomous control system

$$\begin{aligned}
\dot{x}(t) &= f(x(t), u(t)), \quad \text{a.e.} \quad 0 \leq t \leq T \\
x(0) &= x_0, \\
x(T) &= x_T, \\
x(t) &\in X, \\
u(t) &\in U(x(t)), \quad \text{a.e.} \quad 0 \leq t \leq T.
\end{aligned}$$

With the above system we associate the discounted cost criteria

$$J_T(x_0, u(\cdot)) = \int_0^T e^{-\rho t} f_0(x(t), u(t))\, dt,$$

and we assume that $X \subset \mathbb{R}^n$, $U : X \to 2^{\mathbb{R}^m}$ is a given set valued map, $(f_0(\cdot,\cdot), f(\cdot,\cdot)) : \mathbb{R}^n \times \mathbb{R}^m \to \mathbb{R}^{1+n}$ are given functions which are C^2, $(0, x_0)$ and (T, x_T) are fixed, and ρ is a positive discount factor.

With the above problem we associate the following family of mathematical programming problems,

$$\max \quad f_0(x,u) \tag{6.79}$$

s.t.

$$f(x,u) - \rho(x - c) = 0 \tag{6.80}$$

$$x \in X,\ u \in U(x), \tag{6.81}$$

where $c \in \mathbb{R}^n$ is a parameter. For fixed c, we let $(\overline{x}(c), \overline{u}(c))$ denote a solution to the above static optimization problem. We are specifically interested in those parameters c with the property that $\overline{x}(c) = c$. In this situation, the above problem becomes the *Implicit Programming problem*

$$\max \quad f_0(x,u) \tag{6.82}$$

s.t.

$$f(x,u) - \rho(x - \overline{x}(c)) = 0 \tag{6.83}$$

$$x \in X,\ u \in U(x). \tag{6.84}$$

This problem is a well-defined mathematical program with the unique aspect that the state component of the solution, namely $\overline{x} = \overline{x}(c)$, appears in the constraint. Once again we introduce the modified Hamiltonian system

$$\dot{x}(t) = \frac{\partial \mathcal{H}}{\partial q}(x(t), q(t)) \tag{6.17}$$

$$\dot{q}(t) = -\frac{\partial \mathcal{H}}{\partial x}(x(t), q(t)) + \rho q(t), \tag{6.18}$$

where $\mathcal{H}(x,q) = \sup_u \{f_0(x(\cdot), u(\cdot)) + q' f(x(\cdot), u(\cdot)) : u \in U(x)\}$. We assume that $\mathcal{H}$ is concave in x and convex in q. With this notation we have the following.

Theorem 6.5 *Suppose that $\mathcal{H}$, as defined above, is concave in x and convex in q, and assume that $(\overline{x}, \overline{q}) \in X \times \mathbb{R}^n$ is a stationary point of the modified Hamiltonian system* (6.17)–(6.18). *Then $(\overline{x}, \overline{u})$ is a solution to the mathematical programming problem* (6.79)–(6.81) *with $c = \overline{x}$, and $\overline{u}$ is the solution to*

$$\max_u \left\{ f_0(\overline{x}, u) : f(\overline{x}, u) = 0,\ u \in U(\overline{x}) \right\}$$

Proof: The point $(\overline{x},\overline{q})$ being a stationary point for the modified Hamiltonian system implies

$$0 = \frac{\partial \mathcal{H}}{\partial q}(\overline{x},\overline{q})$$

and

$$\rho\overline{q} = \frac{\partial \mathcal{H}}{\partial x}(\overline{x},\overline{q}).$$

An immediate consequence of this fact is that

$$f_0(\overline{x},\overline{u}) = \mathcal{H}(\overline{x},\overline{q}),$$

and so the concavity of $\mathcal{H}$ in x gives us

$$\begin{aligned} f_0(\overline{x},\overline{u}) + \rho\overline{q}'(x-\overline{x}) &= \mathcal{H}(\overline{x},\overline{q}) + \rho\overline{q}'(x-\overline{x}) \\ &\geq \mathcal{H}(x,\overline{q}) \end{aligned}$$

for all $x \in X$. Therefore for all $x \in X$ and $u \in U(x)$ we have,

$$\begin{aligned} f_0(\overline{x},\overline{u}) + \rho\overline{q}'(x-\overline{x}) &\geq \mathcal{H}(x,\overline{q}) \\ &\geq f_0(x,u) + \overline{q}'f(x,u), \end{aligned}$$

from which it follows that

$$f_0(\overline{x},\overline{u}) \geq f_0(x,u)$$

for all $x \in X$ and $u \in U$ satisfying

$$f(x,u) - \rho(x-\overline{x}) = 0.$$

Remark 6.10 *The above result shows that, under the usual concavity hypothesis, any stationary point of the modified Hamiltonian system is a solution of the Implicit Programming problem.*

The next result gives a converse to the above result.

Theorem 6.6 *Let $(\overline{x},\overline{u})$ be a solution to the Implicit Programming problem (i.e., $c = \overline{x}$), and assume H is concave in x, convex in q and continuously differentiable. Further, suppose that $\overline{x} \in X$ and that $\overline{u} \in U(\overline{x})$. Then there exists $\overline{q} \in \mathbb{R}^n$ such that $(\overline{x},\overline{q})$ is a stationary point of the modified Hamiltonian system.*

Proof: We first notice that because $(\overline{x}, \overline{u})$ solves the Implicit Programming problem, there exists a Lagrange multiplier $\overline{q} \in \mathbb{R}^n$ so that, for all $x \in X$ and $u \in U(x)$ we have

$$f_0(\overline{x}, \overline{u}) \geq f_0(x, u) + \overline{q}'[f(x, u) - \rho(x - \overline{x})].$$

Thus,

$$f_0(\overline{x}, \overline{u}) + \rho\overline{q}'(x - \overline{x}) \geq f_0(x, u) + \overline{q}' f(x, u) \quad \text{for all} \quad x \in X,$$

and $u \in U(x)$, so that

$$f_0(\overline{x}, \overline{u}) + \rho\overline{q}'(x - \overline{x}) \geq \mathcal{H}(x, \overline{q}) \text{ for all } x \in X, \tag{6.85}$$

Moreover, we notice that

$$\begin{aligned} \mathcal{H}(\overline{x}, \overline{q}) &= \sup[f_0(\overline{x}, u) + \overline{q}' f(\overline{x}, u) \,:\, u \in U(x)] \\ &\geq f_0(\overline{x}, \overline{u}) + \overline{q}' f(\overline{x}, \overline{u}) \\ &= f_0(\overline{x}, \overline{u}). \end{aligned} \tag{6.86}$$

Combining these results we obtain

$$\mathcal{H}(\overline{x}, \overline{q}) + \rho\overline{q}'(x - \overline{x}) \geq \mathcal{H}(x, \overline{q}) \quad \text{for all} \quad x \in X,$$

which implies

$$\frac{\partial \mathcal{H}}{\partial x}(\overline{x}, \overline{q}) = \rho\overline{q}$$

since $\mathcal{H}(\,\cdot\,, \overline{q})$ is a concave function.

We further notice that since

$$\mathcal{H}(\overline{x}, q) = \sup[f^0(\overline{x}, u) + q' f(\overline{x}, u) \,:\, u \in U(\overline{x})]$$

is continuously differentiable as a function of q and since $\overline{u} \in U(\overline{x})$ we have, that

$$\frac{\partial \mathcal{H}}{\partial q}(\overline{x}, q) = f(\overline{x}, u(q)),$$

where $u(q)$ is a point in $U(\overline{x})$ such that

$$\mathcal{H}(\overline{x}, q) = f_0(\overline{x}, u(q)) + q' f(x, u(q)).$$

Notice that from (6.85) and (6.86) it follows that

$$\mathcal{H}(\overline{x}, \overline{q}) \geq f_0(\overline{x}, \overline{u}) \geq \mathcal{H}(\overline{x}, \overline{q}),$$

and so $\overline{u} = u(\overline{q})$ and thus

$$\frac{\partial \mathcal{H}}{\partial q}(\overline{x}, \overline{q}) = f(\overline{x}, \overline{u}) = 0$$

as desired.

Remark 6.11 *The above theorems are presented in* [75] *under significantly weaker differentiability hypotheses on $\mathcal{H}$. We've presented these results with the above hypotheses partly because of their simplicity and partly because, in the work to follow, we need this regularity anyway.*

The Implicit Programming problem motivates the following Lagrangian function L_ρ : $\mathbb{R}^n \times \mathbb{R}^m \times \mathbb{R}^n \times \mathbb{R}^n \to \mathbb{R} \cup \{-\infty\}$, *defined by*

$$L_\rho(x,u,q,c) = \begin{cases} f_0(x,u) + q'[f(x,u) + \rho(x-c)] & \text{if } x \in X,\ u \in U(x) \\ -\infty & \text{otherwise.} \end{cases}$$

In what follows, we impose the following assumptions on L_ρ.

Assumption 6.4 *Let* $(\overline{x},\overline{u})$ *be a solution to the Implicit Programming problem with* $c = \overline{x}$, *and let* $\overline{q}$ *be a vector corresponding to* $\overline{x}$ *so that* $(\overline{x},\overline{q})$ *denotes a steady-state of the modified Hamiltonian system. We assume that* $L_\rho(\cdot,\cdot,\overline{q},\overline{x})$ *is concave on* $M = \{(x,u) : x \in X,\ u \in U(x)\}$ *and that* L_ρ *is twice continuously differentiable with a negative definite Hessian matrix such that*

$$\sup_{(x,u)} [\lambda(x,u) : (x,u) \in M] = \lambda < 0,$$

where $\lambda(x,u)$ *denotes the largest eigenvalue of the Hessian of* $L_\rho(\cdot,\cdot,\overline{q},\overline{x})$ *evaluated at* (x,u).

We are now ready to begin our treatment of the turnpike property, starting with the following lemma.

Lemma 6.2 (*Strong support Property*). *Assume that Assumption* 6.4 *holds with* $(\overline{x},\overline{u}) \in M$ *and suppose that for* $(x,u) \in M$,

$$f_0(\overline{x},\overline{u}) \geq f_0(x,u) + \overline{q}'[f(x,u) - \rho(x-\overline{x})]. \tag{6.87}$$

Then, for every $\varepsilon > 0$ *and for all* $(x,u) \in M$, *there exists* $\delta_\varepsilon > 0$ *such that if* $\|(x,u) - (\overline{x},\overline{u})\|^2 > \varepsilon$, *we have*

$$f_0(\overline{x},\overline{u}) > f_0(x,u) + \overline{q}'[f(x,u) - \rho(x-\overline{x})] + \delta_\varepsilon. \tag{6.88}$$

Proof: From Taylor's theorem we have that there exists $(\xi, w) \in M$ such that

$$\begin{aligned}
L_\rho(x,u,\overline{q},\overline{x}) &= f_0(x,u) + \overline{q}'[f(x,u) + \rho(x-\overline{x})] \\
&= f_0(\overline{x},\overline{u}) + [\nabla_{(x,u)} L_\rho(\overline{x},\overline{u},\overline{q},\overline{x})]' \begin{bmatrix} x-\overline{x} \\ u-\overline{u} \end{bmatrix} \\
&\quad + \frac{1}{2}\Big[(x-\overline{x})', (u-\overline{u})'\Big] \nabla^2_{(x,u)} L_\rho(\xi,w,\overline{q},\overline{x}) \begin{bmatrix} x-\overline{x} \\ u-\overline{u} \end{bmatrix} \\
&= f_0(\overline{x},\overline{u}) + \frac{1}{2}\Big[(x-\overline{x})', (u-\overline{u})'\Big] \nabla^2_{(x,u)} L_\rho(\xi,w,\overline{q},\overline{x}) \begin{bmatrix} x-\overline{x} \\ u-\overline{u} \end{bmatrix} \\
&\leq f_0(\overline{x},\overline{u}) + \frac{1}{2}\lambda(\xi,w)\|(x,u)-(\overline{x},\overline{u})\|^2 \\
&< f_0(\overline{x},\overline{u}) + \frac{1}{2}\lambda\varepsilon,
\end{aligned}$$

where the above follows from the fact that $\nabla_{(x,u)} L_\rho(\overline{x},\overline{u},\overline{q},\overline{x}) = 0$, and $\lambda(\xi,w) \leq \lambda < 0$. Choosing $\delta_\varepsilon = \frac{1}{2}\lambda\varepsilon > 0$, gives us

$$f_0(\overline{x},\overline{u}) > f_0(x,u) + \overline{q}'[f(x,u) + \rho(x-\overline{x})] + \delta_\varepsilon$$

as desired. ∎

Remark 6.12 *The above strong support property is analogous to Assumption* 4.2 *of Chapter* 4.

From the above lemma and the linear relation between δ_ε and ε we derive the following important corollary.

Corollary 6.3 *Let* $(x(\cdot), u(\cdot))$ *be an admissible pair for the discounted optimal control problem and assume that the hypotheses of Lemma* 6.2 *hold. Then if*

$$\|(x(t),u(t)) - (\overline{x},\overline{u})\|^2 > \varepsilon e^{\rho t},$$

we have,

$$f_0(\overline{x},\overline{u}) > f_0(x(t),u(t)) + \overline{q}'\Big[f(x(t),u(t)) - \rho(x(t)-\overline{x})\Big] + \delta_\varepsilon e^{\rho t}. \tag{6.89}$$

With the above facts we have the following turnpike theorem.

Theorem 6.7 (*Finite horizon turnpike theorem for discounted problems*). *Suppose that Assumption* 6.4 *holds and that* $(\overline{x},\overline{u}) \in M$*, with the support property*

$$f_0(\overline{x},\overline{u}) \geq f_0(x,u) + \overline{q}'[f(x,u) - \rho(x-\overline{x})]$$

for all $(x,u) \in M$ (i.e., the hypothesis of Lemma 6.2 hold). Suppose further that there exist finite times T_1 and T_2 such that the points x_0 and x_T are reachable from $\overline{x}$ in time T_1 and T_2, respectively. Then for any $\varepsilon > 0$, there exists a positive number $\eta(\varepsilon) > 0$ such that for all $T \geq T_1 + T_2$ and optimal solutions $(x^*(\cdot), u^*(\cdot)) : [0,T] \to \mathbb{R}^n \times \mathbb{R}^m$ of the finite horizon, discounted optimal control problem, it happens that

$$\mu\Big[\Big\{t : \|(x^*(t), u^*(t)) - (\overline{x}, \overline{u})\|^2 > \varepsilon e^{\rho t},\ t \in [0,T]\Big\}\Big] < \eta(\varepsilon) \tag{6.90}$$

where $\mu[(\cdot)]$ denotes Lebesgue measure.

Remark 6.13 *As we shall see the number $\eta(\varepsilon)$ above depends only on $(\varepsilon, x_0, x_T, T_1, T_2, \overline{x}, \overline{u}, \overline{q}, \rho)$. In particular we notice that it is independent of the optimal pair $(x^*(\cdot), u^*(\cdot))$ and the terminal time T.*

Proof: From our reachability hypotheses, it is possible to construct an admissible pair $(x(\cdot), u(\cdot))$ which transfers x_0 to x_T in time T for any $T \geq T_1 + T_2$. Specifically, $u(\cdot) : [0,T] \to \mathbb{R}^m$ is given by

$$u(t) = \begin{cases} u_1(t) & 0 \leq t \leq T_1 \\ \overline{u} & T_1 \leq t \leq T - T_2 \\ u_2(t - T + T_2) & T - T_2 \leq t \leq T. \end{cases}$$

If $(x^*(\cdot), u^*(\cdot)) : [0,T] \to \mathbb{R}^n \times \mathbb{R}^m$ is optimal we have,

$$\begin{aligned} \int_0^T e^{-\rho t} f_0(x^*(t), u^*(t))\, dt &\geq \int_0^T e^{-\rho t} f_0(x(t), u(t))\, dt \\ &= \int_0^{T_1} e^{-\rho t} f_0(x_1(t), u_1(t))\, dt + \int_{T_1}^{T-T_2} e^{-\rho t} f_0(\overline{x}, \overline{u})\, dt \\ &\quad + \int_{T-T_2}^{T} e^{-\rho t} f_0(x_2(t - T + T_2), u_2(t - T + T_2))\, dt \end{aligned} \tag{6.91}$$

For $\varepsilon > 0$ let

$$\Sigma = \Sigma(\varepsilon, T) = \Big\{t \in [0,T] : \|(x^*(t), u^*(t)) - (\overline{x}, \overline{u})\|^2 > \varepsilon e^{\rho t}\Big\}.$$

Then for $t \in \Sigma$ we have, from Corollary 6.3, that there is a $\delta_\varepsilon > 0$ such that

$$f_0(\overline{x}, \overline{u}) > f_0(x^*(t), u^*(t)) + \overline{q}'\Big[f(x^*(t), u^*(t)) - \rho(x^*(t) - \overline{x})\Big] + \delta_\varepsilon e^{\rho t},$$

and so on Σ

$$e^{-\rho t} f_0(\overline{x}, \overline{u}) > e^{-\rho t} f_0(x^*(t), u^*(t)) + \overline{q}' \frac{d}{dt}[e^{-\rho t}(x^*(t) - \overline{x})] + \delta_\varepsilon.$$

Therefore, from the support property (6.87) we have

$$\int_0^T e^{-\rho t} f_0(\overline{x},\overline{u})\,dt > \int_0^T e^{-\rho t} f_0(x^*(t),u^*(t))\,dt \qquad (6.92)$$
$$+\overline{q}'[e^{-\rho T}(x_T-\overline{x})-(x_0-\overline{x})]+\delta_\varepsilon\mu(\Sigma).$$

Combining (6.91) and (6.92) we arrive at

$$\int_0^T e^{-\rho t} f_0(\overline{x},\overline{u})dt - \overline{q}'[e^{-\rho T}(x_T-\overline{x})-(x_0-\overline{x})]-\delta_\varepsilon\mu[\Sigma]$$
$$> \int_0^{T_1} e^{-\rho t} f_0(x_1(t),u_1(t))\,dt + \int_{T_1}^{T-T_2} e^{-\rho t} f_0(\overline{x},\overline{u})\,dt$$
$$+\int_{T-T_2}^{T} e^{-\rho t} f_0(x_2(t-T+T_2),u(t-T+T_2))\,dt$$
$$= \int_0^{T_1} e^{-\rho t} f_0(x_1(t),u_1(t))\,dt + \int_{T_1}^{T-T_2} e^{-\rho t} f_0(\overline{x},\overline{u})\,dt$$
$$+e^{-\rho(T-T_2)}\int_0^{T_2} e^{-\rho t} f_0(x_2(t),u_2(t))\,dt,$$

and thus

$$\delta_\varepsilon\mu[\Sigma] < \int_0^{T_1} e^{-\rho t}\Big[f_0(\overline{x},\overline{u})-f_0(x_1(t),u_1(t))\Big]\,dt+\overline{q}'(x_0-\overline{x})$$
$$+e^{-\rho(T-T_2)}\int_0^{T_2} e^{-\rho t}\Big[f_0(\overline{x},\overline{u})-f_0(x_2(t),u_2(t))\Big]\,dt,$$
$$-\overline{q}'e^{-\rho T}(x_T-\overline{x})$$
$$= \int_0^{T_1} e^{-\rho t}\Big[f_0(\overline{x},\overline{u})-f_0(x(t),u(t))\Big]\,dt+\overline{q}'(x_0-\overline{x})$$
$$+e^{-\rho(T-T_2)}\Big\{\int_0^{T_2} e^{-\rho t}\Big[f_0(\overline{x},\overline{u})-f_0(x_2(t),u_2(t))\Big]\,dt$$
$$-\overline{q}'e^{-\rho T_2}(x_T-\overline{x})\Big\}.$$

Now observe that,

$$\int_0^{T_2} \overline{q}\frac{d}{dt}\Big[e^{-\rho t}(x_2(t)-\overline{x})\Big]\,dt = \overline{q}'e^{-\rho T_2}(x_T-\overline{x}),$$

and so

$$\delta_\varepsilon\mu[\Sigma] < \int_0^{T_1} e^{-\rho t}\Big[f_0(\overline{x},\overline{u})-f_0(x(t),u(t))\Big]\,dt+\overline{q}'(x_0-\overline{x})$$

$$
\begin{aligned}
&+e^{-\rho(T-T_2)}\Big\{\int_0^{T_2} e^{-\rho t}\Big[f_0(\overline{x},\overline{u}) - f_0(x_2(t),u_2(t))\\
&-\overline{q}'(f(x_2(t),u_2(t)) - \rho(x_2(t)-\overline{x}))\Big]\,dt\Big\}\\
\le\ &\int_0^{T_1} e^{-\rho t}\Big[f_0(\overline{x},\overline{u}) - f_0(x_1(t),u_2(t))\Big]\,dt + \overline{q}'(x_0-\overline{x})\\
&+e^{-\rho T_1}\int_0^{T_2} e^{-\rho t}\Big[f_0(\overline{x},\overline{u}) - f_0(x_2(t),u_2(t))\\
&-\overline{q}'(f(x_2(t),u_2(t)) - \rho(x_2(t)-\overline{x}))\Big]\,dt
\end{aligned}
$$

where this last inequality follows from the fact that the last integral (over $[0,T_2]$) is nonnegative by the support property (6.87) and that $T_1 \le T - T_2$. Notice that the right hand side is a constant which is independent of $(x^*(\cdot)\,, u^*(\cdot)\,)$ and T. If we denote this constant by K we have,

$$
\mu(\Sigma) < \frac{K}{\delta_\varepsilon} = \nu(\varepsilon),
$$

as desired. ∎

To compare this result with Theorem 4.2 we notice that the set $\Sigma(\varepsilon,T,\rho)$ is the set of times $t\in[0,T]$ for which the optimal pair $(x^*(\cdot)\,, u^*(\cdot)\,)$ satisfies

$$
\|(x^*(t),u^*(t)) - (\overline{x},\overline{u})\|^2 > \varepsilon e^{\rho t}.
$$

If we define the new functions $\xi^*(\cdot)\ :\ [0,T]\to\mathbb{R}^n$ and $w^*(\cdot)\ :\ [0,T]\to\mathbb{R}^m$ by

$$
(\xi^*(t),w^*(t)) = e^{-\frac{1}{2}\rho t}(x^*(t)-\overline{x},u^*(t)-\overline{u})
$$

the above result says that for $t\in\Sigma(\varepsilon,T,\rho)$ we have

$$
\|(\xi^*(t),w^*(t))\|^2 \ = \ e^{-\rho t}\|(x^*(t),u^*(t)) - (\overline{x},\overline{u})\|^2 > \varepsilon.
$$

Thus we see that the above theorem implies that the amount of time $(\xi^*(t),w^*(t))$ stays outside of an ε-tube, centered at $(0,0)$ is bounded. That is $(\xi^*(\cdot)\,, w^*(\cdot)\,)$ satisfies the usual turnpike theorem.

Chapter 7

Turnpike Properties and Existence of Overtaking Optimal Solutions for Classes of Nonautonomous Nonconvex Control Problems

7.1 Introduction

In the previous chapters it has been seen that the turnpike properties of extremal trajectories play an important role in developing existence results for autonomous and discounted infinite horizon optimal control problems. In particular we remind the reader of the sufficient conditions for overtaking optimality of Theorems 2.4, 2.5, and 4.4 as well as the results of Chapter 5 regarding the reduction to finite rewards. We notice that the results of Chapter 2 are given for the general nonautonomous control system where the results given in Chapters 4 and 5 apply only to autonomous problems.

In many of the above mentioned results, most notably in Chapter 4, strong convexity and smoothness hypotheses are required. One of the tasks of this chapter is to present extensions of the theory which, in addition to considering time dependent problems, relies on convexity hypotheses which are more aligned with those utilized in Chapter 5. Such extensions will permit us to consider larger classes of problems which are of interest. The other goal of the chapter is to present a unifying theory of sufficient conditions for overtaking optimality and to establish connections with the classical calculus of variations. In particular we show that these results can be formulated in the framework of a method originally due to Carathéodory [34] and as a consequence lead to a Hamilton-Jacobi theory for infinite horizon optimal control problems. Finally we combine these sufficient conditions with the existence of a strongly optimal solution for an Associated Problem of Lagrange(APL) to obtain existence results for nonautonomous infinite horizon optimal problems.

With these remarks the plan of this chapter is as follows. In Section 7.2 we introduce the notion of a G-supported trajectory as a generalization of the support property given in Theorems 2.4 and 2.5. With this notion we present an extension of the previously presented sufficient conditions with weaker convexity requirements. In Section 7.3 we digress to present a nonsmooth version of Carathéodory's method for fixed end point finite horizon control problems. This motivates the discussion of the next section, Section 7.4, in which we extend Carathéodory's method to the infinite horizon case and overtaking optimality. It is here that connections are made between the reduction to finite rewards presented in Chapter 5, the support properties of Chapter 4, and Carathéodory's method. In Section 7.5 we discuss the relevant growth conditions needed to obtain the relative weak compactness of admissible trajectories. These results are needed to treat the existence of optimal solutions for the the Associated Problem of Lagrange. Section 7.6 we consider the question of existence of a strongly optimal solution and discuss the relevant "upper closure property" needed to address this problem by direct methods. Finally, in Section 7.7 we present the desired existence results for overtaking optimal solutions.

7.2 G-supported trajectories

We begin by once again defining the control system considered. Specifically we consider the ordinary differential equation

$$\dot{x}(t) = f(x(t), u(t), t), \quad \text{a.e. } t \geq 0 \tag{7.1}$$

where, as before, $x \in \mathbb{R}^n$, is the state, $u \in \mathbb{R}^m$, is the control, and $f : \mathbb{R}^n \times \mathbb{R}^m \times [0, \infty) \to \mathbb{R}^n$ is continuous. For each $(x, t) \in \mathbb{R}^n \times [0, \infty)$ we are given a closed subset $U(x, t)$ of $\mathbb{R}^m$, such that the map

$$(x, t) \to U(x, t)$$

is upper semicontinuous. In addition, we impose the following growth condition on the function f:

Assumption 7.1 *There exists a locally integrable function $k(\cdot) : [0, \infty) \to [0, \infty)$ and positive real numbers M and N such that*

$$\|f(x, u, t)\| \leq k(t)(M + N\|x\|) \tag{7.2}$$

$$\limsup_{t \to \infty} k(t) < \infty \tag{7.3}$$

for all $(x, u, t) \in \mathcal{M} \triangleq \{(y, v, s) : y \in \mathbb{R}^n, s \geq 0, v \in U(y, s)\}$.

In addition, we consider the performance index for any $T \geq 0$,

$$J_T(x_0, u(\cdot)) = \int_0^T f_0(x(t), u(t), t)\, dt, \tag{7.4}$$

where $f_0 : \mathbb{R}^n \times \mathbb{R}^m \times [0, \infty) \to \mathbb{R}$ is a continuous map.

Remark 7.1 *The above notation is precisely that considered in Section* 1.5, *but is repeated here for convenience. As in previous chapters a pair of functions* $(x(\cdot), u(\cdot)) : [0, \infty) \to \mathbb{R}^{n+m}$ *will be called admissible if* $x(\cdot)$ *is locally absolutely continuous,* $u(\cdot)$ *is Lebesgue measurable,* $(x(\cdot), u(\cdot))$ *satisfies Equation* (7.1), $u(t) \in U(x(t), t)$, *a.e. for* $t \geq 0$, *and the map* $t \to f_0(x(t), u(t), t)$ *is locally integrable.*

With these notations we introduce the following definition given in [98] and [101].

Definition 7.1 *An admissible pair* $\bar{x}(\cdot)$, *generated by a control* $\bar{u}(\cdot)$, *is said to be* **G-supported** *if there exists a* C^1 *mapping* $G(\cdot) : \mathbb{R}^n \to \mathbb{R}^n$, *with derivative* $g(x)$ *at* x, *and a locally absolutely continuous function* $\bar{p}(\cdot) : [0, \infty) \to \mathbb{R}^n$ *such that for all* $(t, x, u) \in \mathcal{M}$, *we have*

$$\begin{aligned} f_0(\bar{x}(t), \bar{u}(t), t) + \frac{d}{dt}[\bar{p}(t)' G(\bar{x}(t))] \geq \\ f_0(x, u, t) + \bar{p}(t)^T g(x) f(x, u, t) + \dot{\bar{p}}(t)' G(x). \end{aligned} \tag{7.5}$$

For convenience we define the *value-loss function* associated with $(\bar{x}(\cdot), \bar{u}(\cdot))$ by the nonnegative function $\Delta(\cdot, \cdot, \cdot) : \mathcal{M} \to \mathbb{R}$ given by

$$\begin{aligned} \Delta(x, u, t) = f_0(\bar{x}(t), (\bar{u}(t), t) + \frac{d}{dt}[\bar{p}(t)' G(\bar{x}(t))] \\ - f_0(x, u, t) - \bar{p}(t)' g(x) f(x, u, t) - \dot{\bar{p}}(t)' G(x). \end{aligned} \tag{7.6}$$

Remark 7.2 *Observe that if* $G(x)$ *is the identity function, the notion of a G-supported trajectory coincides with the support property given by* (i) *of Theorem* 2.5. *In connection with problems in a discrete time setting we note that when* $G(x)$ *is the identity, the supported trajectory* $\bar{x}(\cdot)$ *is referred to in* [183] *as a* **McKenzie turnpike**.

Before discussing the existence of G-supported trajectories, we indicate their significance, beginning with the following lemma.

Lemma 7.1 *Let* $\bar{x}_1(\cdot)$ *and* $\bar{x}_2(\cdot)$ *be two G-supported trajectories, generated by the controls* $\bar{u}_1(\cdot)$ *and* $\bar{u}_2(\cdot)$, *with support functions* $\bar{p}_1(\cdot)$ *and* $\bar{p}_2(\cdot)$, *respectively. Then for almost all* $t \geq 0$,

$$\frac{d}{dt}\{(\bar{p}_1(t) - \bar{p}_2(t))'[G(\bar{x}_1(t)) - G(\bar{x}_2(t))]\} \geq 0. \tag{7.7}$$

Proof: By using Equation (7.5) twice it is easy to see that for almost all $t \geq 0$,

$$\frac{d}{dt}\{\bar{p}_1(t)'G(\bar{x}_1(t)) - \bar{p}_1(t)'G(\bar{x}_2(t))\}$$
$$\geq f_0(\bar{x}_2(t), \bar{u}_2(t), t) - f_0(\bar{x}_1(t), \bar{u}_1(t), t) \tag{7.8}$$

and

$$\frac{d}{dt}\{\bar{p}_2(t)'G(\bar{x}_2(t)) - \bar{p}_2(t)'G(\bar{x}_1(t))\}$$
$$\geq f_0(\bar{x}_1(t), \bar{u}_1(t), t) - f_0(\bar{x}_2(t), \bar{u}_2(t), t). \tag{7.9}$$

The desired result, Equation (7.7), now follows immediately by adding Equation (7.8) to Equation (7.9). ∎

To investigate the asymptotic stability properties of G-supported trajectories, the above lemma suggests the use of the function

$$V(t) = (\bar{p}_1(t) - \bar{p}_2(t))'[G(\bar{x}_1(t)) - G(\bar{x}_2(t))] \tag{7.10}$$

as a Lyapunov-like function. However as shown by Cass and Shell [37] it is better to consider a function

$$V_\alpha(t) = \alpha(t)V(t), \tag{7.11}$$

where $\alpha : [0, \infty) \to [0, \infty)$ is a nondecreasing and differentiable function. This permits us to prove the following asymptotic convergence result.

Theorem 7.1 *Let $\bar{x}_1(\cdot)$ and $\bar{x}_2(\cdot)$ be to G-supported trajectories with respective support functions $\bar{p}_1(\cdot)$ and $\bar{p}_2(\cdot)$. Assume that for each $\epsilon > 0$ there exists $\delta > 0$, such that for all $t \geq 0$ we have that $\|\bar{x}_1(t) - x\| > \epsilon$ implies*

$$\alpha(t)\Delta_1(x, u, t) > \delta - \dot{\alpha}(t)(\bar{p}_2(t) - \bar{p}_1(t))$$
$$(G(x) - G(\bar{x}_1(t))) \tag{7.12}$$

for all $u \in U(x, t)$, where $\Delta_1, i = 1, 2$ is the value loss function associated with G-supported trajectory $\bar{x}_i(\cdot)$. Further assume that,

$$\limsup_{t\to\infty} \|\bar{x}_i(t)\| < \infty, \quad i = 1, 2 \tag{7.13}$$

and

$$\limsup_{t\to\infty} \alpha(t)\|\bar{p}_i(t)\| < \infty, \quad i = 1, 2 \tag{7.14}$$

both hold. Then

$$\lim_{t\to\infty} \|\bar{x}_1(t) - \bar{x}_2(t)\| = 0. \tag{7.15}$$

Remark 7.3 *For the discounted case (i.e.,* $f_0(x,u,t) = e^{-\rho t} f_0(x,u)$, $\rho > 0$) *an appropriate choice for* α *is* $\alpha(t) = e^{\rho t}$. *Here, we observe that the condition Equation* (7.12), *when* $G(x) = x$, *is effectively the strong support property of Feinstein and Oren* [76] *discussed in Section* 6.7. *Further, in this case, one has* $\lim_{t\to\infty}\left[\frac{\dot\alpha(t)}{\alpha(t)}\right] = \rho$ *and so we see that the last hypothesis places a restriction on the size of the discount rate* ρ.

Proof: The proof is by contradiction, so we assume that Equation (7.15) does not hold. If so, there exists an $\epsilon > 0$ and a sequence $\{t_n\}_{n=1}^{\infty}, t_n \to \infty$, so that

$$\|\bar{x}_1(t_n) - \bar{x}_2(t_n)\| > 2\epsilon \tag{7.16}$$

for all $n = 1, 2, \ldots$. From the growth condition (7.2), (7.3) and the hypotheses (7.13) and (7.14), it follows that each trajectory is a continuous function of bounded variation and so there exists a positive number Δt, such that for all n, and all $t \in [t_n - \Delta t, t_n + \Delta t]$,

$$\|\bar{x}_1(t) - \bar{x}_2(t)\| > \epsilon.$$

This implies that

$$\mu[\{t : \|\bar{x}_1(t) - \bar{x}_2(t)\| > \epsilon\}] = \infty, \tag{7.17}$$

in which $\mu[\cdot]$ denotes Lebesgue measure. Observe now that

$$\begin{aligned} \dot{V}_\alpha(t) &= \frac{d}{dt}[\alpha(t)V(t)] \\ &= \alpha(t)[\dot{V}(t) + \frac{\dot\alpha(t)}{\alpha(t)}V(t)]. \end{aligned} \tag{7.18}$$

In addition, we have

$$\begin{aligned} \dot{V}(t) &= \frac{d}{dt}[(\bar{p}_1(t) - \bar{p}_2(t))'(G(\bar{x}_1(t)) - G(\bar{x}_2(t))] \\ &= \frac{d}{dt}[\bar{p}_1(t)'(G(\bar{x}_1(t)) - G(\bar{x}_2(t))] \\ &\quad - \frac{d}{dt}[\bar{p}_2(t)'(G(\bar{x}_1(t)) - G(\bar{x}_2(t))] \\ &\geq f_0(\bar{x}_1(t), \bar{u}_1(t), t) - f_0(\bar{x}_2(t), \bar{u}_2(t), t) \\ &\quad + \frac{d}{dt}[\bar{p}_1(t)'(G(\bar{x}_1(t)) - G(\bar{x}_2(t))] \\ &= \Delta_1(\bar{x}_2(t), \bar{u}_2(t), t), \end{aligned}$$

and so by multiplying both sides by $\alpha(t)$ we get

$$\alpha(t)\dot{V}(t) \geq \alpha(t)\Delta_1(\bar{x}_2(t), \bar{u}_2(t), t).$$

Therefore, for $t \geq 0$ such that $\|\bar{x}_1(t) - \bar{x}_2(t)\| > \epsilon$, we have from Equations (7.10)–(7.12)

$$\dot{V}_\alpha(t) = \alpha(t)[\dot{V}(t) + \frac{\dot{\alpha}(t)}{\alpha(t)}V(t)] \geq \delta. \tag{7.19}$$

Further,

$$\begin{aligned}
\limsup_{t\to\infty} |V_\alpha(t)| &= \limsup_{t\to\infty}[\alpha(t)|(\bar{p}_1(t) - \bar{p}_2(t))'(G(\bar{x}_1(t)) - G(\bar{x}_2(t)))|] \\
&\leq \limsup_{t\to\infty} \alpha(t)\Big\{\|\bar{p}_1(t)\|[\|G(\bar{x}_1(t))\| + \|G(\bar{x}_2(t))\|] \\
&\quad + \|\bar{p}_2(t)\|[\|G(\bar{x}_1(t))\| + \|G(\bar{x}_2(t))\|]\Big\} \\
&\leq \infty,
\end{aligned}$$

and so, V_α is bounded. On the other hand, Equations (7.17), (7.19) imply that

$$\lim_{t\to\infty} V_\alpha(t) = \infty,$$

an obvious contradiction. Hence Equation (7.15) holds as desired. ■

The above result shows, under appropriate hypotheses, that G-supported trajectories tend to *cluster* at infinity. The next result gives an analogue of the sufficient conditions for overtaking optimality given in Theorem 2.4 for G-supported trajectories.

Theorem 7.2 *Let $x^*(\cdot) : [0,\infty) \to \mathbb{R}^n$ be an admissible pair satisfying $x^*(0) = x_0$ and $x^*(t) \in X$ for $t \geq 0$, where $x_0 \in \mathbb{R}^n$ is given and $X \subseteq \mathbb{R}^n$ is compact, which is a G-supported trajectory whose support function, $p^*(\cdot)$, satisfies*

$$\lim_{t\to\infty} \|p^*(t)\| = 0. \tag{7.20}$$

Then $x^*(\cdot)$ is an overtaking optimal solution over the class of all admissible pairs, $(x(\cdot), u(\cdot))$, emanating from x_0 which satisfy $x(t) \in X$, for $t \geq 0$.

Proof: Let $(x(\cdot), u(\cdot))$ be any admissible pair satisfying $x(0) = x_0$ and $x(t) \in X$ for $t \geq 0$. Then we have for any $T > 0$,

$$\begin{aligned}
J_T(x_0, u^*(\cdot)) - J_T(x_0, u(\cdot)) &= \int_0^T [f_0(x^*(t), u^*(t), t) - f_0(x(t), u(t), t)]\, dt \\
&= \int_0^T \Delta(x(t), u(t), t)\, dt \\
&\quad + \int_0^T \frac{d}{dt}[p^*(t)'(G(x(t)) - G(x^*(t)))]\, dt \\
&= \int_0^T \Delta(x(t), u(t), t)\, dt \\
&\quad + p^*(T)'[G(x(T)) - G(x^*(T))] \\
&\geq p^*(T)'[G(x(T)) - G(x^*(T))],
\end{aligned}$$

and so, the hypotheses placed on $x(\cdot)$ and $x^*(\cdot)$ and Equation (7.20) imply

$$\limsup_{T\to\infty}[J_T(x_0, u^*(\cdot)) - J_T(x_0, u(\cdot))] \geq 0,$$

as desired. ∎

Observe that if one has that Equation (7.14) holds and if in addition,

$$\lim_{t\to\infty} \alpha(t) = \infty,$$

(which does hold in the discounted case), then necessarily Equation (7.20) holds. On the other hand if

$$\lim_{t\to\infty} \alpha(t) < \infty,$$

then without loss of generality we could take $\lim_{t\to\infty} \alpha(t) = 1$ which implies that Equation (7.12) for any fixed p would become

For every $\epsilon > 0$ there exists $\delta > 0$ and $T > 0$ such that for every $t \geq T$

$$\|\bar{x}_1(t) - x\| > \epsilon \Rightarrow \Delta(x, u, t) > \delta. \tag{7.21}$$

This is precisely the assumption needed to weaken the condition given in (7.20).

Theorem 7.3 *If in Theorem* 7.2 *we replace* (7.20) *by* (7.14), *and* $\lim_{t\to\infty} \alpha(t) = 1$, *and assume* (7.21) *holds then* $x^*(\cdot)$ *is overtaking optimal over all admissible pairs* $(x(\cdot), u(\cdot))$ *satisfying* $x(0) = x_0$ *and* $x(t) \in X$, $t \geq 0$ *(recall that* $X \subseteq \mathbb{R}^n$ *is compact).*

Proof: From (7.14) and $\lim_{t\to\infty} \alpha(t) = 1$, we have

$$\limsup_{t\to\infty} \|p^*(t)\| < \infty.$$

If $(x(\cdot), u(\cdot))$ is admissible and satisfies $x(0) = x_0$ and $x(t) \in X$ for $t \geq 0$, then exactly as in the previous proof we have for all $T > 0$

$$J_T(x_0, u^*(\cdot)) - J_T(x_0, u(\cdot)) \geq p^*(T)'[G(x(T)) - G(x^*(T))].$$

Therefore, if $\lim_{t\to\infty} \|x(t) - x^*(t)\| = 0$ we have

$$\liminf_{T\to\infty}[J_T(x_0, u^*(\cdot)) - J_T(x, u)] \geq 0.$$

On the other hand, if $\lim_{t\to\infty} \|x(t) - x^*(t)\| \neq 0$, then for some $\epsilon_o > 0$,

$$\mu[\{t : \|\bar{x}_1(t) - \bar{x}_2(t)\| > \epsilon_o\}] = \infty,$$

which implies that

$$\liminf_{T\to\infty}[J_T(x_0, u^*(\cdot)) - J_T(x_0, u(\cdot))] = \infty.$$

Combining these results leads to the desired conclusion. ■

Having seen the utility of G-supported trajectories for infinite horizon control problems we now present sufficient conditions for the existence of G-supported trajectories. To do this we require the following specializations.

Assumption 7.2

(a) *The mapping* $f(\cdot,\cdot,\cdot) : \mathbb{R}^n \times \mathbb{R}^m \times [0,\infty) \to \mathbb{R}$ *is differentiable with respect to* x *and* u.

(b) *The set* $U(x,t)$ *is defined by*

$$U(x,t) = \{u \in \mathbb{R}^m : h(x,u,t) \geq 0\}, \tag{7.22}$$

where $h(\cdot,\cdot,\cdot) : \mathbb{R}^n \times \mathbb{R}^m \times [0,\infty) \to \mathbb{R}^p$ *satisfies the same hypotheses as the function* f.

(c) *The function* $G(\cdot) : \mathbb{R}^n \to \mathbb{R}^n$ *is one-to-one and twice continuously differentiable.*

With these hypotheses we define the modified Hamiltonian

$$H(x,u,p,t) = f_0(x,u,t) + p'g(x)f(x,u,t), \tag{7.23}$$

the modified Lagrangian

$$L(x,u,p,\mu,t) = H(x,u,p,t) + \mu' h(x,u,t), \tag{7.24}$$

and introduce the following definition.

Definition 7.2 *A trajectory* $\bar{x}(\cdot) : [0,\infty) \to \mathbb{R}^n$ *satisfying* $x(0) = x_0$, *and generated by the admissible control* $\bar{u}(\cdot) : [0,\infty) \to \mathbb{R}^m$ *is called* **G-extremal** *if there exists a locally absolutely continuous function* $\bar{p}(\cdot) : [0,\infty) \to \mathbb{R}^n$ *and a function* $\bar{\mu}(\cdot) : [0,\infty) \to \mathbb{R}^n$ *such that for almost all* $t \geq 0$ *the following conditions are met:*

$$H(\bar{x}(t), \bar{u}(t), \bar{p}(t), t) = \max_{u\in U(x,t)} \{H(\bar{x}(t), u, \bar{p}(t), t)\} \tag{7.25}$$

$$\Big[\frac{d}{dt}\bar{p}(t)\Big]' g(\bar{x}(t)) = -\frac{\partial}{\partial x} L(\bar{x}(t), \bar{u}(t), \bar{p}(t), \bar{\mu}(t), t)' \tag{7.26}$$

$$\bar{p}(t) \geq 0, \qquad \bar{\mu}(t) \geq 0 \tag{7.27}$$

$$\bar{\mu}(t)' h(\bar{x}(t), \bar{u}(t), t) = 0, \tag{7.28}$$

where $\bar{\mu}(t)$ *denotes the Lagrange multiplier associated with the solution of* (7.25).

We now make the following additional assumption.

Assumption 7.3 *The set $\Omega_G(t) \subset \mathbb{R}^n \times \mathbb{R}^m \times \mathbb{R}$, defined for $t \geq 0$ by*

$$\begin{aligned} \Omega_G(t) &= \{(\xi, \eta, \gamma) : \xi \leq G(x), \eta \leq g(x)f(x,u,t), \\ &\quad \gamma \leq f_0(x,u,t), \text{ for some } x \in \mathbb{R}^n \text{ and } u \in U(x,t)\}, \end{aligned} \tag{7.29}$$

is convex and has nonempty interior for each $t \geq 0$.

Theorem 7.4 *If Assumptions 7.2 and 7.3 both hold, then a G-extremal trajectory, $\bar{x}$, is G-supported.*

Proof: For $t \geq 0$ consider the convex set $\pi(t) \subset \mathbb{R}^n \times \mathbb{R}$, defined by

$$\pi(t) = \left\{(\eta, \gamma) : (G(\bar{x}(t)), \eta, \gamma) \in \Omega_G(t)\right\}.$$

The maximum condition (7.25) shows that the vector

$$(\bar{\eta}(t), \gamma(t)) = (g(\bar{x}(t))f(\bar{x}(t), \bar{u}(t), t), f_0(\bar{x}(t), \bar{u}(t), t))$$

is a boundary point of $\pi(t)$ with a supporting hyperplane whose normal is $(\bar{p}(t), 1)$. That is the hyperplane, described by

$$\bar{p}(t)'(\eta - \bar{\eta}(t)) + (\gamma - \bar{\gamma}(t)) = 0.$$

This implies that the point $(G(\bar{x}(t)), \bar{\eta}(t), \bar{\gamma}(t))$ is a boundary point of the set $\Omega_G(t)$, a set with nonempty interior. By appealing to the separation theorem for convex sets it follows that there exists a supporting hyperplane for $\Omega_G(t)$ at this point with normal vector $(M(t), \bar{p}(t), 1)$. Thus for all $(\xi, \eta, \gamma) \in \Omega_G(t)$ we have

$$\begin{aligned} \gamma + \bar{p}(t)'\eta + M(t)'\xi &\leq f_0(\bar{x}(t), \bar{u}(t), t) \\ &\quad + \bar{p}(t)'g(\bar{x}(t))f(\bar{x}(t), \bar{u}(t), t) \\ &\quad + M(t)'G(\bar{x}(t)). \end{aligned} \tag{7.30}$$

It remains to show that $M(t) = \dfrac{d}{dt}\bar{p}(t)$. To this end we observe that (7.30) implies the functional

$$f_0(x,u,t) + \bar{p}(t)'g(x)f(x,u,t) + M(t)'G(x) \tag{7.31}$$

is maximized at $(\bar{x}(t), \bar{u}(t))$ for all (x, u) such that

$$h(x,u,t) \geq 0. \tag{7.32}$$

Taking $x = \bar{x}(t)$ in (7.30) and (7.31) yields the maximization problem given in (7.25) and thus one can take $\bar{\mu}(t)$ as the Lagrange multiplier. This implies that

$$0 = \frac{\partial}{\partial x}\left[f_0(x,u,t) + \bar{p}(t)'g(x)f(x,u,t) + M(t)'G(x) + \bar{\mu}(t)'h(x,u,t)\right]$$

when $(x,u) = (\bar{x}(t), \bar{u}(t))$. This is equivalently expressed as

$$M(t)'g(\bar{x}(t)) = -\frac{\partial}{\partial x}L(\bar{x}(t), \bar{u}(t), \bar{p}(t), \bar{\mu}(t), t)'.$$

The conclusion now follows from (7.26) and the fact that the matrix $g(\bar{x}(t))$ is nonsingular.

∎

To make further comparisons with the previous results, we now restrict our attention to the autonomous, discounted case. Specifically, we now assume that $f_0(x,u,t) = e^{-\rho t}f_0(x,u)$ where $\rho \geq 0$, $f(x,u,t) = f(x,u)$, and $U(x,t) = U(x) = \{u \in \mathbb{R}^m : h(x,u) \geq 0\}$. As in the past, it is convenient to define the current-value modified Hamiltonian

$$\tilde{H}(x,u,q) = f_0(x,u) + q'g(x)f(x,u) \tag{7.33}$$

and the associated current-value modified Lagrangian

$$\tilde{L}(x,u,q,\nu) = \tilde{H}(x,u,q) + \nu'h(x,u). \tag{7.34}$$

Definition 7.3 *A vector $\bar{x} \in \mathbb{R}^n$ is called a* G-extremal steady state *if there exists vectors $\bar{u} \in \mathbb{R}^m$, $\bar{q} \in \mathbb{R}^n$ and $\bar{\nu} \in \mathbb{R}^p$ such that the following five conditions hold:*

$$0 = f(\bar{x}, \bar{u}) \tag{7.35}$$

$$0 \leq h(\bar{x}, \bar{u}), \quad \nu \geq 0 \tag{7.36}$$

$$\tilde{H}(\bar{x}, \bar{u}, \bar{q}) = \max_{u \in U(\bar{x})}\{\tilde{H}(\bar{x}, u, \bar{q})\} \tag{7.37}$$

$$0 = \bar{\nu}'h(\bar{x}, \bar{u}) \tag{7.38}$$

$$0 = \rho\bar{p}'g(\bar{x}) - \frac{\partial}{\partial x}\tilde{L}(\bar{x}, \bar{u}, \bar{q}, \bar{\nu})'. \tag{7.39}$$

Lemma 7.2 *Under Assumption 7.3, a G-extremal steady state, $\bar{x}$, is a G-supported trajectory.*

Proof: Define the function $\tilde{p}(t) = \bar{q}e^{-\rho t}$ and the admissible pair $(\tilde{x}(t), \tilde{u}(t)) \equiv (\bar{x}, \bar{u})$. From (7.35) to (7.39) it follows that $\tilde{x}(\cdot)$ is a G-extremal trajectory. The desired result now is an immediate consequence of Theorem 7.4. ∎

Remark 7.4 *The above definition of a G-extremal steady state can possibly be viewed as a generalization of the implicit programming problem of* [75]. *To see this we observe that if* $(\bar{x}, \bar{u})$ *solves the implicit programming problem*

maximize

$$f_0(x, u) \tag{7.40}$$

subject to

$$\begin{aligned} f(x,u) - \rho(x - \bar{x}) &= 0 \\ h(x,u) &\geq 0, \end{aligned} \tag{7.41}$$

and if certain constraint qualifications are met, there exist multipliers $\bar{q} \in \mathbb{R}^n$ *and* $\bar{\nu} \in \mathbb{R}^p$, $\bar{\nu} \geq 0$, *such that* $\bar{\nu}' h(\bar{x}, \bar{u}) = 0$ *and* $(\bar{x}, \bar{u})$ *is the maximum of the function*

$$f_0(x,u) + \bar{q}'(f(x,u) - \rho(x - \bar{x})) + \bar{\nu}' h(x,u) \tag{7.42}$$

for all (x, u). *Thus,*

$$\frac{\partial f_0}{\partial x}(\bar{x}, \bar{u})' + \bar{q}'\Big(\frac{\partial f}{\partial x}(\bar{x}, \bar{u}) - \rho\Big) + \bar{\nu}' \frac{\partial h}{\partial x}(\bar{x}, \bar{u}) = 0, \tag{7.43}$$

which is precisely condition (7.39) *when we make the identification* $\bar{p} = \bar{q}$ *and* $G(x) = (x - \bar{x})$. *Moreover, we notice that we also have*

$$\tilde{H}(\bar{x}, \bar{u}, \bar{q}) = \max\{\tilde{H}(\bar{x}, u, \bar{q}) : h(\bar{x}, u) \geq 0\}, \tag{7.44}$$

which implies that $\bar{x}$ *is a G-extremal steady state.*

In Feinstein and Luenberger [75] it is shown (under appropriate convexity hypotheses) that if the discount rate $\rho \geq 0$ is sufficiently small then the overtaking optimal trajectories asymptotically approach the solution $\bar{x}$ of the implicit programming problem. To present the analogue of this result for G-supported trajectories we suppose that x_0 is a fixed initial condition and consider an admissible trajectory $x^*(\cdot) : [0, \infty) \to \mathbb{R}^n$, satisfying $x^*(0) = x_0$ and generated by an admissible control $u^*(\cdot) : [0, \infty) \to \mathbb{R}^m$, and a current-value supporting function $q^*(\cdot) : [0, \infty) \to \mathbb{R}^n$ which satisfy the following conditions for almost all $t \geq 0$:

$$h(x^*(t), u^*(t)) \geq 0, \quad \nu \geq 0 \tag{7.45}$$

$$\tilde{H}(x^*(t), u^*(t), q^*(t)) = \max_{u \in U(x^*(t))} \tilde{H}(x^*(t), u, q^*(t)) \tag{7.46}$$

$$\bar{\nu}' h(x^*(t), u^*(t)) = 0 \tag{7.47}$$

$$(\dot{q}^*(t) - \rho q^*(t))' g(x^*(t)) = -\frac{\partial}{\partial x} \tilde{L}(x^*(t), u^*(t), q^*(t), \bar{\nu})' \tag{7.48}$$

$$\lim_{t \to \infty} q^*(t) = \bar{q}. \tag{7.49}$$

If such a solution exists, then $x^*(\cdot)$ is clearly G-extremal (with $p^*(t) = e^{-\rho t} q^*(t)$).

Assumption 7.4 *The G-extremal steady state $\bar{x}$ and the supporting vector $\bar{q}$ are such that $\forall\epsilon \geq 0$, $\exists\,\delta \geq 0$ such that $\forall x \in \mathbb{R}^n$, $\|x-\bar{x}\| \geq \epsilon$ implies for all $t \geq 0$*

$$f_0(\bar{x},\bar{u}) - f_0(x,u) - \bar{q}'g(x)f(x,u) \geq \delta - \rho\bar{q}'(G(x) - G(\bar{x})) \tag{7.50}$$

for all $u \in U(x)$.

Theorem 7.5 *Under Assumptions 7.2 and 7.4, if the trajectory $x^*(\cdot)$ satisfying (7.45) to (7.49) satisfies $x^*(t) \in X$, $t \geq 0$, where $X \subset \mathbb{R}^n$ is compact, then $x^*(\cdot)$ is overtaking optimal over all admissible trajectories $x(\cdot)$, satisfying $x(0) = x_0$ and $x(t) \in X$ for $t \geq 0$, provided $\rho > 0$ is sufficiently close to zero. Moreover we also have*

$$\lim_{t\to\infty} x^*(t) = \bar{x}. \tag{7.51}$$

Proof: The fact that $x^*(\cdot)$ is overtaking optimal follows from Theorem 7.2 since

$$\lim_{t\to\infty} \|p^*(t)\| = \lim_{t\to\infty} e^{-\rho t}\|q^*\| = 0.$$

From Assumption 7.4 and the compactness of it follows that for t sufficiently large, as $q^*(t) \to \bar{q}$, (7.12), (7.13), (7.14) all hold with $\bar{x}_1(t) = \bar{x}$, $\bar{p}_1(t) = \bar{q}e^{-\rho t}$, and $\alpha(t) = e^{\rho t}$, $\bar{p}_2(t) = p^*(t)$ and $\bar{x}_2(t) = x^*(t)$. Then by Theorem 7.1 we have

$$\lim_{t\to\infty} \|x^*(t) - \bar{x}\| = 0.$$

■

We conclude our discussion of G-supported trajectories by considering the following example appearing in [101] (see also [98]).

Example 7.1 This example arises from studying the optimal harvest rates of a predator and a prey in mutual dependence. Specifically the dynamics of this system is described by the control system

$$\begin{aligned}
\dot{x}(t) &= x(t)[f_1(x(t),y(t)) - u(t)] && (7.52)\\
\dot{y}(t) &= y(t)[f_2(x(t),y(t)) - v(t)] && (7.53)\\
0 \leq u(t) &\leq h_1(k_1(t),x(t)) && (7.54)\\
0 \leq v(t) &\leq h_2(k_2(t),y(t)) && (7.55)\\
0 < a \leq x(t),\ & 0 < a \leq y(t) && (7.56)\\
0 \leq k_i(t),\ & i = 1,2. && (7.57)
\end{aligned}$$

In the above description we assume $a > 0$ is fixed and we interpret x as the prey, y as the predator, k_i as the stock of physical capital used for the exploitation of each species, and u and v denote the per capita harvest rates of the two species. The functions $f_1(\cdot,\cdot)$

and $f_2(\cdot,\cdot)$ describe the birth rates of the two species in interaction and the functions $h_1(\cdot,\cdot)$ and $h_2(\cdot,\cdot)$ represent production functions of the harvesting activity. From economic and biological considerations we can assume that the functions f_1, f_2, h_1, and h_2 are all concave. The performance criterion we consider is given by

$$J_T(x_0, y_0, u(\cdot), v(\cdot)) = \int_0^T e^{-\rho t} f_0(x(t), y(t), u(t), v(t), k_1(t), k_2(t))\, dt \tag{7.58}$$

where f_0 is an instantaneous utility function and $\rho > 0$ is a discount rate.

For this model, we consider the function $G : \mathbb{R}^2 \to \mathbb{R}^2$ defined by

$$G(x, y) = (\log(x), \log(y)) \tag{7.59}$$

which is well defined on $(0, \infty) \times (0, \infty)$. Its derivative g at (x, y) is given by the matrix

$$g(x, y) = \begin{bmatrix} 1/x & 0 \\ 0 & 1/y \end{bmatrix} \tag{7.60}$$

In this situation, the set $\Omega_G(t)$ is given by

$$\begin{aligned} \Omega_G(t) = \{(\xi_1, \xi_2, \eta_1, \eta_2, \gamma) \;:\; & \xi_1 \le \log(x), \xi_2 \le \log(y), \\ & \eta_1 \le f_1(x, y) - u, \;\; \eta_2 \le f_2(x, y) - v, \\ & \gamma \le e^{-\rho t} f_0(x, y, u, v, k_1, k_2), \\ & \text{for } \;(x, y, u, v, k_1, k_2) > 0 \;\text{ satisfying} \\ & u \le h_1(k_1, x) \;\text{ and }\; v \le h_2(k_2, y)\}. \end{aligned} \tag{7.61}$$

As a consequence of our concavity assumptions on f_1, f_2, h_1, h_2, and f_0 it is easy to see that $\Omega_G(t)$ is a convex set and that the continuity of these functions guarantee that it has nonempty interior. Therefore it is possible to determine a G-supported trajectory by looking for a G-extremal steady state. If such a steady state exists then the results discussed previously would be applicable.

7.3 Carathéodory's method for finite horizon optimal control problems

In the calculus of variations the Hamilton-Jacobi theory plays an important role in establishing sufficient conditions for optimality. Our goal here is to present an extension of this theory, first exposed in [43], to finite horizon optimal control problems, with fixed endpoints. The approach given here extends the concept of equivalent variational problems originally due to Carathéodory [34]. These results also contain earlier approaches due to Bridgeland [25] and Snow [170] who both assume stronger smoothness hypotheses than considered here.

To compare the results given here with those of Carathéodory we pose our problem as a nonsmooth calculus of variations problem. We have seen this technique applied to the autonomous infinite horizon problem considered previously. Specifically, we consider the problem of maximizing the integral functional

$$J(x(\cdot)) = \int_{t_o}^{t_1} \mathcal{L}(x(t), \dot{x}(t), t)\, dt \tag{7.62}$$

over all functions $x(\cdot) : [t_o, t_1] \to \mathbb{R}^n$ which are absolutely continuous, satisfy $J(x)$ is finite, and which satisfy the fixed end conditions

$$x(t_o) = x_0 \text{ and } x(t_1) = x_1. \tag{7.63}$$

The function $\mathcal{L}(\cdot,\cdot) : \mathbb{R}^n \times \mathbb{R}^n \times [t_o, t_1] \to \overline{\mathbb{R}}$ is an extended real valued function.

Throughout this discussion it will be necessary to deal with several different, though similar, measure spaces defined for a product set of the form $\mathbb{R}^k \times I$, in which $I = [t_o, t_1]$ and k is a positive integer. These are defined as follows.

Definition 7.4 *A set $B \subset \mathbb{R}^k \times I$ will be called* $(\mathbb{R}^k \times I)$-**measurable** *if B is measurable with respect to the σ-algebra generated by products of Borel measurable subsets of $\mathbb{R}^k$ with Lebesgue measurable subsets of I. Correspondingly, a function $f : \mathbb{R}^k \times I \to \mathbb{R}^p$ will be called a* $(\mathbb{R}^k \times I)$-**measurable function** *if each of its components is measurable with respect to this σ-algebra.*

Assumption 7.5 *We assume throughout that the integrand $\mathcal{L}$ satisfies the following:*

$$\mathcal{L}(\cdot,\cdot,t) \quad \textit{is upper semicontinuous for almost all} \quad t \in I \tag{7.64}$$

$$\mathcal{L}(\cdot,\cdot,\cdot) \quad \textit{is} \quad (\mathbb{R}^{2n} \times I)\textit{-measurable} \tag{7.65}$$

$$\mathcal{L}(x,\cdot,t) \quad \textit{is concave for all} \quad (x,t) \in \mathbb{R}^k \times I. \tag{7.66}$$

We also require $\mathcal{L}$ to satisfy the following basic growth condition.

Assumption 7.6 *We say that $\mathcal{L}(\cdot,\cdot,\cdot)$ satisfies the basic growth condition if for each fixed $p \in \mathbb{R}^n$ and bounded set $K \subset \mathbb{R}^n$ there exists a Lebesgue integrable function $Q(\cdot)$: $[t_o, t_1] \to \mathbb{R}$ such that the maximized Hamiltonian satisfies,*

$$\mathcal{H}(x,p,t) = \sup_{z \in \mathbb{R}^n} [p'z + \mathcal{L}(x,z,t)] \le Q(t) \tag{7.67}$$

for almost all (x,t).

Remark 7.5 *This growth condition has been shown to be equivalent to a growth condition originally due to Cesari, LaPalm and Nishiura* [49] *(see also* [47]*) used later in this chapter, to guarantee that admissible trajectories are weakly compact in the space of locally absolutely continuous function. In addition we remark that this growth condition is more general than those used previously.*

Under Assumption 7.6, the Hamiltonian is $(\mathbb{R}^{2n} \times I)$-measurable and in addition is such that $\mathcal{H}(x,\cdot,t)$ is upper semicontinuous and convex. Moreover, we also have the reciprocal relation

$$\mathcal{L}(x,z,t) = \inf_{p \in \mathbb{R}^n} [\mathcal{H}(x,p,t) - p'z] \tag{7.68}$$

With these preliminary facts we now present Carathéodory's method.

The basic idea is to define a new integrand $\mathcal{T} : \mathbb{R}^n \times \mathbb{R}^n \times I \to \mathbb{R}^n \cup \{-\infty\}$ by the formula

$$\mathcal{T}(x,z,t) = \mathcal{L}(x,z,t) + S_t(t,x) + \nabla_x S(t,x)'z, \tag{7.69}$$

in which $S(\cdot,\cdot) : \mathbb{R}^n \times I \to \mathbb{R}$ is a given differentiable function. Observe that for each trajectory $x(\cdot) : I \to \mathbb{R}^n$ which is feasible for the optimization problem (7.62) and (7.63) we have that

$$\begin{aligned} \tilde{J}(x)(\cdot) &\triangleq \int_{t_o}^{t_1} \mathcal{T}(x(t),\dot{x}(t),t)\,dt \\ &= J(x(\cdot)) + [S(x_1,t_1) - S(x_0,t_o)], \end{aligned} \tag{7.70}$$

from which it can be seen that the minimizers of $\tilde{J}$ are the same as those of J. Carathéodory called the above two variational problems *equivalent.* As a consequence of the arbitrary nature of the function S we are led to a family of equivalent variational problems. This however fails to be useful unless we can choose an appropriate function S. In particular we search for two functions $S(\cdot,\cdot) : \mathbb{R}^n \times I \to \mathbb{R}$ and $\tilde{p}(\cdot,\cdot) : \mathbb{R}^n \times I \to \mathbb{R}^n$ satisfying the following properties:

$$\mathcal{L}(x,z,t) + \frac{\partial S}{\partial t}(x,t) + \nabla_x S(x,t)'z \leq 0 \tag{7.71}$$

for all (x,z) and almost all t and

$$\mathcal{L}(x,\tilde{p}(x,t),t) + \frac{\partial S}{\partial t}(x,t) + \nabla_x S(x,t)'\tilde{p}(x,t) = 0 \tag{7.72}$$

for all (x,t). If such a pair of functions can be found, then a solution to the optimization problem described by (7.62) and (7.63) solves the two-point boundary value problem

$$\dot{x}(t) = \tilde{p}(x(t),t) \quad \text{a.e.} \quad t \in I \tag{7.73}$$

$$x(t_o) = x_0 \quad \text{and} \quad x(t_1) = x_1. \tag{7.74}$$

As a consequence of this observation, the remainder of this section concerns the existence of the pair of functions $S(\cdot,\cdot)$ and $\tilde{p}(\cdot,\cdot)$. The first result we present provides necessary conditions for (7.71) and (7.72) to hold.

Theorem 7.6 *Let* $\mathcal{L}(\cdot,\cdot,\cdot) : \mathbb{R}^n \times \mathbb{R}^n \times I \to \overline{\mathbb{R}}$ *satisfy Assumption* 7.5, *let* $\tilde{p}(\cdot,\cdot) : \mathbb{R}^n \times I \to \mathbb{R}^n$ *be* $(\mathbb{R}^n \times I)$-measurable, *and let* $S(\cdot,\cdot) : \mathbb{R}^n \times I \to \mathbb{R}$ *be differentiable. Then a necessary condition for both* (7.71) *and* (7.72) *to hold is that* $S(\cdot,\cdot)$ *be a solution of the Hamilton-Jacobi equation*

$$\frac{\partial S}{\partial t}(x,t) + \mathcal{H}(x, \nabla_x S(x,t), t) = 0 \tag{7.75}$$

and that $\tilde{p}(\cdot,\cdot)$ *be related to* $S(\cdot,\cdot)$ *through the inclusion*

$$\nabla_x S(x,t) \in \partial_z \mathcal{L}(x, \tilde{p}(t,x), t) \tag{7.76}$$

where "∂_z*" denotes the (concave) subgradient of* $\mathcal{L}(\cdot,\cdot,\cdot)$ *with respect to its second argument.*

Proof: We first notice that (7.71) and (7.72) imply, for all (x,z) and almost all t, that the vector $\tilde{p}(x,t)$ maximizes the concave funtion $\mathcal{T}(x,\cdot,t)$ given by (7.69). Necessary conditions for this to hold are that

$$0 \in \partial_z \mathcal{T}(x, \tilde{p}(x,t), t),$$

where "∂_z" denotes the (concave) subgradient of $\mathcal{T}$ with respect to z, which implies,

$$\nabla_x S(x,t) \in \partial_z \mathcal{L}(x, \tilde{p}(x,t), t),$$

and so (7.76) holds. To prove (7.75) we observe that for almost all (x,z,t) we have

$$\nabla_x S(x,t)'z + \mathcal{L}(x,z,t) \le -\frac{\partial S}{\partial t}(x,t),$$

so that

$$\mathcal{H}(x, \nabla_x S(x,t), t) \le -\frac{\partial S}{\partial t}(x,t).$$

On the other hand,

$$\begin{aligned} -\frac{\partial S}{\partial t}(x,t) &= \nabla_x S(x,t)'\tilde{p}(x,t) + \mathcal{L}(x, \tilde{p}(x,t), t) \\ &\le \mathcal{H}(x, \nabla_x S(x,t), t). \end{aligned}$$

Combining these results gives (7.75). ∎

Remark 7.6 *The above result coincides exactly with those of Carathéodory* [34], *when one assumes that* $\mathcal{L}(\cdot,\cdot,t)$ *is* C^2.

Remark 7.7 *The hypothesis that $\mathcal{L}(x,\cdot,t)$ is concave in the above result can be relaxed to merely requiring $\mathcal{L}(x,\cdot,t)$ to be locally Lipschitz. In this case the proof remains the same provided $\partial_z\mathcal{L}(x,\cdot,t)$ is interpreted as a generalized gradient of Clarke* [56].

The function $\tilde{p}(\cdot,\cdot)$ given above could be called a *generalized slope function* since its role is analogous to the slope function found in the classical theory. Pursuing these ideas further we give the following definition.

Definition 7.5 *Let $\tilde{p}(\cdot,\cdot) : \mathbb{R}^n \times I \to \mathbb{R}^n$ be a given* ($\mathbb{R}^n \times I$)-measurable *function. We say $\tilde{p}(\cdot,\cdot)$ is a* slope function *for the integrand $\mathcal{L}(\cdot,\cdot,\cdot)$ if there exists a normal integrand $\tilde{q}(\cdot,\cdot) : \mathbb{R}^n \times I \to \mathbb{R}^n$ with*

$$\tilde{q}(x,t) \in \partial_z\mathcal{L}(x,\tilde{p}(x,t),t), \quad \text{in } \mathbb{R}^n \times I \tag{7.77}$$

and such that for all absolutely continuous curves $C = \{(\phi(t),t) : t \in [\tau_o,\tau_1]\}$ lying in $\mathbb{R}^n \times I$ the Hilbert invariant line integral

$$\begin{aligned}\int_C [\mathcal{L}(x,\tilde{p}(x,t),t) \; &- \; \tilde{q}(x,t)'\tilde{p}(x,t)]dt + \tilde{q}(x,t)'dx \\ &= \int_{\tau_o}^{\tau_1} [\mathcal{L}(\phi(t),\tilde{p}(\phi(t),t),t) \\ &\quad -\tilde{q}(\phi(t),t)'\tilde{p}(\phi(t),t)\;]dt + \tilde{q}(\phi(t),t)'\dot{\phi}'(t)\; dt\end{aligned} \tag{7.78}$$

exists and depends only on the endpoints of the curve C.

Using this notion we have the following analogue of the Weierstrass sufficiency theorem.

Theorem 7.7 *Let $\mathcal{L}(\cdot,\cdot,\cdot) : \mathbb{R}^n \times \mathbb{R}^n \times I \to \overline{\mathbb{R}}^1$ satisfy Assumption* 7.5 *and suppose that $\tilde{p}(\cdot,\cdot)$ is a slope function for $\mathcal{L}(\cdot,\cdot,\cdot)$. Then if $x^*(\cdot) : I \to \mathbb{R}^n$ is an absolutely continuous function which satisfies the two-point boundary value problem*

$$\dot{x}(t) = \tilde{p}(x(t),t)\,, \quad \text{a.e. } \; t \in I$$

$$x(t_o) = x_0 \;\; \text{and} \;\; x(t_1) = x_1,$$

it solves the variational equation given by (7.62)–(7.63).

Proof: Let $x(\cdot)$ be any admissible trajectory and $x^*(\cdot)$ be as above. Then since $\tilde{p}$ is a slope function there exists a function $\tilde{q}$ such that (7.77) and (7.78) both hold. This implies,

$$\int_{t_o}^{t_1} \mathcal{L}(x(t),\dot{x}(t),t) \;-\; \mathcal{L}(x^*(t),\dot{x}^*(t),t)dt$$

$$= \int_{t_o}^{t_1} [\mathcal{L}(x(t),\dot{x}(t),t) - \mathcal{L}(x(t),\tilde{p}(x(t),t)$$
$$+\tilde{q}(x(t),t)'(\dot{x}(t) - \tilde{p}(x(t),t))]\; dt$$
$$\leq 0,$$

because $\mathcal{L}(x(t),\cdot,t)$ is a concave function for almost all $t \in I$ and since $\tilde{q}$ satisfies (7.77). Thus we obtain

$$\int_{t_o}^{t_1} \mathcal{L}(x(t),\dot{x}(t),t)\; dt \leq \int_{t_o}^{t_1} \mathcal{L}(x^*(t),\dot{x}^*(t),t)\; dt.$$

as desired. ∎

We now investigate the existence of a slope function.

Theorem 7.8 *If* $\tilde{p}(\cdot,\cdot) : \mathbb{R}^n \times I \to \mathbb{R}^n$ *is a slope function for an integrand* $\mathcal{L}(\cdot,\cdot,\cdot)$ *satisfying Assumption 7.5 then*

$$\mathcal{H}(x,\tilde{q}(x,t),t) + \mathcal{L}(x,\tilde{p}(x,t),t) = \tilde{q}(x,t)'\tilde{p}(x,t) \tag{7.79}$$

holds for all $(x,t) \in \mathbb{R}^n \times I$*, where* $\tilde{q}(\cdot,\cdot)$ *is as in Definition 7.4.*

Conversely, if $\tilde{p}(\cdot,\cdot) : \mathbb{R}^n \times I \to \mathbb{R}^n$ *is a Carathéodory function and* $S(\cdot,\cdot) : \mathbb{R}^n \times I \to \mathbb{R}$ *is a solution of the Hamilton-Jacobi equation (7.75) whose gradient has components which are normal integrands which are bounded on compact subsets of* $\mathbb{R}^n \times I$ *and is such that the inclusion (7.76) holds, then* $\tilde{p}(\cdot,\cdot)$ *is a slope function.*

Proof: The condition (7.79) is an easy consequence of the inclusion (7.77) and the concavity of $\mathcal{L}(x,\cdot,t)$.

On the other hand, if $\tilde{p}(\cdot,\cdot)$ and $S(\cdot,\cdot)$ are as above then $\nabla_x S(\cdot,\cdot)$ is $(\mathbb{R}^n \times I)$-measurable and satisfies (7.76). Moreover for any absolutely continuous curve $\mathcal{C}$ lying in $(\mathbb{R}^n \times I)$ we have, since $S(\cdot,\cdot)$ is a solution of the Hamilton-Jacobi equation,

$$\int_{\mathcal{C}} [\mathcal{L}(x,\tilde{p}(x,t),t) \;+\; \nabla_x S(x,t)'\tilde{p}(x,t)]\; dt + \nabla_x S(t,x)'dx$$

$$= \int_{\mathcal{C}} \frac{\partial S}{\partial t}\; dt + \nabla_x S(x,t)'dx$$

$$= \int_{\tau_o}^{\tau_1} \frac{d}{dt} S(\phi(t),t)\; dt$$

$$= S(\phi(\tau_1),\tau_1) - S(\phi(\tau_o),\tau_o)\;,$$

in which $t \to (\phi(t), t)$ for $t \in [\tau_o, \tau_1]$ denotes any parametrization of $\mathcal{C}$. From the above it now follows that $\tilde{p}(\cdot,\cdot)$ is a slope function with $\tilde{q}(\cdot,\cdot) = \nabla_x S$. ∎

The above theorem shows that if we have a sufficiently smooth solution of the Hamilton-Jacobi equation then one can obtain a slope function provided we can solve the inclusion

$$\nabla_x S(x,t) \in \partial_z \mathcal{L}(x,p,t)$$

for a Carathéodory function $p(\cdot,\cdot) : \mathbb{R}^n \times I \to \mathbb{R}^n$. To give sufficient conditions for the solvability of this inclusion we require more regularity than previously assumed. In particular we require $\mathcal{L}(x,z,t)$ to be continuous and have a Lipschitz continuous gradient $\partial_z \mathcal{L}(x,z,t)$. This smoothness is needed in order to apply the implicit function theorem found in Clarke [56, p. 256].

We now give the following result which provides sufficient conditions for the existence of a slope function. In this theorem $\pi_z \partial F(x,p,t)$ denotes the set of all $n \times n$ matrices $\mathcal{M}$ such that for some $n \times (n+1)$ matrix $\mathcal{N}$, the $n \times (2n+1)$ matrix $[\mathcal{N}, \mathcal{M}]$ belongs to the generalized Jacobian, $\partial F(x,p,t)$ (see [56] for specific details concerning this notation).

Theorem 7.9 *Let $\mathcal{L}(\cdot,\cdot,\cdot) : \mathbb{R}^n \times \mathbb{R}^n \times I \to \mathbb{R}$ be continuous with $\mathcal{L}(x,\cdot,t)$ concave and such that the gradient $\nabla_z \mathcal{L} : \mathbb{R}^n \times \mathbb{R}^n \times I \to \mathbb{R}^n$ is Lipschitzian. Further assume that $S(\cdot,\cdot) : \mathbb{R}^n \times I \to \mathbb{R}^n$ is a solution of the Hamilton-Jacobi equation* (7.75) *which is differentiable with a Lipschitz continuous gradient $(\frac{\partial S}{\partial t}, \nabla_x S)$. Then if $(\hat{x}, \hat{z}, \hat{t}) \in \mathbb{R}^n \times \mathbb{R}^n \times I$ is such that*

$$F(\hat{x},\hat{z},\hat{t}) = \nabla_x S(\hat{x},\hat{t}) - \nabla_z \mathcal{L}(\hat{x},\hat{z},\hat{t}) = 0$$

and in addition if $\pi_z \partial F(\hat{x},\hat{z},\hat{t})$ has maximal rank, there exists a neighborhood $\mathcal{O}$ of $(\hat{x},\hat{t})$ and a Lipschitz continuous function $\tilde{p}(\cdot,\cdot) : \mathcal{O} \to \mathbb{R}^n$ such that $\tilde{p}(\hat{x},\hat{t}) = \hat{z}$ and

$$\nabla_x S(x,t) = \nabla_z \mathcal{L}(x, \tilde{p}(x,t), t)$$

for all $(x,t) \in \mathcal{O}$. In particular, this implies that $\tilde{p}(\cdot,\cdot)$ is a slope function for $\mathcal{L}(\cdot,\cdot,\cdot)$ restricted to $\{(x,z,t) : (x,t) \in \mathcal{O}, z \in \mathbb{R}^n\}$.

Proof: The proof of this result is established by first observing that the function $F(\cdot,\cdot,\cdot) : \mathbb{R}^n \times \mathbb{R}^n \times I \to \mathbb{R}^n$ satisfies the hypotheses needed to apply the implicit function theorem found in [56]. This implies that there exists a neighborhood, say $\mathcal{O}$, of $(\hat{x},\hat{t})$ and a Lipschitz continuous function $\tilde{p}(\cdot,\cdot)$ satisfying the (7.76) on $\mathcal{O}$. The conclusion that $\tilde{p}$ is a slope function now follows immediately from Theorem 7.8. ∎

Combining the above results we can now state the following corollary which brings the essential facts together.

Corollary 7.1 *Let $\mathcal{L}(\cdot,\cdot,\cdot)$ and $S(\cdot,\cdot)$ satisfy the hypotheses of Theorem 7.9 and let $(\hat{x},\hat{z},\hat{t})$ be as indicated above. Then there exists an open set $\mathcal{O}$ of $(\hat{x},\hat{t})$ and a Lipschitz continuous slope function $\tilde{p}(\cdot,\cdot):\mathcal{O}\to\mathbb{R}^n$. Moreover for any two points (x_0,t_0) and (x_1,t_1) in $\mathcal{O}$, if $x^*:[t_0,t_1]\to\mathbb{R}^n$ solves the two-point boundary value problem (7.73)–(7.74), then $x^*(\cdot)$ is a solution of the variational problem described by (7.62)–(7.63) over all absolutely continuous arcs $x(\cdot):[t_o,t_1]\to\mathbb{R}^n$ whose graphs lie in $\mathcal{O}$ (i.e., $(x(t),t)\in\mathcal{O}$).*

Remark 7.8 *The above corollary represents a nonsmooth version of* Carathéodory's *method* [34, pp. 227–239].

To see how the results are applicable in optimal control theory we consider the Lagrange problem of optimal control which consists of maximizing

$$J(x_0,u(\cdot)) = \int_{t_o}^{t_1} f_0(x(t),u(t),t)\,dt$$

over all admissible pairs $(x(\cdot),u(\cdot)):I\to\mathbb{R}^{n+m}$ satisfying

$$\begin{aligned}\dot{x}(t) &= f(x(t),u(t),t) \quad \text{a.e. } on\ I\\ x(t_o) &= x_0 \text{ and } x(t_1)=x_1\\ (x(t),t) &\in A \text{ on } I\\ u(t) &\in U(x(t),t) \quad \text{a.e. on } I.\end{aligned}$$

It is well known (see e.g., Rockafellar [158]) that the above problem can be reformulated as an equivalent nonsmooth calculus of variations problem by defining $\mathcal{L}(\cdot,\cdot,\cdot):\mathbb{R}^n\times\mathbb{R}^n\times I\to\overline{\mathbb{R}}$ through the formula

$$\mathcal{L}(x,z,t)=\begin{cases}\sup\{f_0(x,u,t):\ z=f(x,u,t),\ u\in U(x,t)\} & \text{if } (x,t)\in A\\ -\infty & \text{otherwise.}\end{cases}$$

Therefore the previous results are applicable. Moreover, if $\tilde{p}(\cdot,\cdot)$ is a slope function, the Kuratowski, Ryll-Nardzewski selection theorem can be applied to assert the existence of an $(\mathbb{R}^n\times I)$-measurable function $u(\cdot,\cdot):\mathbb{R}^n\times I\to\mathbb{R}^m$ satisfying $\mathcal{L}(x,\tilde{p}(x,t),t)=f_0(x,u(x,t),t)$ on A and $\tilde{p}(x,t)=f(x,u(x,t),t)$ on A. This implies that $u(\cdot,\cdot):\mathbb{R}^n\times I\to\mathbb{R}^m$ is a measurable optimal feedback control whenever Corollary 7.1 can be applied.

We conclude our discussion of Carathéodory's method for finite horizon optimization problems with the following simple example.

Example 7.2 The example we consider consists of maximizing

$$J(x_0,u(\cdot)) = \int_{t_o}^{t_1}\left(cx^2(t)-u^2(t)\right)dt$$

over all pairs of functions $(x(\cdot), u(\cdot)) : [t_o, t_1] \to \mathbb{R}^2$ satisfying

$$\dot{x}(t) = ax(t) + bu(t) \quad \text{a.e. } t_o \leq t \leq t_1$$

$$x(t_o) = x_0 \text{ and } x(t_1) = x_1,$$

where a, b, and c are constants with $b \neq 0$. Observe that $J(x(\cdot), u(\cdot))$ is convex in $x(\cdot)$ and concave in $u(\cdot)$ with $x(\cdot)$ satisfying fixed initial and terminal constraints. Thus, this problem is not the classical linear-quadratic optimal control problem. For this problem it is easy to see that $\mathcal{L}(\cdot,\cdot) : \mathbb{R} \times \mathbb{R} \to \mathbb{R}$ is given by

$$\mathcal{L}(x,z) = \left(c - \frac{a^2}{b^2}\right)x^2 + \frac{2a}{b^2}xz - \frac{1}{b^2}z^2,$$

and the Hamiltonian is given by

$$\mathcal{H}(x,p) = \frac{b^2}{4}p^2 + axp + cx^2.$$

Therefore the Hamilton-Jacobi equation is

$$\frac{\partial S}{\partial t}(x,t) + \frac{b^2}{4}\left(\frac{\partial S}{\partial x}(x,t)\right)^2 + ax\frac{\partial S}{\partial x}(x,t) = -cx^2.$$

We seek a solution of the form $S(x,t) = At + B(x)$, where A is an arbitrary constant and B is smooth. From the above partial differential equation we obtain

$$\frac{d}{dx}B(x) = \frac{2}{b^2}\left[-ax \pm \left((a^2+b^2c)x^2 - b^2A\right)^{\frac{1}{2}}\right].$$

For this example the slope function $\tilde{p}(\cdot)$ must satisfy

$$\tilde{p}(x) = \frac{b^2}{2}\frac{d}{dx}B(x) + ax$$

and so

$$\tilde{p}(x) = \pm\left((a^2+b^2c)x^2 - b^2A\right)^{\frac{1}{2}}.$$

Therefore, an optimal solution is obtained by solving the differential equation

$$\dot{x}(t) = \pm\left((a^2+b^2c)x^2 - b^2A\right)^{\frac{1}{2}} \quad \text{a.e. } I.$$

This yields the solution

$$x(t) = \left(\frac{b^2A}{(a^2+b^2c)^{\frac{1}{2}}}\right)\cosh\left((a^2+b^2c)^{\frac{1}{2}}t + K\right)$$

where K is a constant of integration. Hence for a given pair of boundary data, (x_0, t_o) and (x_1, t_1) we obtain a unique solution. To see that we do indeed have a solution to our variational problem we observe that

$$\mathcal{L}(x,z) + \frac{\partial S}{\partial t}(x,t) + \frac{\partial S}{\partial x}(x,t)z = -\left[\frac{1}{2}b\frac{d}{dx}B(x) + \left(\frac{ax-z}{b}\right)\right]^2$$

which is negative for all (x,z), except for $z = \tilde{p}(x)$ where the above expression is identically zero. In addition we observe that the optimal feedback control is given by

$$\begin{aligned} u(x) &= \frac{1}{b}\left(\tilde{p}(x) - ax\right) \\ &= \frac{1}{2}b\frac{d}{dx}\,B(x) \\ &= \pm\left((a^2 + b^2c)x^2 - b^2A\right)^{\frac{1}{2}} - ax. \end{aligned}$$

7.4 Carathéodory's method for infinite horizon optimal control problems

In this section, we extend the results of the previous section to the infinite horizon optimal control problem. These results were first reported in [41]. By appropriate modifications to Carathéodory's method we arrive at a general theorem which gives sufficient conditions for overtaking optimal solutions. This result encompasses most of the previously presented results. In keeping with the notation of the previous section we consider the nonsmooth cost functional

$$J_T(x(\cdot)) \triangleq \int_0^T \mathcal{L}(x(t), \dot{x}(t), t)\, dt \tag{7.80}$$

where $x(\cdot) : [0,\infty) \to \mathbb{R}^n$ is locally absolutely continuous and satisfies the fixed initial condition $x(0) = x_0$. Once again we assume $\mathcal{L} : \mathbb{R}^n \times \mathbb{R}^n \times I \to \overline{\mathbb{R}}$, where now $I = [0,\infty)$, satisfies Assumption 7.5. The basic growth condition given in Assumption 7.6 must be modified to account for the unbounded interval $[0,\infty)$. Thus we have the following assumption.

Assumption 7.7 *The infinite horizon optimization problem associated with the functional* (7.80) *satisfies the basic growth condition if for every $p \in \mathbb{R}^n$ and $T \geq 0$ there exists an integrable function $Q_T : [0,T] \to \mathbb{R}$ such that*

$$\mathcal{H}(x,p,t) = \sup_{z\in\mathbb{R}^n} \left[p'z + \mathcal{L}(x,z,t)\right] \leq Q_T(t) \tag{7.81}$$

for all $x \in \mathbb{R}^n$ and almost all $t \in [0,T]$.

Remark 7.9 *Assumption* 7.7 *merely requires that Assumption* 7.6 *holds on every interval* $[0,T]$, $T \geq 0$.

In Carathéodory's method, as presented in Section 7.3, the fixed end conditions are essential. Therefore, to extend these ideas to the infinite horizon requires a slightly different notion of equivalent problem. Specifically, we need to place some asymptotic conditions on the difference $S(\cdot, x(\cdot)) - S(\cdot, y(\cdot))$ in which $x(\cdot)$ and $y(\cdot)$ are two different admissible trajectories. With this observation we have the following definition.

Definition 7.6 *Suppose that $\mathcal{L}(\cdot,\cdot,\cdot) : \mathbb{R}^n \times \mathbb{R}^n \times I \to \mathbb{R}^n \cup \{-\infty\}$ satisfies Assumption 7.5 and that $S(\cdot,\cdot) : \mathbb{R}^n \times I \to \mathbb{R}$ is a Lipschitz continuous function. Define the integrand $\mathcal{T}(\cdot,\cdot,\cdot) : \mathbb{R}^n \times \mathbb{R}^n \times I \to \mathbb{R}^n \cup \{-\infty\}$ by*

$$\mathcal{T}(x,z,t) \triangleq \mathcal{L}(x,z,t) + \frac{\partial S}{\partial t}(x,t) + \nabla_x S(x,t)'z \tag{7.82}$$

and consider the cost functional

$$\tilde{J}_T(x) \triangleq \int_0^T \mathcal{T}(x(t),\dot{x}(t),t)\, dt, \tag{7.83}$$

where $x(\cdot)$ is absolutely continuous with $x(0) = x_0$. We now say that the optimization problem described by (7.83) *is equivalent in the sense of Carathéodory to the original optimization problem* (7.80) *if $S(\cdot,\cdot)$ meets the following requirements:*

1. *For every absolutely continuous function $x(\cdot) : [0,+\infty) \to \mathbb{R}^n$, satisfying $x(0) = x_0$,*

$$\liminf_{t\to\infty} S(x(t),t) > -\infty\ ; \tag{7.84}$$

2. *There exists a constant C such that for every admissible trajectory $x(\cdot)$, $x(0) = x_0$, for which $\lim_{T\to+\infty} \tilde{J}_T(x)$ is finite, it happens that*

$$\lim_{t\to\infty} S(x(t),t) = C\ ; \tag{7.85}$$

3. $$\mathcal{T}(x,z,t) \le 0 \textit{ for all } (x,z) \in \mathbb{R}^{2n} \textit{ for almost all } t \ge 0. \tag{7.86}$$

Our choice for the above definition is justified by the following theorem.

Theorem 7.10 *Let $\mathcal{L}(\cdot,\cdot,\cdot)$ satisfy Assumptions* 7.5 *and* 7.7 *and assume that there exists a function $S(\cdot,\cdot) : \mathbb{R}^n \times I \to \mathbb{R}$ such that the conditions given in Definition* 7.5 *are satisfied. Then if $x^*(\cdot) : [0,\infty) \to \mathbb{R}$, $x^*(0) = x_0$, is a* strongly optimal *solution of the optimization problem described by the cost functional* (7.83)*, then $x^*(\cdot)$ is an overtaking optimal solution for the optimization problem described by the cost functional* (7.80).

Proof: The strong optimality of $x^*(\cdot)$ implies $\lim_{T\to\infty} \tilde{J}_T(x^*(\cdot))$ is finite and therefore from (7.85) we have

$$\lim_{t\to\infty} S(x^*(t),t) = C\ .$$

To prove the above assertion we let $x(\cdot)\ :\ [0,\infty) \to \mathbb{R}^n$ be absolutely continuous such that $x(0) = x_0$, and divide our consideration into two cases:

Case 1. $\lim_{T\to+\infty} \tilde{J}_T(x)$ is finite. In this case we have $\lim_{t\to+\infty} S(x(t),t) = C$ and so for any $T > 0$,

$$\int_0^T \Big\{\mathcal{L}(x^*(t),\dot{x}^*(t),t) - \mathcal{L}(x(t),\dot{x}(t),t)\Big\}\, dt$$
$$= \int_0^T \Big\{\mathcal{T}(x^*(t),\dot{x}^*(t),t) - \mathcal{T}(x(t),\dot{x}(t),t)\Big\}\, dt$$
$$+ \int_0^T \frac{d}{dt}\left[S(x(t),t) - S(x^*(t),t)\right]\, dt.$$

The optimality of $x^*(\cdot)$ implies that for every $\epsilon > 0$ there exists $T_1 > 0$ so that for all $T > T_1$ we have

$$\tilde{J}_T(x^*(\cdot)) - \tilde{J}_T(x(\cdot)) \geq -\frac{\epsilon}{2}\ .$$

Moreover, since $\lim_{T\to\infty} \tilde{J}_T(x)$ is finite, it follows (from (7.85)) that there exists $T_2 > 0$ so that for all $T \geq T_2$,

$$S(x(T),T) - S(x^*(T),T) > -\frac{\epsilon}{2}\,.$$

Therefore for all $T \geq \max\{T_1, T_2\}$ we have

$$\int_0^T \mathcal{L}(x^*(t),\dot{x}^*(t),t) - \mathcal{L}(x(t),\dot{x}(t),t)\ dt > -\epsilon$$

which implies $x^*(\cdot)$ overtakes $x(\cdot)$.

Case 2. $\lim_{T\to+\infty} \tilde{J}_T(x(\cdot)) = -\infty$. In this case we have for any T,

$$\int_0^T \mathcal{L}(x^*(t),\dot{x}^*(t),t) - \mathcal{L}(x(t),\dot{x}(t),t)\ dt$$
$$= \left[\int_0^T \mathcal{T}(x^*(t),\dot{x}^*(t),t)\ dt - S(x^*(T),T)\right]$$
$$+ \left[S(x(T),T) - \int_0^T \mathcal{T}(x(t),\dot{x}(t),t)\ dt\right]\ .$$

From our hypotheses the first term has a finite limit as $T \to \infty$. Further, the second term tends to infinity as $T \to \infty$ since $\tilde{J}_T(x) \to -\infty$ as $T \to \infty$ and since $\liminf_{T\to\infty} S(x(T),T) > -\infty$. Therefore we can conclude

$$\liminf_{T\to\infty} [J_T(x^*(\cdot)) - J_T(x(\cdot))] \geq 0 .$$

Combining both cases we obtain the desired conclusion. ∎

As a consequence of the previous result it is easy to see that it would be useful to obtain a function $S(\cdot,\cdot)$ for which all the conditions of Definition 7.5 hold. Moreover, it would also be particularly useful if a function $\tilde{p}(\cdot,\cdot) : \mathbb{R}^n \times I \to \mathbb{R}^n$ could be found such that

$$\mathcal{T}(x, \tilde{p}(x,t), t) = 0 , \tag{7.87}$$

for then a strongly optimal solution for the functional $\tilde{J}_T$ could be obtained by solving the initial value problem

$$\dot{x}(t) = \tilde{p}(x(t),t) \quad \text{a.e. } t \geq 0 \tag{7.88}$$

$$x(0) = x_0. \tag{7.89}$$

We already know (from Theorem 7.6) that necessary conditions for (7.86) and (7.87) to hold are that $S(\cdot,\cdot)$ solve the Hamilton-Jacobi equation and that $\tilde{p}$ be related to S through the inclusion

$$\nabla_x S(x,t) = \partial_z \mathcal{L}(x, \tilde{p}(x,t), t) .$$

With these observations it is easy to establish the following set of sufficient conditions for overtaking optimality.

Theorem 7.11 *Let $\mathcal{L}(\cdot,\cdot,\cdot) : \mathbb{R}^n \times \mathbb{R}^n \times I \to \overline{\mathbb{R}}$ satisfy Assumptions 7.5 and 7.7 and let $S(\cdot,\cdot) : \mathbb{R}^n \times I \to \mathbb{R}$ be a Lipschitz continuous solution of the Hamilton-Jacobi equation*

$$\frac{\partial S}{\partial t}(x,t) + \mathcal{H}(x, \nabla_x S(x,t), t) = 0$$

satisfying (7.84) and (7.85). Further suppose that there exists a continuous function $\tilde{p}(\cdot,\cdot) : \mathbb{R}^n \times I \to \mathbb{R}^n$ such that

$$\nabla_x S(x,t) \in \partial_z \mathcal{L}(x, \tilde{p}(x,t), t)$$

for all $x \in \mathbb{R}^n$ and for almost all $t \geq 0$. Then if $x^*(\cdot) : [0,\infty) \to \mathbb{R}^n$ solves the initial value problem

$$\dot{x}(t) = \tilde{p}(x(t),t)$$

$$x(0) = x_0 ,$$

it is an overtaking optimal solution for the optimization problem described by the integral functional (7.80).

Proof: From our previous discussions it is clear that $x^*(\cdot)$ as indicated above is a strongly optimal solution to the equivalent problem described by (7.83). Therefore an application of Theorem 7.10 provides us with the desired conclusion. ■

Remark 7.10 *The above discussion forms an infinite horizon analogue of Carathéodory's method for overtaking optimal solutions.*

As in the finite horizon case, the utility of this method rests on being able to determine appropriate functions $S(\cdot,\cdot)$ and $\tilde{p}(\cdot,\cdot)$ meeting the desired requirements. This can possibly be realized by solving the corresponding Hamilton-Jacobi equation. With regards to this, the works of Lions [127], Crandall and Lions [58] and of Barbu and DaPrato [14] as well as others concerning the Hamilton-Jacobi equation could be perhaps utilized to obtain an appropriate function S.

Probably the most difficult criterion to satisfy in Definition 7.5 is the condition given by (7.85). The next result provides conditions under which this difficulty might be overcome.

Theorem 7.12 *Let* $\mathcal{L}(\cdot,\cdot,\cdot) : \mathbb{R}^n \times \mathbb{R}^n \times I \to \overline{\mathbb{R}}$ *satisfy Assumptions* 7.5 *and* 7.7 *and let* $S(\cdot,\cdot) : \mathbb{R}^n \times I \to \overline{\mathbb{R}}$ *be a Lipschitz continuous function which satisfies conditions* (7.84) *and* (7.86) *as well as*

(i) *Either*

$$\limsup_{t\to\infty} \frac{d}{dt} S(x(t),t) < +\infty \tag{7.90}$$

or

$$\liminf_{t\to\infty} \frac{d}{dt} S(x(t),t) > -\infty \tag{7.91}$$

holds for every admissible trajectory $x(\cdot) : [0,\infty) \to \mathbb{R}^n$, $x(0) = x_0$.

(ii) *For every* $\epsilon > 0$ *and admissible trajectory* $x(\cdot)$, *with* $x(0) = x_0$, *there exists* $\delta = \delta(\epsilon, x) > 0$ *such that* $|S(x(t),t)| > \epsilon$ *implies*

$$\mathcal{T}(x(t),\dot{x}(t),t) < -\delta\,, \tag{7.92}$$

where $\mathcal{T}(\cdot,\cdot,\cdot)$ *is given by* (7.82). *Then if* $x(\cdot)$ *is an admissible trajectory emanating from* x_0 *such that*

$$\int_0^\infty \mathcal{T}(x(t),\dot{x}(t),t)\, dt > -\infty\,,$$

it happens that

$$\lim_{t\to\infty} S(x(t),t) = 0\,.$$

Proof: Let $\mu[\cdot]$ denote Lebesgue measure on $\mathbb{R}$; and for $\epsilon > 0$, let

$$I(\epsilon) = \{t \geq 0 : |S(x(t), t)| > \epsilon\} .$$

From (7.92), there exists $\delta = \delta(\epsilon, x) > 0$ such that for all $t \in I(\epsilon)$,

$$\mathcal{T}(x(t), \dot{x}(t), t) < -\delta .$$

This implies

$$0 < \mu\,[I(\epsilon)] < -\frac{1}{\delta} \int_0^{+\infty} \mathcal{T}(x(t), \dot{x}(t), t)\, dt < \infty ,$$

so that $\mu\,[I(\epsilon)]$ is finite.

Now suppose that $\lim_{t\to\infty} S(x(t), t) \neq 0$. Then there exists $\epsilon_o > 0$ and a sequence $\{t_n\}_{n=1}^{\infty}$ of nonnegative numbers tending to infinity such that

$$|S(x(t_n), t_n)| > \epsilon_o .$$

The continuity of S and the fact that $\mu\,[I(\epsilon)]$ is finite allows us to assert the existence of sequences $\{t_n^1\}_{n=1}^{\infty}$ and $\{t_n^2\}_{n=1}^{\infty}$ satisfying:

(i) $t_n^1 < t < t_n^2$

(ii) $|S(x(t_n^i), t_n^i)| > \frac{1}{3}\,\epsilon_o \;,\; i = 1, 2$

(iii) $\lim_{n\to\infty} (t_n^i - t_n) = 0 \;,\; i = 1, 2$

(iv) $|S(x(t_n^i), t_n^i) - S(x(t_n), t_n)| > \frac{1}{2}\,\epsilon_o \;,\; i = 1, 2.$

We now consider four cases.

Case 1: $S(x(t_n), t_n) > \epsilon_o$ and condition (7.90) holds.

From (7.90) it follows that there exists $\eta > 0$ and $\tau > 0$ such that for all $t \geq \tau$,

$$\frac{d}{dt}\,[S(x(t), t)] < \eta .$$

Therefore there exists an integer $N > 0$ such that $t_n^1 \geq \tau$ for all $n \geq N$ implying

$$\frac{S(x(t_n), t_n) - S(x(t_n^1), t_n^1)}{t_n - t_n^1} \leq \eta .$$

On the other hand, our assumptions concerning the sequence $\{t_n^1\}_{n=1}^{\infty}$ implies

$$\eta \geq \frac{S(x(t_n), t_n) - S(x(t_n^1), t_n^1)}{t_n - t_n^1} > \frac{\epsilon_o}{2(t_n - t_n^1)}$$

since otherwise $\mu\,[I(\epsilon_o)] = \infty$. Clearly this leads to a contradiction since *(iii)* holds.

Case 2: $S(x(t_n), t_n) > \epsilon_o$ and condition (7.91) holds.

In this case, (7.91) implies there exists $\eta \geq 0$ and $T > 0$ such that for all $t \geq T$,

$$\frac{d}{dt}\left[S(x(t), t)\right] > -\eta \, .$$

Hence there exists $N > 0$ such that $t_n \geq T$ for all $n \geq N$ which yields

$$\frac{S(x(t_n^2), t_n^2) - S(x(t_n), t_n)}{t_n^2 - t_n} > -\eta \, .$$

As above, our assumptions on $\{t_n^2\}_{n=1}^{\infty}$ imply

$$-\eta < \frac{S(x(t_n^2), t_n^2) - S(x(t_n), t_n)}{t_n^2 - t_n} < -\frac{\epsilon_o}{2(t_n^2 - t_n)} \, ,$$

which again leads to a contradiction.

The remaining two cases:

Case 3: $S(x(t_n), t_n) < -\epsilon_o$ and (7.90) holds.

Case 4: $S(x(t_n), t_n) < -\epsilon_o$ and (7.91) holds.

are contradicted in an indentical manner. Therefore, we must have

$$\lim_{t \to \infty} S(x(t), t) = 0 \, . \qquad \blacksquare$$

We now compare Carathéodory's method for overtaking optimal solutions with previous results. We begin by showing that the notion of G-supported trajectories and Yano's *McKenzie turnpike* ideas are special cases of the above theory. This is easily seen since, in the previous theories, we can take

$$S(x, t) = \bar{p}(t)' \left(G(x) - G(\bar{x}(t))\right) \, ,$$

where $\bar{p} : [0, \infty) \to \mathbb{R}^n$ is absolutely continuous, $G : \mathbb{R}^n \to \mathbb{R}^n$ is continuously differentiable, and $\bar{x}(\cdot) : [0, \infty) \to \mathbb{R}^n$ is a fixed admissible trajectory. In the following result we present an extension of these ideas.

Theorem 7.13 *Let* $\mathcal{L}(\cdot, \cdot, \cdot) : \mathbb{R}^n \times \mathbb{R}^n \times [0, \infty) \to \overline{\mathbb{R}}$ *satisfy Assumptions* 7.5 *and* 7.7, *let* $\pi(\cdot, \cdot) : \mathbb{R}^n \times [0, \infty) \to \mathbb{R}$ *be a given differentiable function and assume that* $\bar{x}(\cdot) : [0, \infty) \to \mathbb{R}^n$ *is a given admissible trajectory (with* $\bar{x}(0)$ *not necessarily* x_0*) which satisfies the following conditions:*

(i) $\mathcal{L}(x,z,t)+\dot{\pi}(x,z,t) \leq \mathcal{L}(\bar{x}(t),\dot{\bar{x}}(t),t)+\dot{\pi}(\bar{x}(t),\dot{\bar{x}}(t),t)$, *a.e. on* $[0,\infty)$, $(x,z) \in \mathbb{R}^{2n}$, *where*
$\dot{\pi}(x,z,t) = \frac{\partial \pi}{\partial t}(x,t) + \nabla_x \pi(x,t)'z$

(ii) $\liminf_{t\to\infty} \pi(\bar{x}(t),t) > -\infty$ *and* $\limsup_{t\to\infty} \pi(x(t),t) < \infty$, *for each admissible trajectory* $x(\cdot)$, *satisfying* $x(0) = x_0$;

(iii) *either*

$$(a) \qquad \limsup_{t\to\infty} \dot{\pi}(\bar{x}(t),\dot{\bar{x}}(t),t) < \infty \text{ and } \liminf_{t\to\infty} \dot{\pi}(x(t),\dot{x}(t),t) > -\infty$$

or

$$(b) \qquad \liminf_{t\to\infty} \dot{\pi}(\bar{x}(t),\dot{\bar{x}}(t),t) > -\infty \text{ and } \limsup_{t\to\infty} \dot{\pi}(x(t),\dot{x}(t),t) < \infty \,;$$

(iv) *for every* $\epsilon > 0$ *and admissible trajectory* $x(\cdot)$, *with* $x(0) = x_0$, *there exists* $\delta = \delta(\epsilon, x) > 0$ *such that* $|\pi(x(t),t) - \pi(\bar{x}(t),t)| > \epsilon$ *implies*

$$\mathcal{L}(x(t),\dot{x}(t),t) - \mathcal{L}(\bar{x}(t),\dot{\bar{x}}(t),t) + \frac{d}{dt}[\pi(x(t),t) - \pi(\bar{x}(t),t)] < -\delta\,.$$

If $x^*(\cdot) : [0,\infty) \to \mathbb{R}^n$, $x^*(0) = x_0$, *is a strongly optimal solution associated with the performance criterion*

$$\tilde{J}(x(\cdot)) = \int_0^\infty \left\{\mathcal{L}(x(t),\dot{x}(t),t) - \mathcal{L}(\bar{x}(t),\dot{\bar{x}}(t),t) + \frac{d}{dt}[\pi(x(t),t) - \pi(\bar{x}(t),t)]\right\} dt$$

with respect to all admissible trajectories $x(\cdot) : [0,\infty) \to \mathbb{R}^n$, $x(0) = x_0$, *then it is also an overtaking optimal solution for the original infinite horizon optimization problem.*

Proof: We begin by defining the functions $\mathcal{T}(\cdot,\cdot,\cdot) : \mathbb{R}^n \times \mathbb{R}^n \times [0,\infty) \to \overline{\mathbb{R}}$ and $S(\cdot,\cdot) : \mathbb{R}^n \times [0,\infty) \to \mathbb{R}$ by the formulas

$$\mathcal{T}(\xi,\gamma,t) = \mathcal{L}(\xi + \bar{x}(t), \gamma + \dot{\bar{x}}(t), t) - \mathcal{L}(\bar{x}(t),\dot{\bar{x}}(t),t) \tag{7.93}$$

and

$$S(\xi,t) = \pi(\xi + \bar{x}(t),t) - \pi(\bar{x}(t),t)\,. \tag{7.94}$$

From the hypotheses it is easy to see that for almost all $(\xi,\gamma,t) \in \mathbb{R}^n \times \mathbb{R}^n \times [0,\infty)$

$$\begin{aligned} &\mathcal{T}(\xi,\gamma,t) + \frac{\partial S}{\partial t}(\xi,t) + \nabla_\xi S(\xi,t) \\ &= [\mathcal{L}(\xi + \bar{x}(t), \gamma + \dot{\bar{x}}(t), t) + \dot{\pi}(\xi + \bar{x}(t), \gamma + \dot{\bar{x}}(t), t)] \\ &\quad - [\mathcal{L}(\bar{x}(t),\dot{\bar{x}}(t),t) + \dot{\pi}(\bar{x}(t),\dot{\bar{x}}(t),t)] \\ &\leq 0\,. \end{aligned}$$

Furthermore, it is equally easy to see that $\mathcal{T}(\cdot,\cdot,\cdot)$ and $S(\cdot,\cdot)$ satisfy the hypotheses of Theorem 7.12. Thus it follows that we can apply Theorem 7.11 to the problem of obtaining overtaking optimal solutions for the problem associated with

$$\hat{J}_T(\xi(\cdot)) = \int_0^T \mathcal{T}(\xi(t), \dot{\xi}(t), t)\, dt \; ; \quad T > 0 \, , \tag{7.95}$$

where now we consider those feasible trajectories $\xi(\cdot) : [0,\infty) \to \mathbb{R}^n$ which satisfy the initial condition

$$\xi(0) = x_0 - \bar{x}(0) \, . \tag{7.96}$$

We observe that if $x(\cdot) : [0,\infty) \to \mathbb{R}^n$ is an admissible trajectory for the original problem, then $\xi(\cdot) : [0,\infty) \to \mathbb{R}^n$ defined by

$$\xi(t) = x(t) - \bar{x}(t) \tag{7.97}$$

is admissible for the above problem. Moreover, the converse is also true. Finally, we notice that the continuous function $\xi^*(\cdot) : [0,\infty) \to \mathbb{R}^n$ defined by

$$\xi^*(t) = x^*(t) - \bar{x}(t)$$

is a strongly optimal solution for the problem of maximizing

$$\int_0^\infty [\mathcal{T}(\xi(t), \dot{\xi}(t), t) + \frac{\partial S}{\partial t}(\xi(t), t) + \nabla_\xi S(\xi(t), t)'\dot{\xi}(t)]\, dt$$

over all trajectories $\xi(\cdot)$, with $\xi(0) = x_0 - \bar{x}(0)$. In fact, we have

$$\begin{aligned} &\int_0^\infty [\mathcal{T}(\xi^*(t), \dot{\xi}^*(t), t) + \frac{\partial S}{\partial t}(\xi^*(t), t) + \nabla_\xi S(\xi^*(t), t)'\dot{\xi}^*(t)]\, dt \\ &= \int_0^\infty \mathcal{L}(x^*(t), \dot{x}^*(t), t) - \mathcal{L}(\bar{x}(t), \dot{\bar{x}}(t), t) \\ &\quad + \frac{d}{dt}\left[\pi(x^*(t), t) - \pi(\bar{x}(t), t)\right]\, dt. \end{aligned}$$

Hence by Theorem 7.11, $\xi^*(\cdot)$ is overtaking optimal for the infinite horizon optimal control problem described by (7.93) and (7.94). The desired conclusions now follows as a direct consequence of the optimality of $\xi^*(\cdot)$ and the one-to-one correspondence described by (7.97). ∎

Remark 7.11 *By defining $\pi(x,t) = \bar{p}(t)'G(x)$ it is easy to see that the hypotheses assumed in Theorem 7.3, imply the conditions given in the above theorem since in Theorem 7.3 we impose the condition that all admissible trajectories are confined to a compact set X.*

To compare the results presented in this section with those of Section 5.8 we begin by observing that Equation (5.52)

$$\begin{aligned} \mu &= \max_{z\in \mathbb{R}^n} \{\mathcal{L}(x,z) + z'\nabla p(x)\} \\ &= \mathcal{H}(x, \nabla p(x)). \end{aligned}$$

Thus if we are given a good solution, say $(p(\cdot)\,,\mu)$, of Equation (5.52) (see Definition 5.6), with $x \to p(x)$ differentiable, we see that the function $S : [0,\infty) \times \mathbb{R}^n \to \mathbb{R}$ defined by the formula

$$S(x,t) = \mu t - p(x),$$

satisfies the Hamilton-Jacobi equation (7.75). At this point Theorems 7.10 and 7.11 are directly applicable only when $\mu \geq 0$ since in this case (7.84) holds (recall in Section 5.8 the admissible trajectories are required to lie in a compact set K). However, if we define the associated problem of maximizing

$$\hat{J}(x(\cdot)\,) = \int_0^\infty \{\mathcal{L}(x(t), \dot{x}(t) + \mu\}\; dt$$

over all arcs $x(\cdot) \in AC_{loc}$ with $x(0) = x_0$, we see that the function $\hat{S} : \mathbb{R}^n \to \mathbb{R}^n$given by

$$\hat{S}(x) = -p(x)$$

solves the Hamilton-Jacobi equation

$$\frac{\partial S}{\partial t}(x,t) + \mathcal{H}_\mu(x, \nabla_x S(x,t)) = 0,$$

in which

$$\mathcal{H}_\mu(x,p) = \sup_{z\in \mathbb{R}^n} \left[\mathcal{L}(x,z) + \mu + z'p\right].$$

Further, we observe that since all admissible trajectories are required to lie in a compact set K it is clear that (7.84) holds for this associated problem and now, if $\hat{S}$ satisfies the appropriate additional hypotheses, the results of this section are indeed applicable. In fact we see that the work presented in Section 5.8 is essentially a more refined version of Carathéodory's method for autonomous infinite horizon optimal control problems.

7.5 The growth condition (γ) and the compactness of the set of admissible trajectories

In this section we concern ourselves with a control system having the form

$$\dot{x}(t) = f(x(t), u(t), t), \text{ a.e. on } [0,\infty) \tag{7.98}$$

$$x(0) = x_0 \text{ and } (x(t),t) \in A \quad \text{for } t \in [0,\infty) \tag{7.99}$$

$$u(t) \in U(x(t), t) \text{ a.e. in } [0,\infty) \tag{7.100}$$

where we assume that $(x(\cdot), u(\cdot)) : [0,\infty) \to \mathbb{R}^n \times \mathbb{R}^m$, $x_0 \in \mathbb{R}^n$ is given, $A \subset \mathbb{R}^n \times [0,\infty)$ is a given closed set, $U(\cdot,\cdot) : A \to 2^{\mathbb{R}^m}$ is a given set-valued mapping with closed graph $M = \{(x,u,t) : (x,t) \in A,\ u \in U(x,t)\}$, and where $f(\cdot,\cdot,\cdot) : M \to \mathbb{R}^n$ is a given vector-valued function which satisfies the usual Carathéodory conditions (i.e., it is continuous in (x,u) for almost all $t \geq 0$ and Lebesgue measurable in t for each (x,u)).

Associated with the above control system is a cost criterion described by the definite integral

$$J_T(x_0, u(\cdot)) = \int_0^T f_0(x(t), u(t), t)\, dt, \tag{7.101}$$

where we assume that $f_0(\cdot,\cdot,\cdot) : M \to \mathbb{R}$ is a given function with the properties that $f_0(t,\cdot,\cdot)$ is upper semicontinuous for almost all $t \geq 0$ and that $f_0(\cdot,\cdot,\cdot)$ is measurable with respect to the σ-algebra generated by products of Lebesgue measurable subsets of $[0,\infty)$ with the Borel measurable subsets of $\mathbb{R}^{n+m}$.

With the above notation we give a precise definition of an admissble pair.

Definition 7.7 *A pair of function $(x(\cdot), u(\cdot)) : [0,\infty) \to \mathbb{R}^n \times \mathbb{R}^m$ will be called an admissible pair if $x(\cdot)$ is absolutely continuous on every compact subset of $[0,\infty)$ (henceforth denoted $x(\cdot) \in AC_{loc}$) if $u(\cdot)$ is Lebesgue measurable, the pair $(x(\cdot), u(\cdot))$ satisfies the control system* (7.98)–(7.100), *and if the map $t \to f_0(x(t), u(t), t)$ is locally Lebesgue integrable on $[0,\infty)$.*

The above notations and hypotheses give a precise description of the control systems we treat in this chapter. We observe that the smoothness requirements described above are more general than those considered in previous chapters, and therefore the systems considered before are contained in the above system as a special case. In order to achieve the desired results it will become necessary to impose further conditions on the above model. At present we refrain from giving these hypotheses preferring instead to present them separately in subsequent sections so as to investigate their importance.

We now begin our discussion of the compactness of the set of admissible trajectories. The compactness of the set of admissible trajectories played a crucial role in the existence discussions given in Chapters 4 and 5. The requisite compactness conditions were assured by the result of Varaiya, Theorem 4.6, in Chapter 4 and by a coercivity condition in Chapter 5. In our previous treatment this result assured us that the set of all admissible trajectories formed a compact subset of the space of continuous functions on $[0,\infty)$ endowed with the topology of uniform convergence on compact subsets of $[0,\infty)$. For the results presented here we choose a weaker topology placed on the space AC_{loc} instead of the space of continuous functions. As we shall see, this permits us to weaken both the growth conditions and the convexity requirements of Varaiya. We begin by giving a precise definition of our topology.

Definition 7.8 *A sequence $\{x_n(\cdot)\}_{n=1}^{\infty} \subset AC_{loc}$ converges weakly to $x(\cdot) \in AC_{loc}$ if for every subset compact S of $[0,\infty)$ every vector $a \in \mathbb{R}^n$ and every function $b(\cdot) \in L^{\infty}(S)$, the space of essentially bounded functions from S into $\mathbb{R}^n$, we have*

$$\lim_{n\to\infty}\left[a'\left(x_n(0)-x(0)\right)+\int_S b(t)'\left(\dot{x}_n(t)-\dot{x}(t)\right)\,dt\right]=0\,. \tag{7.102}$$

The above definition clearly shows that the notion of weak convergence in AC_{loc} corresponds to requiring pointwise convergence of the initial values $\{x_n(0)\}_{n=1}^{\infty}$ and the weak convergence of the derivatives $\{\dot{x}_n(\cdot)\}_{n=1}^{\infty}$ in $L^1(S)$ for every compact subset S of $[0,\infty)$. It is thus not surprising that conditions for weak relative compactness in AC_{loc} can be obtained by applying the corresponding result for $AC[0,T]$ for each $T > 0$ in conjunction with a standard diagonalization argument. With this remark we give the following result.

Theorem 7.14 *A subset $B \subset AC_{loc}$ is relatively weakly sequentially compact if and only if*

(i) *$\{x(0) : x(\cdot) \in B\}$ is bounded, and*

(ii) *For every set $S \subset [0,\infty)$, the sets $\dot{B}_S = \{\dot{x}|_S(\cdot) : x(\cdot) \in B\}$, where $\dot{x}|_S(\cdot)$ denotes the restriction of $\dot{x}(\cdot)$ to the set S, satisfy any one of the following three conditions*

(a) *For every $\epsilon > 0$ there exists a $\delta > 0$ such that for all measurable subsets $R \subset S$ with Lebesgue measure $\mu(R) \leq \delta$,*

$$\left|\int_R y(t)\ dt\right| < \epsilon, \text{ (equiabsolutely continuous integrals)}$$

for each $y(\cdot) \in \dot{B}|_S$.

(b) *For every $\epsilon > 0$, there exists $\lambda > 0$ such that*

$$\int_{\{t:|y(t)|\geq\lambda\}} |y(t)|\,dt < \epsilon \text{ (uniform integrability)}$$

for each $y(\cdot) \in \dot{B}|_S$.

(c) *There exists a positive Borel measurable function $\phi(\cdot) : [0,\infty) \to [0,\infty)$ such that $\lim_{z\to\infty} \phi(z)/z = \infty$ and*

$$\sup_{y\in\dot{B}|_S}\left\{\int_S \phi(|y(t)|)\,dt\right\} < \infty\,.$$

Proof: Let $\{x_n(\cdot)\}_{n=1}^{\infty}$ be a sequence in B. Then from (i) above it follows that there exists a subsequence, say still $\{x_n(\cdot)\}_{n=1}^{\infty}$, such that $\lim_{n\to\infty} x_n(0) = \bar{x}_0$ exists. Moreover if any of the conditions (a) to (c) hold on $[0, N]$ for each integer N we have that the sequence $\{\dot{x}_n(\cdot)|_{[0,N]}\}_{n=1}^{\infty}$ is weakly sequentially compact in $L^1[0, N]$ (see e.g., Cesari [47]). Therefore, proceeding successively on $[0, N]$ for each N we can extract a subsequence, say still $\{\dot{x}_n(\cdot)\}_{n=1}^{\infty}$, which converges weakly in $L^1[0, N]$ for each N to some function $y(\cdot) \in \mathcal{L}^1_{loc}[0, \infty)$. The desired conclusion now follows by defining $\bar{x}(\cdot) : [0, \infty) \to \mathbb{R}^n$ by

$$\bar{x}(t) = \bar{x}_0 + \int_0^t y(t)\, dt.$$

■

To be able to apply the above result to the set of admissible trajectories we must place growth restrictions on the functions $f(\cdot,\cdot,\cdot)$ and $f_0(\cdot,\cdot,\cdot)$. The most general of these is apparently the growth condition (γ) given below

Definition 7.9 *We say the functions $f(\cdot,\cdot,\cdot) : M \to \mathbb{R}^n$ and $f_0(\cdot,\cdot,\cdot) : M \to \mathbb{R}^n$ satisfy the growth condition (γ) if for every $\epsilon > 0$ and $T \geq 0$ there exists a nonnegative Lebesgue integrable function $\psi_{\epsilon,T}(\cdot) : [0, T] \to \mathbb{R}$ and a constant $\mathcal{L}_T \geq 0$ such that*

$$f_0(x, u, t) \geq \mathcal{L}_T \tag{7.103}$$

and

$$|f(x, u, t)| \leq \psi_{\epsilon,T}(t) - \epsilon f_0(x, u, t) \tag{7.104}$$

both hold for almost all $t \in [0, T]$, $(x, u, t) \in M$.

Remark 7.12 *The growth condition (γ) given above is a modification of the growth condition (γ) used in Cesari, LaPalm, and Nishiura [49] to establish weak relative compactness of the derivatives of admissible trajectories for finite horizon optimal control problems. In Cesari [47] or Cesari and Suryanarayana [50] it has been shown that this growth condition is equivalent to the growth condition used in the work of Rockafellar [158] for the same purpose.*

With the above growth conditions we have the following compactness result.

Theorem 7.15 *If the functions $f(\cdot,\cdot,\cdot) : M \to \mathbb{R}^n$ and $f_0(\cdot,\cdot,\cdot) : M \to \mathbb{R}$ satisfy the growth condition (γ), then the set of all admissible trajectories $x(\cdot) : [0, \infty) \to \mathbb{R}^n$ forms a relatively weak sequentially compact subset of AC_{loc}.*

Proof: Let $(x(\cdot)\,, u(\cdot)\,)$ denote any admissible pair and let $T > 0$ be given. Then for any $\epsilon > 0$, there exists a nonnegative integrable function $\psi_{\epsilon,T}(\cdot) \,:\, [0,T] \to \mathbb{R}$ and a constant $\mathcal{L}_T \geq 0$ such that for almost all $t \in [0,T]$ we have

$$f_0(x(t), u(t), t) \geq -\mathcal{L}_T$$

and

$$|\dot{x}(t)| \leq \psi_{\epsilon,T}(t) - \epsilon f_0(x(t), u(t), t) \leq \psi_{\epsilon T}(t) + \epsilon \mathcal{L}_T.$$

Choose $\delta > 0$ such that for any $S \in [0,T]$ Lebesgue measure $\mu(S) \leq \delta$ we have

$$\int_S \psi_{\epsilon,T}(t)\, dt < \frac{\epsilon}{2}$$

and

$$\mu(S) < \frac{1}{2(\mathcal{L}_T + 1)}.$$

Notice that for any $S \subset [0,T]$ with $\mu(S) < \delta$ we have

$$\begin{aligned}
\left|\int_S \dot{x}(t)\, dt\right| &\leq \int_S |\dot{x}(t)|\, dt \\
&\leq \int_S \psi_{\epsilon,T}(t)\, dt + \epsilon \int_S \mathcal{L}_T\, dt \\
&< \frac{\epsilon}{2} + \epsilon\Big(\frac{\mathcal{L}_T}{2(\mathcal{L}_T + 1)}\Big) \\
&< \frac{\epsilon}{2} + \frac{\epsilon}{2} \\
&= \epsilon.
\end{aligned}$$

This proves $\{\dot{x}(\cdot)\,|_{[0,T]} : x \text{ is an admissible trajectory}\}$ forms an equiabsolutely integrable family of functions in $L^1[0,T]$ (i.e., Condition (ii a) of Theorem 7.14 holds). The desired conclusion now follows from Theorem 7.14 since for every admissible trajectory, $x(\cdot)$, we have $x(0) = x_0$. ∎

The above result shows that the growth condition is sufficient to insure that the set of admissible trajectories forms a relatively weakly compact subset of AC_{loc} and therefore every sequence of admissible trajectories has a subsequence which weakly converges to some function in AC_{loc}. It does not however insure that this limit trajectory is also an admissible trajectory. In Varaiya's result this was guaranteed by assuming that the velocity set

$$\Phi(x,t) = \{f(x,u,t) : u \in U(x,t)\}$$

is convex for each $(x,t) \in A$. As we shall see, in the next section, a similar condition will be used here.

7.6 Upper closure and the existence of strongly optimal solutions

All theorems concerning the existence of a maximum for a functional, $J(\cdot)$, over some set, contains two components. The first of these is the compactness of the set of admissible elements. In our situation we require the set of admissible trajectories to form a weakly compact subset of AC_{loc}. From the previous section we see that the growth condition (γ) is sufficient to insure only relative weak compactness. The missing ingredient for obtaining compactness in AC_{loc} is closedness of the set of admissible trajectories. This is one of the topics discussed in this section. The second component we need to discuss is the upper semicontinuity of the reward. For the case of strong optimality this is described by the improper integral

$$J(x_0, u(\cdot)) = \int_0^\infty f_0(x(t), u(t), t)\, dt, \tag{7.105}$$

which is assumed to be finite for some subset of all admissible pairs. Both upper semicontinuity of J and the closedness of admissible trajectories is insured by establishing an *upper closure theorem*. Such a result usually comes under the heading of a lower closure theorem. This is because in most discussions of the existence of optimal solutions the existence of a minimum is treated, whereas in our discussion we are investigating the existence of a maximum.

To give this result it is necessary to require certain convexity and seminormality conditions be placed on our model. For optimal control problems these conditions are most easily described in terms of the set-valued map $\tilde{Q}(\cdot,\cdot) : A \to 2^{\mathbb{R}^{1+n}}$ defined for $(x,t) \in A$ by

$$\tilde{Q}(x,t) = \{(z_0, z) : z_0 \leq f_0(x,u,t),\ z = f(x,u,t),\ u \in U(x,t)\}. \tag{7.106}$$

Specifically we require the sets $\tilde{Q}(x,t)$ to be convex and satisfy the upper semicontinuity condition property $(\mathcal{K})$ described by the following equality

$$\tilde{Q}(x,t) = \bigcup_{\delta>0} \text{cl}\ \left[\bigcap\left\{\tilde{Q}(y,t) : |x-y| < \delta\right\}\right] \tag{7.107}$$

for almost all $t \geq 0$, $(x,t) \in A$. To see how these ideas apply to the optimal control model we observe that if $(x(\cdot), u(\cdot))$ is an admissible pair we have that $(x(t), \eta(t), \xi(t))$, where for $t \geq 0$

$$\eta(t) = f_0(x(t), u(t), t), \tag{7.108}$$

$$\xi(t) = f(x(t), u(t), t) \tag{7.109}$$

satisfy the following orientor field equations

$$(t, x(t)) \in A \text{ and } (\eta(t), \xi(t)) \in \tilde{Q}(x(t), t) \text{ a.e. } t \geq 0. \tag{7.110}$$

Conversely, we will see that if $(x(\cdot), \eta(\cdot), \xi(\cdot))$ is a solution to the orientor field equations (7.110) there exists a measurable function $u(\cdot) : [0, \infty) \to \mathbb{R}^m$ so that for almost all $t \geq 0$

$$\xi(t) = f(x(t), u(t), t), \tag{7.111}$$

$$\eta(t) \leq f_0(x(t), u(t), t), \tag{7.112}$$

and

$$(t, x(t)) \in A. \tag{7.113}$$

In this way we will show it is possible to define an admissible pair $(x(\cdot), u(\cdot)) : [0, \infty) \to \mathbb{R}^n \times \mathbb{R}^m$.

With this motivation in mind we give the following definition of upper closure.

Definition 7.10 *Let $\xi(\cdot) : [0, \infty) \to \mathbb{R}^n$, $x(\cdot) : [0, \infty) \to \mathbb{R}^n$ and let $\eta_k(\cdot) : [0, \infty) \to \mathbb{R}$, $\xi_k(\cdot) : [0, \infty) \to \mathbb{R}^n$, and $x_k(\cdot) : [0, \infty) \to \mathbb{R}^n$ for $k = 1, 2, \ldots$ be given measurable functions satisfying the orientor field equations (7.110) a.e. on $[0, \infty)$. That is,*

$$(t, x_k(t)) \in A \text{ and } (\eta_k(t), \xi_k(t)) \in \tilde{Q}(x_k(t), t) \tag{7.114}$$

and such that

(i) *$x_k(t) \to x(t)$ pointwise in $\mathbb{R}^n$ as $k \to \infty$*

(ii) *$\xi_k(\cdot) \to \xi(\cdot)$ weakly in $\mathcal{L}^1_{loc}[0, \infty)$ as $k \to \infty$*

(iii) *$\eta_k(\cdot) \in L^1[0, \infty)$ for each $k = 1, 2, \ldots$*

We say the sequence $\{(x_k(\cdot), \eta_k(\cdot), \xi_k(\cdot))\}_{n=1}^{\infty}$ has the upper closure property *if there exists $\eta(\cdot) \in \mathcal{L}^1[0, \infty)$ such that the triple $(x(\cdot), \eta(\cdot), \xi(\cdot))$ satisfies the orientor field equations (7.110) a.e. on $[0, \infty)$ and in addition,*

$$-\infty < \limsup_{k \to \infty} \int_0^\infty \eta_k(t)\, dt \leq \int_0^\infty \eta(t)\, dt. \tag{7.115}$$

With this definition we give the following upper closure theorem. For technical convenience and for brevity we state this result without proof, referring the reader to Bates [15], Carlson [40] or to Balder [12] for the necessary details.

Theorem 7.16 *Let $A \subset \mathbb{R}^n \times [0, \infty)$ be given and assume that for almost all $t \geq 0$ the sets $A(t) = \{x : (x, t) \in A\}$ are closed and nonempty. Let $\tilde{Q}(\cdot, \cdot) : A \to 2^{\mathbb{R}^{1+n}}$ be a given set valued map such that for almost all $(x, t) \in A$ the sets $\tilde{Q}(x, t)$ are closed, convex and enjoy the upper semicontinuity condition $(\mathcal{K})$ as defined by (7.107). Further let*

$\xi(\cdot)$, $x(\cdot)$, $x_k(\cdot)$, $\eta_k(\cdot)$, $\xi_k(\cdot)$, $\lambda_k(\cdot)$, $t \geq 0$ be given measurable functions with $\xi(\cdot), x_k(\cdot) \in \mathcal{L}^1_{loc}([0,\infty);\mathbb{R}^n)$, $\eta_k \in \mathcal{L}^1([0,\infty);\mathbb{R})$, $x_k(\cdot) \to x(\cdot)$ pointwise a.e. on $[0,\infty)$, $\xi_k(\cdot) \to \xi(\cdot)$ weakly in $\mathcal{L}^1_{loc}([0,\infty);\mathbb{R}^n)$, the triples $(x_k(\cdot), \eta_k(\cdot), \xi_k(\cdot))$ satisfy the orientor field equations (7.110) a.e. on $[0,\infty)$,

$$-\infty < l = \limsup_{k\to\infty} \int_0^\infty \eta_k(t)\,dt < \infty, \tag{7.116}$$

$$\eta_k(t) \leq \lambda_k(t), \quad \lambda(\cdot),\ \lambda_k(\cdot) \in \mathcal{L}^1([0,\infty);\mathbb{R}), \quad \lambda_k(\cdot) \to \lambda(\cdot) \tag{7.117}$$

weakly in $\mathcal{L}^1([0,\infty);\mathbb{R})$. Then there exists $\eta \in \mathcal{L}^1([0,\infty);\mathbb{R})$ such that $(x(\cdot), \eta(\cdot), \xi(\cdot))$ satisfies the orientor field equations (7.110) and

$$l \leq \int_0^\infty \eta(t)\,dt \tag{7.118}$$

(i.e., the sequence $(x_k(\cdot), \eta_k(\cdot), \xi_k(\cdot))$ has the upper closure property).

Remark 7.13 *To appreciate the hypotheses listed above we define the Lagrangian $\mathcal{L}(\cdot,\cdot,\cdot) : \mathbb{R}^n \times \mathbb{R}^n \times [0,\infty) \to \mathbb{R}$ by*

$$\mathcal{L}(x,\xi,t) = \begin{cases} \sup\left\{\eta : (\eta,\xi) \in \tilde{Q}(x,t)\right\} & \text{if } \tilde{Q}(x,t) \neq \emptyset \\ -\infty & \text{otherwise} \end{cases} \tag{7.119}$$

As a consequence of the hypotheses placed on the set-valued map $\tilde{Q}$ it follows that $\mathcal{L}(x,\cdot,t)$ is concave, $\mathcal{L}(\cdot,\cdot,t)$ is upper semicontinuous and $\mathcal{L}(\cdot,\cdot,\cdot)$ is measurable with respect to the σ-algebra generated by products of Borel measurable subsets of $\mathbb{R}^{2n}$ with the Lebesgue measurable subsets of $[0,\infty)$. In particular we observe that if $\tilde{Q}$ is time independent, so is $\mathcal{L}(\cdot,\cdot,\cdot)$ and the regularity conditions satisfied by $\mathcal{L}(\cdot,\cdot,\cdot)$ agree with those placed on $\mathcal{L}$ in Chapters 4 and 5. In addition we observe that the conclusion of the above theorem applied to the sequence $(x_k(\cdot), \eta_k(\cdot), \xi_k(\cdot)) = (x_k(\cdot), \mathcal{L}(x_k(\cdot), \xi_k(\cdot), t), \xi_k(\cdot))$, where $(x_k(\cdot), \xi_k(\cdot))$ are as above yields

$$-\infty < \limsup_{k\to\infty} \int_0^\infty \mathcal{L}(x_k(t),\xi_k(t),t)\,dt \leq \int_0^\infty \mathcal{L}(x(t),\xi(t),t)\,dt < \infty.$$

That is, the functional

$$\mathcal{J}(x(\cdot), \xi(\cdot)) = \int_0^\infty \mathcal{L}(x(t),\xi(t),t)\,dt \tag{7.120}$$

is upper semicontinuous with respect to the topology of pointwise convergence in $x(\cdot)$ and weak convergence in $L^1([0,\infty);\mathbb{R}^n)$ in ξ. An analogous result played a crucial role in establishing the existence of a strongly optimal solution for the associated problem of Lagrange in Chapters 4 and 5.

With the above remarks we now give a result which insures the existence of a strongly optimal solution for the infinite horizon optimal control problem described by (7.98)–(7.101).

Theorem 7.17 Existence of Strongly Optimal Solutions. *Let $A \subset \mathbb{R}^n \times [0,\infty)$ be such that for almost all $t \geq 0$ the sets $A(t) = \{x : (x,t) \in A\}$ are closed and nonempty. Let $U(\cdot,\cdot) : A \to 2^{\mathbb{R}^m}$ be a given set valued map with closed graph $M = \{(x,u,t) : (x,t) \in A,\ u \in U(x,t)\}$. Let $f_0(\cdot,\cdot,\cdot) : M \to \mathbb{R}$ and $f(\cdot,\cdot,\cdot) : M \to \mathbb{R}^n$ satisfy the growth condition (γ). Further assume that the following three conditions hold:*

(i) *There exists $\alpha \in \mathbb{R}$ such that the set Ω_α, given by*

$$\Omega_\alpha = \{(x(\cdot), u(\cdot)) : (x(\cdot), u(\cdot)) \text{ is an admissible pair and } \int_0^\infty f_0(x(t),u(t),t)\,dt \geq \alpha\}, \tag{7.121}$$

is nonempty.

(ii) *The set valued map $\tilde{Q}(\cdot,\cdot) : A \to 2^{\mathbb{R}^{1+n}}$ defined by (7.106) is closed, convex and enjoys property $(\mathcal{K})$.*

(iii) *There exists $\lambda(\cdot) : [0,\infty) \to \mathbb{R}$, Lebesgue integrable, such that for almost all $(x(\cdot), u(\cdot)) \in \Omega_\alpha$,*

$$f_0(x(t),u(t),t) \leq \lambda(t) \text{ a.e. on } [0,\infty). \tag{7.122}$$

Then the infinite horizon optimal control problem (7.98)–(7.101) has a strongly optimal solution $(x^*(\cdot), u^*(\cdot))$.

Proof: The hypotheses (i) and (iii) above insure

$$\alpha \leq l = \sup\left\{\int_0^\infty f_0(x(t),u(t),t)\,dt\right\} \leq \int_0^\infty \lambda(t)\,dt < \infty,$$

where the supremum is taken over all admissible pairs $(x(\cdot), u(\cdot))$. Therefore there exists a sequence $\{(x_k(\cdot), u_k(\cdot))\}_{k=1}^\infty$ of admissible pairs such that

$$\lim_{k\to\infty} \int_0^\infty f_0(x_k(t),u_k(t),t)\,dt = l. \tag{7.123}$$

As a consequence of the growth condition (γ) we can assume, without loss of generality, that the corresponding sequence of admissible trajectories, $\{x_k(\cdot)\}_{k=1}^{\infty}$, converges weakly to a function $x^*(\cdot) \in AC_{loc}$. For each $k = 1, 2, \ldots$, define $\eta_k(t) = f_0(x_k(t), u_k(t), t)$, $\xi_k(t) = \dot{x}_k(t)$, and $\lambda_k(t) = \lambda(t)$. Observe that the triples $(x_k(\cdot), \eta_k(\cdot), \xi_k(\cdot))$ satisfy the hypotheses of the upper closure theorem, Theorem 7.16, with $\lambda_k(\cdot) \equiv \lambda(\cdot)$. Therefore, there exists $\eta(\cdot) \in L^1[0,\infty)$ such that

$$(x^*(t), t) \in A \text{ and } (\eta(t), \dot{x}^*(t)) \in \tilde{Q}(x^*(t), t) \text{ a.e. } t \geq 0$$

and

$$l \leq \int_0^\infty \eta(t)\,dt \leq \int_0^\infty \lambda(t)\,dt. \tag{7.124}$$

To conclude the proof define the set valued map $\Gamma(\cdot) : [0,\infty) \to 2^{\mathbb{R}^m}$ by

$$\begin{aligned} \Gamma(t) \triangleq \{u \in U(x^*(t), t) :& \\ &\eta(t) \leq f_0(x^*(t), u, t) \leq \lambda(t), \\ &\dot{x}^*(t) = f(x^*(t), u, t)\}. \end{aligned} \tag{7.125}$$

Since $(x^*(\cdot), \eta(\cdot), \dot{x}^*(\cdot))$ satisfy the orientor field equations, it follows that $\Gamma(\cdot)$ is a closed valued, nonempty, measurable set valued map, and therefore by the Kuratowski, Ryll-Nardzewski measurable selection theorem (see [112]). There exists a Lebesgue measurable function $u^*(\cdot) : [0,\infty) \to \mathbb{R}^m$ such that

$$u^*(t) \in \Gamma(t) \text{ a.e. on } t \geq 0. \tag{7.126}$$

From the definition of $\Gamma(\cdot)$ it therefore follows that

$$\begin{aligned} \dot{x}^*(t) &= f(x^*(t), u^*(t), t) \text{ a.e. on } [0,\infty) \\ x^*(0) &= x_0,\ (x^*(t), t) \in A \text{ on } [0,\infty) \\ u^*(t) &\in U(x^*(t), t) \text{ a.e. on } [0,\infty), \end{aligned}$$

and

$$l \leq \int_0^\infty f_0(x^*(t), u^*(t), t)\,dt \leq \int_0^\infty \lambda(t)\,dt,$$

from which we obtain that $(x^*(\cdot), u^*(\cdot))$ is indeed an admissible pair with

$$\sup\left\{\int_0^\infty f_0(x(t), u(t), t)\,dt\right\} = \int_0^\infty f_0(x^*(t), u^*(t), t)\,dt,$$

where the supremum is taken over all admissible pairs, as desired.

Remark 7.14 *The above result was first presented for free problems in the calculus of variations in 1940 by Cinquini* [54] *with extensions given in Faedo* [70], [71], [72], *Darbo* [59], *and again Cinquini* [53]. *In the context of optimal control this result first appears in 1976 in Baum* [16] *with updated versions given in Bates* [15] *culminating in the above result given in Balder* [12]. *Other results along these lines appear in the economics literature (e.g., Takekuma* [172], *Magill* [132], [133], *Thisse and Weiserbs* [173], *and Maruyama* [137],[138]) *all of which are included as special cases of the above result. A similar result appears in Toman* [174].

We conclude this section by considering a simple example used to model the optimal harvesting of a renewable resource.

Example 7.3 In this example we let $x(t)$ denote the quantity of renewable resource available at time $t \geq 0$. The goal here is to determine a harvest rate which maximizes total revenue. The dynamics of the resource, without harvesting is given by the differential equation

$$\dot{x}(t)) = F(x(t)) \qquad \text{a.e.} \quad t \geq 0$$
$$x(0) = x_0, \quad 0 \leq x(t) \leq K \quad \text{for} \quad t \geq 0$$

where $F(\cdot) : [0, K) \to \mathbb{R}^n$ is a given continuous function, satisfying $F(0) = F(K) = 0$. Here K represents the carrying capacity of the population. The control parameter, u, denotes the amount of effort applied by the harvester and leads to the following harvesting model

$$\begin{aligned} \dot{x}(t) &= F(x(t)) - qx(t)u(t) \qquad \text{a.e.} \quad t \geq 0 \\ x(0) &= x_0, \quad 0 \leq x(t) \leq K \quad \text{for} \quad t \geq 0 \\ & 0 \leq u(t) \leq E \qquad \text{a.e.} \quad t \geq 0 \end{aligned} \tag{7.127}$$

where q is a positive constant (in fishery economics q is called the "catchability coefficient" (see Clark [55] for a detailed discussion)) and E denotes an upper bound on the amount of effort which can be applied by the harvester.

The accumulated total revenue depends on the price, p, obtained through selling the harvested resource, the cost of unit effort $c(u)$, and the harvest rate. For the purposes of the example we consider a discounting factor, $e^{-\delta t}$, to provide a subjective rate of time preference. With this notation the accumulated profit at time $T > 0$ is given by the definite integral

$$J_T(x_0, u(\cdot)) = \int_0^T e^{-\delta t} \left[pqx(t)u(t) - c(u(t))\right] dt. \tag{7.128}$$

Therefore we seek a stongly optimal solution for the control problem described by (7.127)–(7.128). The function $c(\cdot) : [0, M] \to \mathbb{R}$ is assumed to be a strictly increasing, continuous, convex function with $c(0) = 0$

We now show that the hypotheses of Theorem 7.17 are satisfied. We begin by observing the following:

(i) The set $A = [0, K] \times [0, \infty)$ clearly satisfies the condition that the sets $A(t) = \{x : (x, t) \in A\} = [0, K]$ are closed and nonempty.

(ii) The set valued map $U(\cdot, \cdot) : A \to 2^{\mathbb{R}}$ is given by $U(x, t) = [0, E]$ with closed graph $M = [0, K] \times [0, E] \times [0, \infty)$.

(iii) The functions $f_0(\cdot, \cdot, \cdot) : M \to \mathbb{R}$ and $f(\cdot, \cdot, \cdot) : M \to \mathbb{R}$ are give respectively by $f_0(x, u, t) = e^{-\delta t}[pqxu - c(u)]$ and $f(x, u, t) = F(x) - qxu$ are both continuous and therefore satisfy the requisite regularity hypotheses.

(iv) To see that the growth condition (γ) holds we notice that for $(x, u) \in [0, E] \times [0, K]$, there exists positive constants α and β such that both

$$|F(x) - qxu| \leq \alpha$$

and

$$|c(u) - pqxu| \leq \beta$$

hold. Therefore for any $\epsilon > 0$ we have

$$e^{-\delta t}(pqxu - c(u)) \leq e^{-\delta t}\beta \leq \beta$$

and

$$|F(x) - qxu| \leq \left(\alpha - \epsilon e^{-\delta t}\beta\right) - \epsilon\,(pqxu - c(u))\,e^{-\delta t}$$

which is precisely the growth condition (γ) (i.e., $\mathcal{L}_T = \beta$ and $\psi_{\epsilon T}(t) = \left(\alpha - \epsilon e^{-\delta t}\beta\right)$).

(v) Since $c(0) = 0$ it follows that the natural growth of the resource (i.e., $u(t) \equiv 0$) provides an admissible pair which gives zero profit. Therefore for any $\alpha \leq 0$, the set Ω_α described by (7.121) is nonempty.

(vi) Since $(x, u) \in [0, K] \times [0, E]$ we also have that

$$e^{-\delta t}(pqxu - c(u)) \leq \beta e^{-\delta t}$$

and so (7.123) holds as well.

(vii) The only hypotheses left unverified concern the set valued map $\tilde{Q}(\cdot, \cdot) : [0, K] \times [0, \infty) \to 2^{\mathbb{R}^2}$ defined by

$$\begin{aligned} \tilde{Q}(x, t) &= \{(z_0, z) : z_0 \leq e^{-\delta t}(pqxu - c(u)), \\ z &= F(x) - qxu,\ u \in [0, E]\}. \end{aligned}$$

The convexity of the function $c(\cdot)$ easily implies that these sets are convex. It remains to prove the upper semicontinuity property $(\mathcal{K})$. To this end we fix (x,t) and let

$$(z_0, z) \in \bigcup_{\delta>0} \text{cl} \left[\bigcap \left\{\tilde{Q}(y,t) : |y-x| < \delta, \ y \in [0,K]\right\}\right].$$

This inclusion implies there exists a null sequence $\{\delta_k\}_{k=1}^{\infty}$, and sequences $\{x_k\}_{k=1}^{\infty}$ and $\{(z_{0k}, z_k)\}_{k=1}^{\infty}$ such that

(a) $|x_k - x| < \delta_k$, $k = 1, 2, \ldots$

(b) $(z_{0k}, z_k) \in \tilde{Q}(x_k, t)$, $k = 1, 2, \ldots$

(c) $\lim_{k\to\infty} (z_{0k}, z_k) = (z_0, z)$.

As a consequence of (b) for each $k = 1, 2, \ldots$ there exists $u_k \in [0, E]$ such that

$$z_{0k} \leq e^{-\delta t} \left(pqx_k u_k - c(u_k)\right)$$

and

$$z_k = F(x_k) - qx_k u_k.$$

Without loss of generality we assume the $\lim_{k\to\infty} u_k = u \in [0, E]$ exists and thus

$$z_0 \leq \lim_{k\to\infty} \left(pqx_k u_k - c(u_k)\right) = pqxu - c(u)$$

and

$$z = \lim_{k\to\infty} \left(F(x_k) - qx_k u_k\right) = F(x) - qxu,$$

implying that $(z_0, z) \in \tilde{Q}(x,t)$ as desired.

We have just shown that the above example meets all the hypotheses of Theorem 7.17 and therefore has a strongly optimal solution.

Remark 7.15 *Clearly the most difficult hypothesis to verify in the above example was property $(\mathcal{K})$. We chose to show this condition by applying the definition directly in order to indicate how to apply the definition. Known conditions assuming property $(\mathcal{K})$ could have been utilized. For a detailed treatment of such results we direct the reader to Cesari* [47, Chapter 6].

Remark 7.16 *We observe that no hypothesis concerning the convexity of F was made. As a consequence of this fact the velocity set considered in Chapter 4 is not necessarily convex and thus the joint convexity hypotheses made in many of the previous results are not applicable.*

7.7 The existence of overtaking optimal solutions

In this section we combine the sufficient conditions given by Carathéodory's method (Theorem 7.10) with the existence result, Theorem 7.17, for strongly optimal solutions to obtain nonautonomous analogues of the existence results given in Chapters 4 and 5. These results were first reported in Carlson [42]. After presenting these results we illustrate their utility with two examples. We begin by stating our first existence result.

Theorem 7.18 *Let $A \subset \mathbb{R}^n \times [0, \infty)$ be a given set such that for almost all $t \geq 0$, $A(t) = \{x : (x,t) \in A\}$ is closed and nonempty. Let $U(\cdot,\cdot) : A \to 2^{\mathbb{R}^m}$ be a given set valued mapping with closed graph $M = \{(x,u,t) : (x,t) \in A,\ u \in U(x,t)\}$ and let $f_0(\cdot,\cdot,\cdot) : M \to \mathbb{R}$ and $f(\cdot,\cdot,\cdot) : M \to \mathbb{R}^n$ be given functions satisfying the regularity hypotheses given in Section 7.6. In addition, let $S(\cdot,\cdot) : A \to \mathbb{R}$ be a given differentiable function whose gradient $\left(\frac{\partial S}{\partial t}(\cdot,\cdot), \nabla_x S(\cdot,\cdot)\right)$ satisfies the Carathéodory conditions and moreover satisfies the conditions*

(a) *For $(x,u,t) \in M$ the function $h(\cdot,\cdot,\cdot) : M \to \mathbb{R}$ given by*

$$h(x,u,t) = f_0(x,u,t) + \frac{\partial S}{\partial t}(x,t) + \nabla_x S(x,t)' f(x,u,t) \tag{7.129}$$

is nonpositive.

(b) *For every admissible trajectory $x(\cdot)$, we have*

$$\liminf_{t \to \infty} S(x(t),t) > -\infty \tag{7.130}$$

(c) *There exists a constant C such that for every admissible pair $(x(\cdot), u(\cdot))$ with the property that whenever*

$$J^*(x_0, u(\cdot)) = \int_0^\infty h(x(t),u(t),t)\,dt$$

is finite we have

$$\lim_{t \to \infty} S(x(t),t) = C. \tag{7.131}$$

Then if the following conditions (i) *to* (iii) *hold;*

(i) *There exists $\alpha \leq 0$ such that the set Ω_α, given by*

$$\Omega_\alpha = \{(x(\cdot), u(\cdot)) \text{ is an admissible pair: } J^*(x_0, u(\cdot)) \geq \alpha\}, \tag{7.132}$$

is nonempty.

(ii) *The set valued map* $\tilde{Q}(\cdot,\cdot) : A \to 2^{\mathbb{R}^{1+n}}$, *defined by* (7.106), *is convex valued and enjoys property* $(\mathcal{K})$ *with respect to* x *on* A.

(iii) *The functions* $f_0(\cdot,\cdot,\cdot)$ *and* $f(\cdot,\cdot,\cdot)$ *satisfy the growth condition* (γ) *given in Definition* 7.9;

there exists an admissible pair $(x^*(\cdot), u^*(\cdot))$ *which is an overtaking optimal solution for the optimal control problem described by* (7.98)-(7.101).

Proof: The conditions placed on the function S (i.e., (a), (b) and (c) above) are precisely those given in Definition 7.5. Therefore, as a result of Theorem 7.10, it is sufficient to prove that the optimal control problem consisting of minimizing the integral functional $J^*(x_0, u(\cdot))$ over all admissible paris $(x(\cdot), u(\cdot))$ has a strongly optimal solution. To do this we apply Theorem 7.17. Observe that for the problem under consideration, the conditions (7.121) and (7.123) are seen to be satisfied respectively by (7.129) and (7.131) (here $\lambda(t) \equiv 0$). The only hypotheses of Theorem 7.17 not immediately met concern the set valued map

$$\tilde{Q}^*(x,t) = \{(z_0, z) : \; z_0 \leq h(x,u,t), \; z = f(x,u,t), \; u \in U(x,t)\} \tag{7.133}$$

defined for $(x,t) \in A$. However we notice that for fixed $(x,t) \in A$ the map $\Gamma_{x,t}(\cdot,\cdot) : \tilde{Q}^*(x,t) \to \tilde{Q}(x,t)$ by

$$\Gamma_{x,t}(z_0, z) = \left(z_0 - \frac{\partial S}{\partial t}(x,t) - \nabla_x S(x,t)'z, z\right),$$

defines a bijection from which it is easy to deduce that $\tilde{Q}^*(\cdot,\cdot)$ enjoys the same properties as $\tilde{Q}(\cdot,\cdot)$. Thus all the hypotheses of Theorem 7.17 and Theorem 7.10 are satisfied implying the existence of an overtaking optimal solution for the control problem described by (7.98)-(7.101). ∎

We now present a useful corollary of the above result which establishes the conditions for the existence of overtaking optimal solutions when it is know that either a G-supported trajectory or McKenzie turnpike exists.

Theorem 7.19 *Let* A, $U(\cdot,\cdot) : A \to 2^{\mathbb{R}^m}$, $f_0(\cdot,\cdot,\cdot) : M \to \mathbb{R}$, *and* $f(\cdot,\cdot,\cdot) : M \to \mathbb{R}^n$ *satisfy the hypotheses of Theorem* 7.18. *Let* $q : A \to \mathbb{R}$ *be a given continuously differentiable function and let* $(\bar{x}(\cdot), \bar{u}(\cdot))$ *be an admissible pair with* $\bar{x}(0)$ *not necessarily equal to* x_0, *satisfying the following properties:*

(a) *For all* $(x,u,t) \in M$,

$$\begin{aligned} f_0(x,u,t) \; &+ \; \frac{\partial q}{\partial t}(x,t) + \nabla_x q(x,t)'f(x,u,t) \\ &\leq \; f_0(\bar{x}(t), \bar{u}(t), t) + \frac{d}{dt}q(\bar{x}(t), t). \end{aligned} \tag{7.134}$$

(b)

$$\liminf_{t\to\infty} q(\bar{x}(t),t) > -\infty \text{ and } \limsup_{t\to\infty} q(x(t),t) < \infty \tag{7.135}$$

for all admissible trajectories $x : [0,\infty) \to \mathbb{R}^n$.

(c) *Either*

$$\limsup_{t\to\infty} \frac{d}{dt} q(\bar{x}(t),t) < \infty \quad \text{and} \quad \liminf_{t\to\infty} \frac{d}{dt} q(x(t),t) > -\infty$$

or

$$\liminf_{t\to\infty} \frac{d}{dt} q(\bar{x}(t),t) > -\infty \quad \text{and} \quad \limsup_{t\to\infty} \frac{d}{dt} q(x(t),t) < \infty \tag{7.136}$$

for all admissible trajectories $x : [0,\infty) \to \mathbb{R}^n$.

(d) *For every* $\epsilon > 0$ *and admissible pair* $(x(\cdot), u(\cdot))$, *there exists a constant* $\delta = \delta(\epsilon, x(\cdot), u(\cdot)) > 0$ *such that* $|q(x(t),t) - q(\bar{x}(t),t)| > \epsilon$ *implies,*

$$f_0(x(t),u(t),t) - f_0(\bar{x}(t),\bar{u}(t),t) + \frac{d}{dt}\left[q(x(t),t) - q(\bar{x}(t),t)\right] < -\delta. \tag{7.137}$$

Then if the conditions (i) *to* (iii) *hold;*

(i) *There exists* $\alpha \leq 0$ *such that the set* Ω_α , *given by*

$$\Omega_\alpha = \{(x(\cdot), u(\cdot)) \text{ admissible: } \int_0^\infty f_0(x(t),u(t),t)$$
$$-f_0(\bar{x}(t),\bar{u}(t),t) + \frac{d}{dt}\left[q(x(t),t) - q(\bar{x}(t),t)\right] dt \leq \alpha\}, \tag{7.138}$$

is nonempty.

(ii) *The set valued map* $\tilde{Q}(\cdot,\cdot) : A \to 2^{\mathbb{R}^{1+n}}$, *defined by* (7.106), *is convex valued and satisfies property* $(\mathcal{K})$ *with respect to* x *on* A.

(iii) *The functions* $f_0(\cdot,\cdot,\cdot)$ *and* $f(\cdot,\cdot,\cdot)$ *satisfy the growth condition* (γ);

there exists an overtaking optimal solution for the optimal control problem described by (7.98)–(7.101).

Proof: This result can be established directly by an argument identical to that of the previous theorem. Alternatively one can utilize the transformations found in the proof of Theorem 7.13 and apply Theorem 7.18 directly. ∎

Remark 7.17 *In [66] an existence result, which is similar to the above, is obtained for* $q(x,t) = \bar{p}(t)'(x - \bar{x}(t))$ *where for almost all* $t \geq 0$, *there exists an admissible control* $\bar{u}(t)$ *such that*

$$f_0(\bar{x}(t), \bar{u}(t), t) = \max\{f_0(x,u,t) : 0 = f(x,u,t),\ u \in U(x),\ x \in X\}.$$

We conclude this section with two examples. The first of these is purely mathematical and utilizes Theorem 7.18 while the second is an undiscounted version of the optimal harvesting model considered in Example 7.3 to which we apply Theorem 7.19.

Example 7.4 In this example we consider the problem of finding an overtaking optimal solution for the control problem whose reward is described by

$$J_T(x_0, u(\cdot)) = \int_0^T \left[(\cos(t)+1)(1 - \sqrt{x(t)+1}\,) - \sin(t)u(t)\right] dt$$

over all pairs of functions $(x(\cdot), u(\cdot)) : [0,\infty) \to \mathbb{R} \times \mathbb{R}$ satisfying

$$\begin{aligned} \dot{x}(t) &= 2\sqrt{x(t)+1}\,u(t) \quad \text{a.e. on } [0,\infty) \\ x(0) &= x_0, \quad 0 \leq x(t) \leq 100 \quad \text{on } [0,\infty) \\ -1 \leq u(t) &\leq 1 \quad \text{a.e. on } [0,\infty). \end{aligned}$$

For this problem we define $S(\cdot,\cdot) : [0,100] \times [0,\infty) \to \mathbb{R}$ by

$$S(x,t) = \sin(t)(\sqrt{x+1} - 1).$$

Observe that $h(\cdot,\cdot,\cdot) : [0,100] \times [-1,1] \times [0,\infty) \to \mathbb{R}$ defined by

$$\begin{aligned} h(x,u,t) &= (\cos(t)+1)(1 - \sqrt{x+1} - \sin(t)u \\ &\quad + \frac{\partial S}{\partial t}(x,t) + \frac{\partial S}{\partial x}(x,t)(2\sqrt{x+1}\,)u \\ &= 1 - \sqrt{x+1} \\ &\leq 0, \end{aligned}$$

since $x \geq 0$. Further, for any admissible pair $(x(\cdot), u(\cdot))$ we have

$$S(x(t),t) \leq \sqrt{x(t)+1} - 1 \leq \sqrt{101} + 1$$

and

$$\frac{d}{dt}S(x(t),t) = \cos(t)(\sqrt{x(t)+1} - 1) + \sin(t)u \leq \sqrt{101}.$$

Therefore for any admissible pair $(x(\cdot), u(\cdot))$,

$$\limsup_{t\to\infty} S(x(t),t) < \infty$$

and

$$\limsup_{t\to\infty} \frac{d}{dt} S(x(t),t) < \infty.$$

Also, for any $\epsilon > 0$ and admissible pair $(x(\cdot), u(\cdot))$ we have

$$h(x(t),u(t),t) = 1 - \sqrt{x(t)+1} < -\epsilon,$$

wherever

$$|S(x(t),t)| = |\sin(t)(\sqrt{x(t)+1} - 1)| > \epsilon$$

Thus $S(\cdot,\cdot)$ meets the requirements of Theorem 7.12 and therefore, for any admissible pair $(x(\cdot), u(\cdot))$ for which

$$\int_0^\infty h(x(t),u(t),t)\,dt > -\infty,$$

we have

$$\lim_{t\to\infty} S(x(t),t) = 0.$$

This analysis shows $S(\cdot,\cdot)$ satisfies the conditions of Theorem 7.18.

We further observe that the functions $(y(\cdot), v(\cdot)) : [0,\infty) \to \mathbb{R} \times \mathbb{R}$ defined by

$$(y(t),v(t)) = \begin{cases} \left((\sqrt{x_0+1} - t)^2 - 1, -1\right) & \text{if } 0 \le t \le \sqrt{x_0+1} - 1 \\ (0,0) & \text{if } \sqrt{x_0+1} - 1 \le t \end{cases}$$

satisfies

$$\begin{aligned} \int_0^\infty h(y(t),v(t),t)\,dt &= \int_0^\infty (1 - \sqrt{y(t)+1}\,)\,dt \\ &= \int_0^{\sqrt{x_0+1}-1} \left(1 - (\sqrt{x_0+1} - t)\right)\,dt \\ &= -(1 - \sqrt{x_0+1})^2 + \frac{1}{2}(1 - \sqrt{x_0+1})^2 \\ &= -\frac{1}{2}(1 - \sqrt{x_0+1})^2 \end{aligned}$$

Therefore, if we choose $\alpha \le -\frac{1}{2}(1 - \sqrt{x_0+1})^2$ the set Ω_α given by (7.131) is clearly nonempty. In addition since $f_0(x,u,t) = (\cos(t)+1)(1 - \sqrt{x+1}) - \sin(t)u$ and $f(x,u,t) = 2\sqrt{x+1}\,u$ are bounded on $[0,100] \times [-1,-1] \times [0,\infty)$, it is easy to see that the growth conditon (γ) holds. The remaining hypotheses of Theorem 7.18 concern the sets, $\tilde{Q}(x,t)$, defined here for $(x,t) \in [0,100] \times [0,\infty)$ by

$$\begin{aligned} \tilde{Q}(x,t) = \{(z^o, z) :\; & z^o \le (\cos(t)+1)\left(1 - \sqrt{x+1}\right) - \sin(t)u, \\ & z = 2\sqrt{x+1}\,u,\; -1 \le u \le -1\}. \end{aligned}$$

These sets are clearly convex and moreover property $(\mathcal{K})$ can be verified easily form the definition. Thus all the hypotheses of Theorem 7.18 are met and we conclude that there exists an overtaking optimal solution.

Example 7.5 The undiscounted analogue of Example 7.3 consists of seeking overtaking optimal solutions for the optimal control problem described by

$$
\begin{aligned}
J_T(x_0, u(\cdot)) &= \int_0^T [pqx(t)u(t) - c(u(t))]\, dt \\
\dot{x}(t) &= F(x(t)) - qx(t)u(t) \qquad \text{a.e. } t \geq 0 \\
x(0) = x_0, &\quad 0 \leq x(t) \leq K \qquad \text{on } [0, \infty) \\
0 &\leq u(t) \leq E \qquad \text{a.e. } t \geq 0.
\end{aligned}
$$

We will assume the same basic hypotheses as in Example 7.3 except now we assume that E is chosen sufficiently large to insure that

$$F(x) = qxu \tag{7.139}$$

admits at least one solution in $[0, E]$ for each fixed $x \in [0, K]$.

In addition we assume there exists a unique solution, $(\overline{x}, \overline{u})$ of the optimal steady state maximization problem

$$\max_{(x,u)\in[0,K]\times[0,E]} \{[pqxu - c(u)\} : 0 = F(x) - qxu\}. \tag{7.140}$$

Our goal is to show that $(\overline{x}, \overline{u})$ is the required admissible pair needed in Theorem 7.19 and that an appropriate function $q(x,t) = q(x)$ can be chosen to be any antiderivative of

$$-\frac{1}{qx}\left[\frac{d}{du}\, c\Big(\frac{F(x)}{qx}\Big)\right] + p, \tag{7.141}$$

which we will assume is a continuous function of x.

Remark 7.18 *A special case for this last assumption to hold is when $c(u) = u^\beta$ and $F(x) = rx^\alpha\left(1 - \frac{x}{K}\right)$ where α and β are two fixed constants satisfying $\alpha > 1$, $\beta > 1$ and $\alpha \geq \beta(\beta - 1)^{-1}$.*

To show the admissible pair $(\overline{x}, \overline{u})$ and the function $q(x)$ are as desired we let $x \in (0, K]$ be arbitrary (but fixed) and let $\tilde{u} \in [0, E]$ be such that

$$0 = F(x) - qx\tilde{u}. \tag{7.142}$$

Then for any $u \in [0, E]$ we have

$$
\begin{aligned}
&[pqxu - c(u)] - [pqx\tilde{u} - c(\tilde{u})] \\
&\quad = c(\tilde{u}) - c(u) + pqx(u - \tilde{u}) \\
&\quad \leq -\frac{d}{du}c(u)\Big|_{u=\tilde{u}}(u - \tilde{u}) + pqx(u - \tilde{u}) \\
&\quad = \left[-\frac{d}{du}c(u)\Big|_{u=\tilde{u}} + pqx\right](u - \tilde{u}),
\end{aligned}
$$

where the above inequality is a consequence of the convexity of the function c. Now since $(x, \tilde{u})$ satisfies (7.142) and $x \neq 0$, the above becomes

$$\begin{aligned}
&[pqxu - c(u)] - [pqx\tilde{u} - c(\tilde{u})] \\
&\leq \Big[-\frac{d}{du}c\Big(\frac{F(x)}{qx}\Big) + pqx\Big]\Big(u - \frac{F(x)}{qx}\Big) \\
&= -\Big[-\frac{1}{qx}\frac{d}{du}c\Big(\frac{F(x)}{qx}\Big) + p\Big](F(x) - qxu) \\
&= \Big(\frac{d}{dx}q(\cdot)\Big|_{x=\overline{x}}\Big)(F(x) - qxu),
\end{aligned}$$

or equivalently,

$$pqxu - c(u) + q'(x) = \Big(\frac{d}{dx}q(\cdot)\Big|_{x=\overline{x}}\Big)(F(x) - qxu) \leq pqx\tilde{u} - c(\tilde{u}).$$

Moreover, the optimality of $(\overline{x}, \overline{u})$ gives us

$$pqx\tilde{u} - c(\tilde{u}) \leq pq\overline{x}\,\overline{u} - c(\overline{u}),$$

and so for all $(x, u) \in (0, K] \times [0, E]$ we have

$$\begin{aligned}
&pqxu - c(u)\Big(\frac{d}{dx}q(\cdot)\Big|_{x=\overline{x}}\Big)(x)(F(x) - qxu) \\
&\leq pq\overline{x}\,\overline{u} - c(\overline{u}) \qquad (7.143)\\
&= pq\overline{x}\,\overline{u} - c(\overline{u})\Big(\frac{d}{dx}q(\cdot)\Big|_{x=\overline{x}}\Big)(F(\overline{x}) - q\overline{x}\,\overline{u}).
\end{aligned}$$

This is precisely (7.134) in Theorem 7.19. We further notice that if $x = 0$, the hypothesis $F(0) = 0$ implies $(u, 0)$ is a feasible point for the optimal steady state problem, and so the inequality (7.143) holds for $x = 0$, $u \in [0, E]$ as well. Further because of the smoothness hypothesis on q (i.e., q and $\frac{d}{dx}q(\cdot)$ are continuous) and the bounds on $(x, u) \in [0, K] \times [0, E]$, it is easy to conclude that conditions (7.135) and (7.136) of Theorem 7.19 both hold. The only hypothesis concerning the function q left to treat is (7.137). We show this condition holds by contradiction. That is, we will assume there exists $\epsilon_0 > 0$, and admissible pair $(x(\cdot), u(\cdot)) : [0, \infty) \to \mathbb{R} \times \mathbb{R}$ such that for each integer n there exists $t_n \geq 0$ such that $|q(x(t_n)) - q(\overline{x})| > \epsilon_0$ and

$$\begin{aligned}
0 \geq\ &(pqx(t_n)u(t_n) - c(u(t_n))) - (pq\overline{x}\,\overline{u} - c(\overline{u})) \qquad (7.144)\\
&+ \frac{d}{dx}q(\cdot)\,|_{x=x(t_n)}\,[F(x(t_n)) - qx(t_n)u(t_n)] > -\frac{1}{n}.
\end{aligned}$$

Since the sequence $\{x(t_n)\}_{n=1}^{\infty} \subset [0, K]$ we assume without loss of generality that $\lim_{n\to\infty} x(t_n) = \tilde{x}$. For each integer n choose $\tilde{u}_n \in [0, E]$ so that

$$F(x(t_n)) = qx(t_n)\tilde{u}_n,$$

and recall that if $x(t_n) \neq 0$ we have from above

$$\begin{aligned}[pqx(t_n)u(t_n) - c(u(t_n))] \;-\; & [pqx(t_n)\tilde{u} - c(\tilde{u})] \\ & \leq \frac{d}{dx}q(\cdot)\,|_{x=x(t_n)}\,[F(x(t_n) - qx(t_n)u(t_n)]\,.\end{aligned} \tag{7.145}$$

Combining (7.144) and (7.145) yields, if $x(t_n) \neq 0$,

$$-\frac{1}{n} < [pqx(t_n)\tilde{u}_n - c(\tilde{u}_n)] - [pq\overline{x}\,\overline{u} - c(\overline{u}\,)] \leq 0,$$

where the last inequality follows from the optimality of $(\overline{x}, \overline{u}\,)$. Therefore we get

$$\lim_{n\to\infty} [pqx(t_n)\tilde{u}_n - c(\tilde{u}_n)] = [pq\overline{x}\,\overline{u} - c(\overline{u}\,)].$$

By definition if $\tilde{x} \neq 0$,

$$\lim_{n\to\infty} \tilde{u}_n = \lim_{n\to\infty} \frac{F(x(t_n))}{qx(t_n)} = \frac{F(\tilde{x})}{q\tilde{x}} \triangleq \tilde{u}$$

and so we get that $(\tilde{x}, \tilde{u})$ is admissible for the optimal steady state problem and moreover

$$pq\tilde{x}\tilde{u} - c(\tilde{u}) = pq\overline{x}\,\overline{u} - c(\overline{u}\,).$$

This implies $(\tilde{x}, \tilde{u}) = (\overline{x}, \overline{u}\,)$ by our uniqueness hypothesis for the solution of the steady state problem. However, on the other hand

$$0 < \epsilon_0 \leq \lim_{n\to\infty} |q(x(t_n)) - q(\overline{x}\,)| = 0,$$

which is clearly a contradiction. Therefore we must assume $\tilde{x} = 0$. In this case $\overline{x} \neq 0$ since $\lim_{n\to\infty} |q(x(t_n)) - q(\overline{x}\,)| \geq \epsilon_0$. From our assumptions (in particular (7.144)) it is easy to see that

$$\begin{aligned}&\lim_{n\to\infty} [pqx(t_n)u(t_n) - c(u(t_n))] \\ &= \lim_{n\to\infty} [-c(u(t_n))] \\ &= pq\overline{x}\,\overline{u} - c(\overline{u}\,) > 0\end{aligned}$$

where the last inequality follows since $(0, 0)$ is admissible for the optimal steady state and $\overline{x} \neq 0$. However by hypothesis $c(u) \geq 0$ so we get $\lim_{n\to\infty} [-c(u(t_n))] \leq 0$, clearly a contradiction. From the above analysis we must therefore conclude that (7.137) holds.

The remaining hypotheses of Theorem 7.19 are now verified as in Example 7.3 and we conclude that there exists an overtaking optimal solution to the undiscounted harvesting model.

Chapter 8

Control of Systems with Integrodifferential Equations

8.1 Introduction

It has long been recognized that time delays are important in formulating economic models. This was observed as early as 1935 when Kalecki [109] introduced a class of such models described by linear differential-difference equations. These models were further developed by Leontief [125] and others. To quote Gandolfo's 1971 text [82]:

> "The fundamental reason (that we discuss such systems) is that we think that mixed differential-difference equations are much more suitable than differential equations alone or difference equations alone, for an adequate treatment of dynamic economic phenomena."

More recently, Volterra integral and integrodifferential equations have been used to model optimal dynamic advertising models with continuous lags in the finite time horizon. In particular we would mention the works of Pauwels [147], Hartl [94], and Hartl and Sethi [95] in which their primary concern was that of obtaining necessary conditions and sufficient conditions for optimality via the method of dynamic programming. In addition Haurie et. al. [104] proposed a similar model for the purpose of investigating social service policies directed to large populations (e.g., the elderly, the handicapped, or other dependents). Once again, this work is primarily concerned with finite horizon models, however it is remarked that "a complete discussion of these theories (with regards to infinite horizon control models) to the case of optimal control of distributed parameters is not yet completed".

In this chapter we investigate the existence of overtaking optimal solutions for a class of models in which the states are governed by a nonlinear autonomous Volterra integrodifferential equation with infinite delay in which the delay occurs only with respect to the state variable. These results were first reported in Carlson [38] and are most closely aligned with the results presented in Chapter 4. As we shall see, the theory we develop here is obtained under essentially the same hypotheses as those found in Chapter 4. With these brief remarks, the format of this chapter is as follows. In Section 8.2 we introduce the model considered and indicate the basic hypotheses assumed throughout. Section 8.3 summarizes several technical results concerning linear hereditary operators due to Marcus and Mizel [136] as well as presenting a general upper closure theorem which is similar in spirit to Theorem 7.16. These results will of course be utilized in Section 8.4 where we develop the associated problem of Lagrange, present the requisite asymptotic turnpike property, and present the existence result. We conclude our discussion with Section 8.5 in which several examples are presented. These examples include a version of the classical economic growth model with distributed time delay as well as a model which investigates the optimal exploitation of a renewable resource.

8.2 The basic model

We consider a system described by a Volterra integrodifferential equation of the form

$$\dot{x}(t) = f(x(t), u(t)) + \int_0^{\infty} g(t-s)h(x(s))\,ds \quad \text{a.e. } t \geq 0, \tag{8.1}$$

where $x(\cdot) : (-\infty, \infty) \to \mathbb{R}^n$ is a bounded continuous function that is locally absolutely continuous on $[0, \infty)$ and satisfies the prescribed initial condition

$$x(s) = x_0(s) \qquad \text{for all } s < 0, \tag{8.2}$$

where $x_0(\cdot) : (-\infty, 0] \to \mathbb{R}^n$ is a given bounded continuous function, as well as the state constraints

$$x(t) \in X \qquad \text{for } t \in (-\infty, \infty), \tag{8.3}$$

in which X is a closed subset of n-dimensional Euclidean space $\mathbb{R}^n$. The control function $u(\cdot) : [0, \infty) \to \mathbb{R}^n$ is assumed to be Lebesgue measurable and satisfies the feedback control constraints

$$u(t) \in U(x(t)) \qquad \text{a.e. } 0 \leq t, \tag{8.4}$$

where $U(\cdot) : X \to 2^{\mathbb{R}^m}$ is a point to set mapping with closed graph $M = \{(x, u) : x \in X$ and $u \in U(x)\}$.

As regards the functions $f(\cdot,\cdot)$, $g(\cdot)$, and $h(\cdot)$ we assume that $f(\cdot,\cdot) : M \to \mathbb{R}^n$ and $h(\cdot) : X \to \mathbb{R}^p$ are both continuous and that $g(\cdot) = (g_{ij}(\cdot))_{n \times p}$ is an $n \times p$ matrix function defined for $t \geq 0$ with entries satisfying

$$\text{(i)} \qquad \int_0^{\infty} |g_{ij}(t)|\,dt < \infty,$$

$$\text{(ii)} \qquad \int_0^\infty t\ |g_{ij}(t)|\, dt < \infty, \tag{8.5}$$

$$\text{(iii)} \qquad \sum_{m=1}^{\infty} \|g_{ij}\|_m^\infty < +\infty,$$

where $\|g_{ij}\|_m^\infty$ is the essential supremum of g_{ij} restricted to the interval $[m-1, m]$, $m = 1, 2, \ldots$. We remark that the assumption (8.5(iii)) given above implies (8.5(i)), but we have included both for definiteness. The performance of the above control system is described for any positive T by the integral functional

$$J_T(x_0(\cdot)\,, u(\cdot)\,) = \int_0^T f_0(x(t), u(t))\, dt, \tag{8.6}$$

where $f_0(\cdot,\cdot) : M \to \mathbb{R}^n$ is a given upper semicontinuous function.

With this notation we give the following definition.

Definition 8.1 *A bounded continuous function $x(\cdot) : (-\infty, \infty) \to \mathbb{R}^n$ will be called a* trajectory *if $x(\cdot)$ is locally absolutely continuous on $[0, \infty)$ and if there exists a Lebesgue measurable function (referred to as a* control*) $u(\cdot) : [0, \infty) \to \mathbb{R}^m$ such that the pair $(x(\cdot)\,, u(\cdot)\,)$ satisfies* (8.1), (8.3), (8.4), *and the map $t \to f_0(x(t), u(t))$ is locally Lebesgue integrable on $[0, \infty)$. If in addition, the trajectory $x(\cdot)$ satisfies the prescribed initial condition* (8.2), *we will call $x(\cdot)$ an admissible trajectory and $u(\cdot)$ an admissible control.*

For brevity we let A denote the set of all trajectory-control pairs $(x(\cdot)\,, u(\cdot)\,)$ and let $A_0 \subset A$ be the set of all admissible pairs.

As is usual in our discussion concerning the existence of overtaking optimal solutions we require certain convexity and growth hypotheses on the model. These conditions are analogous to those of Chapter 4 and Chapter 7. In particular we require

Assumption 8.1 *For each $x \in X$, the set $\tilde{Q}(x)$ given by*

$$\tilde{Q}(x) = \{(z^o, z) :\ z^o \leq f_0(x, u),\ z = f(x, u),\ u \in U(x)\} \tag{8.7}$$

is a nonempty closed, convex set that satisfies the upper semicontinuity condition property $(\mathcal{K})$ given as

$$\tilde{Q}(x) = \bigcap_{\delta > 0} \text{cl}\ \left[\bigcup \left\{\tilde{Q}(y) :\ \|y - x\| < \delta\right\}\right]. \tag{8.8}$$

Assumption 8.2 *We assume that for each $\epsilon > 0$ there exists $c_\epsilon > 0$ such that for all $(x, u) \in M$ we have*

$$|f(x, u)| + |h(x)| \leq c_\epsilon - \epsilon f_0(x, u). \tag{8.9}$$

Remark 8.1 *The conditions placed on $\tilde{Q}(x)$ are identical with those found in Sections* 7.6 *and* 7.7 *and these conditions will be utilized in a similar way to insure the appropriate upper semicontinuity of the functional $(x)(\cdot), u(\cdot)) \to J_T(x_0(\cdot), u(\cdot))$ with respect to the weak topology on $AC([0,T]; \mathbb{R}^n)$ when it is regarded as a functional defined on the set of admissible trajectories. Further, the growth condition found in Assumption* 8.2 *provides for the equiabsolute integrability of the derivatives of the admissible trajectories on $[0,T]$ along a minimizing sequence, and consequently gives us the requisite compactness conditions we require. This growth condition is similar to the growth condition (γ) found in Chapter* 7. *Moreover, in the ordinary differential equation case (i.e., $h(x) \equiv 0$) this growth condition is weaker than the growth condition used in Section* 4.9, *but is equivalent to the classical growth condition of Nagumo and Tonelli, used in* [119] *(see also Chapter* 5 *Section* 5.8*). For further information regarding these growth conditons the reader is referred to Cesari* [47, Section 10.4].

To conclude our description of the model, we introduce the optimal steady-state problem. As is typical of our discussions in Chapter 4 we will establish that the overtaking optimal solutions we obtain will converge to the optimal steady state. We now describe this steady-state problem

Assumption 8.3 *We assume that the optimal steady-state problem (henceforth OSSP) described as*

maximize

$$f_0(x, u)$$

subject to

$$\begin{aligned} 0 &= f(x,u) + \Big(\int_0^\infty g(s)\,ds\Big)h(x), \qquad (8.10) \\ x &\in X \\ u &\in U(x) \end{aligned}$$

has a solution $(\bar{x}, \bar{u}) \in \mathbb{R}^{n+m}$, with $\bar{x}$ uniquely determined.

We further assume that there exists $\bar{p} \in \mathbb{R}^n$ such that the upper semicontinuous function

$$\mathcal{L}_0(x,u) = f_0(x,u) - f_0(\bar{x},\bar{u}) + \langle \bar{p}, f(x,u) + \Big(\int_0^\infty g(s)ds\Big)h(x)\rangle \qquad (8.11)$$

is nonpositive, where $\langle \cdot, \cdot \rangle$ denotes the usual inner product in $\mathbb{R}^n$.

Remark 8.2 *The OSSP described above plays an analogous role for the nondelay infinite horizon optimal control problem described by* *maximize*

$$\int_0^\infty f_0(x(t), u(t))\, dt$$

subject to

$$\dot{x}(t) = f(x(t), u(t)) + \Big(\int_0^\infty g(s)\, ds\Big) h(x(t)) \quad a.e. \quad t \geq 0,$$

$$x(0) = x^o(0),$$

$$x(t) \in X \quad for \quad t \in (0, \infty),$$

and

$$u(t) \in U(x(t)) \quad a.e. \quad t \geq 0.$$

In fact, under essentially the same hypotheses as those utilized here, it can easily be seen that the results of Chapter 4 and Chapter 5 can be used to establish the existence of an overtaking optimal solution, say $(\hat{x}(\cdot), \hat{u}(\cdot))$, for this nondelay optimal control problem. Moreover, we also obtain that

$$\lim_{t \to \infty} \hat{x}(t) = \bar{x}.$$

As we shall see in Section 8.5 this property also holds for the delay case considered here. In this way we see that for large $t > 0$,the optimal trajectories for both systems are close together.

Concerning the function $\mathcal{L}_0(\cdot, \cdot)$ we have the following elementary result.

Proposition 8.1 *Under the hypotheses placed on X, $U(\cdot)$, $f_0(\cdot, \cdot)$, $f(\cdot, \cdot)$, $h(\cdot)$, $g(\cdot)$, if the sets $\tilde{Q}(x)$, given by (8.7), satisfy the conditions outlined in Assumption 8.1 we have that the sets $\tilde{Q}_{\mathcal{L}}(x)$ defined for $x \in X$ by*

$$\tilde{Q}_{\mathcal{L}}(x) = \{(z_0, z) : \; z_0 \leq \mathcal{L}_0(x, u), \; z = f(x, u), \; u \in U(x)\}$$

enjoy the same properties. Furthermore, if $f(\cdot, \cdot)$, $h(\cdot)$, and $f_0(\cdot, \cdot)$ satisfy the growth condition, Assumption 8.2, the same holds for $f(\cdot, \cdot)$, $h(\cdot)$, and $\mathcal{L}_0(\cdot, \cdot)$. That is, for each $\epsilon > 0$ there exists $c_\epsilon > 0$ so that

$$\|f(x, u)\| + \|h(x)\| \leq c_\epsilon - \epsilon\, \mathcal{L}_0(x, u)$$

for all $(x, u) \in M$.

Proof: The convexity and upper semicontinuity properties of the sets $\tilde{Q}_{\mathcal{L}}(x)$ are an easy consequence of the fact that for each $x \in X$ the affine mapping $\Gamma_x : \tilde{Q}(x) \to \tilde{Q}_{\mathcal{L}}(x)$ defined by

$$\Gamma_x(z_0, z) = \left(z_0 - f_0(x,u) + \left\langle \bar{p}, z + \left(\int_0^\infty g(s)\,ds\right)h(x)\right\rangle, z\right)$$

is one-to-one and onto. We leave the details of this argument to the reader.

To establish the growth condition we let $\epsilon > 0$ be given, let

$$K = \max\left[\|\bar{p}\|, \left\|\left(\int_0^\infty g(s)\,ds\right)\bar{p}\right\|\right],$$

and observe that for each $(x,u) \in M$ we have

$$\begin{aligned} f_0(x,u) &= \mathcal{L}_0(x,u) + f_0(\bar{x},\bar{u}) - \left\langle \bar{p}, z + \left(\int_0^\infty g(s)\,ds\right)h(x)\right\rangle \\ &\geq \mathcal{L}_0(x,u) + f_0(\bar{x},\bar{u}) - K\left(\|f(x,u)\| + \|h(x)\|\right). \end{aligned}$$

Let $\eta > 0$ be chosen so that $\eta < \min\left\{\frac{\epsilon}{1+\epsilon K}, \frac{1}{K}\right\}$ if $K \neq 0$ and $\eta = \epsilon$ if $K = 0$. From Assumption 8.2 there exists $c_\eta > 0$ such that (8.11) holds with $\epsilon = \eta$. This implies that for all $(x,u) \in M$,

$$\begin{aligned} &\|f(x,u)\| + \|h(x)\| \\ &\leq c_\eta - \eta f_0(x,u) \\ &\leq c_\eta - \eta\left[\mathcal{L}_0(x,u) + f_0(\bar{x},\bar{u}) - \left\langle \bar{p}, z + \left(\int_0^\infty g(s)\,ds\right)h(x)\right\rangle\right] \\ &\leq c_\eta - \eta\left[\mathcal{L}_0(x,u) + f_0(\bar{x},\bar{u}) - K\left[\|f(x,u)\| + \|h(x)\|\right]\right] \end{aligned}$$

or equivalently,

$$(1-\eta K)\left(\|f(x,u)\| + \|h(x)\|\right) \leq \left(c_\eta - \eta f_o(\bar{x},\bar{u})\right) - \eta\mathcal{L}_o(x,u).$$

From our choice of η it follows that $\dfrac{\eta}{(1-\eta K)} < \epsilon$ and that $1 - \eta K > 0$. Therefore we obtain

$$\begin{aligned} |f(x,u)| + |h(x)| &\leq \frac{1}{1-\eta K}\left(c_\eta - \eta f_0(\bar{x},\bar{u})\right) - \frac{\eta}{1-\eta K}\mathcal{L}_0(x,u) \\ &\leq \frac{1}{1-\eta K}\left(c_\eta - \eta f_0(\bar{x},\bar{u})\right) - \epsilon\mathcal{L}_o(x,u). \end{aligned}$$

The desired conclusion follows by choosing $c_\epsilon > \max\left\{0, \frac{1}{1-\eta K}\left(c_\eta - \eta f_0(\bar{x},\bar{u})\right)\right\}$. ∎

8.3 Linear hereditary operators and an upper closure theorem

To establish our results we will need certain properties of the linear integral operator G defined by

$$(Gy(\cdot))(t) = \int_{-\infty}^{t} g(t-s)y(s)\,ds = \int_{0}^{\infty} g(s)y(t-s)\,ds, \quad -\infty < t < \infty, \tag{8.12}$$

in which $g(\cdot)$ is an $n \times p$ matrix function satisfying condition (8.5(iii)). Clearly the operator G is not well defined for all choices of functions $y(\cdot)$. However, operators of this type have been thoroughly examined in Marcus and Mizel [136], who chose as a domain for this operator the space $N_1((-\infty,\infty);\mathbb{R}^p)$ consisting of all locally integrable functions $y(\cdot) \in L^1_{loc}((-\infty,\infty);\mathbb{R}^p)$ with the following property

$$\|y(\cdot)\|^* = \sup_{-\infty<s<\infty} \left\{ \int_{s-1}^{s} \|y(t)\|\,dt \right\} < \infty. \tag{8.13}$$

It is easy to see that $\|\cdot\|^*$ defines a norm on the space $N_1((-\infty,\infty);\mathbb{R}^p)$. With this notation we summarize the results from [136] that we require in the following theorem.

Theorem 8.1 *Let $g(\cdot)$ be an $n \times p$ matrix function on $[0,\infty)$ satisfying (8.5(iii)) and define the operator G on $N_1((-\infty,\infty);\mathbb{R}^p)$ by (8.12). The following properties hold.*

(A) *For every $y(\cdot) \in N_1((-\infty,\infty);\mathbb{R}^p)$ the integral $(Gy(\cdot))(t)$, defined by (8.12) for each $t \in (-\infty,\infty)$, exists as a Lebesgue integral.*

(B) *For every $y(\cdot) \in N_1((-\infty,\infty);\mathbb{R}^p)$, the function $(Gy(\cdot))(\cdot) : (-\infty,\infty) \to \mathbb{R}^n$ is continuous.*

(C) *If $\{y_k(\cdot)\}_{k=1}^{\infty} \subset N_1((-\infty,\infty);\mathbb{R}^p)$ is a norm bounded sequence (i.e., there exists a real number $C > 0$ so that $\|y_k(\cdot)\|^* \le C$ for all k) such that for some $y(\cdot) \in N_1((-\infty,\infty);\mathbb{R}^p)$ the sequence $\{y_k(\cdot)|_{(\alpha,\beta)}\}_{k=1}^{\infty}$ converges strongly in $\mathcal{L}^1(\alpha,\beta)$ to $y(\cdot)$ for all reall numbers α, β, $\alpha < \beta$; then the sequence $\left\{(Gy_k(\cdot))(\cdot)\right\}_{k=1}^{\infty}$ converges uniformly on compact subsets of $(-\infty,\infty)$ to $(Gy(\cdot))(\cdot)$.*

(D) *If D is a norm bounded subset of $N_1((-\infty,\infty);\mathbb{R}^p)$ such that the elements of D are is uniformly integrable on $(-T,T)$ for every $T > 0$, then the set $G(D) = \{(Gy(\cdot))(\cdot) : y(\cdot) \in D\}$ is precompact in $C[(-T,T);\mathbb{R}^n)]$, the space of continuous functions on $[-T,T]$, for all $T > 0$.*

Proof: The proof of the above results are found in Lemmas 7.5 and 7.10 of [136]. ∎

Remark 8.3 *In* [136] *the kernel* $g(\cdot)$ *is assumed to satisfy stronger hypotheses than those indicated above (see* [136, p.21]*). However, a careful examination of the proofs of the results given above shows that this additional hypothesis is not required. On the other hand, it is needed to prove all of the results given in* [136, Lemma 7.5]. *We also note that the proof of part (D) above, given in* [136, Lemma 7.10] *is only concerned with* $D \subset N_1((-\infty, 0); \mathbb{R}^p)$ *but a straightforward modification of their argument gives the result for the case considered here.*

In addition to the above theorem we will need the following upper closure theorem. For the applications of this result, the sets $\mathcal{G}_k$ below are subsets of the time axis $(-\infty, \infty)$.

Theorem 8.2 *Let* $(\mathcal{G}, \mu)$ *be a* σ*-finite measure space,* $\mathcal{G} = \bigcup_{k=1}^{\infty} \mathcal{G}_k$ *where* $\mathcal{G}_k \subset \mathcal{G}_{k+1}$ *and* $\mu[\mathcal{G}_k] < \infty$*, let* $X \subset \mathbb{R}^n$ *be closed, and let* $\mathcal{R} : X \to 2^{\mathbb{R}^{1+r}}$ *be a given set valued map that is closed and convex valued and satisfies the upper semicontinuity property* $(\mathcal{K})$ *given by* (8.10). *Furthermore, assume that there exists measurable functions* $\eta_k : \mathcal{G}_k \to \mathbb{R}$, $\lambda_k : \mathcal{G}_k \to \mathbb{R}$, $\lambda : \mathcal{G} \to \mathbb{R}$, $\xi_k : \mathcal{G}_k \to \mathbb{R}$, $\xi : \mathcal{G} \to \mathbb{R}^n$, $x_k : \mathcal{G}_k \to \mathbb{R}^n$, *and* $x : \mathcal{G} \to \mathbb{R}^n$, $k = 1, 2, \ldots$ *satisfying the following conditions:*

(i) *For each index* k*, and almost all* $t \in \mathcal{G}_k$,

$$x_k(t) \in \mathcal{G}_k \quad \text{and} \quad (\eta_k(t), \xi_k(t)) \in \mathcal{R}(x_k(t)).$$

(ii) $x_k(t) \to x(t)$ *pointwise,* $\xi_k(\cdot) \to \xi(\cdot)$ *weakly in* $L^1_{loc}(\mathcal{G}; \mathbb{R}^n)$*, and* $\lambda_k(\cdot) \to \lambda(\cdot)$ *weakly in* $L^1_{loc}(\mathcal{G}; \mathbb{R}^n)$ *as* $k \to \infty$.

(iii) $\lambda(\cdot) \in L^1(\mathcal{G}; \mathbb{R})$, $\eta_k(t) \leq \lambda_k(t)$ *almost everywhere in* $\mathcal{G}_k$*, and*

$$-\infty < \limsup_{k \to \infty} \int_{\mathcal{G}_k} \eta_k(t)\, dt < \infty.$$

Then there exists $\eta(\cdot) \in L^1(\mathcal{G}; \mathbb{R})$ *such that*

$$x(t) \in X \quad \text{and} \quad (\eta(t), \xi(t)) \in \mathcal{R}(x(t)) \quad \text{a.e.} \quad t \in \mathcal{G}$$

and

$$\limsup_{k \to \infty} \int_{\mathcal{G}_k} \eta_k(t)\, dt \leq \int_{\mathcal{G}} \eta(t)\, dt < \infty.$$

Proof: See Carlson [40, Theorem 3.3]. ∎

Remark 8.4 *The above theorem is a more general version of the upper closure theorem, Theorem 7.16, presented earlier.*

8.4 Existence of overtaking optimal solutions

8.4.1 Support property assumptions

With the notation and hypotheses given above we concern ourselves with the existence of overtaking optimal solutions for the integrodifferential equation model considered in this chapter. We begin by investigating the "asymptotic turnpike property" of certain admissible trajectories. Following the ideas presented in Section 4.9 we let $\mathcal{F}$ denote the set of all trajectories $x(\cdot) : \mathbb{R} \to \mathbb{R}^n$ satisfying

$$\mathcal{L}_0(x(t), u(t)) = 0 \quad a.e.\ t \geq 0 \tag{8.14}$$

where $u(\cdot) : [0, \infty) \to \mathbb{R}^m$ is a measurable control function corresponding to the trajectory $x(\cdot)$ (see Definition 8.1). We observe that the optimal steady state $x(t) = \bar{x}$ is such a trajectory so that $\mathcal{F} \neq \emptyset$. As regards to $\mathcal{F}$ we make the following assumption.

Assumption 8.4 *For each $\epsilon > 0$ there exists $t_\epsilon > 0$ such that for all $t \geq t_\epsilon$ and all $x(\cdot) \in \mathcal{F}$, $|x(t) - \bar{x}| < \epsilon$.*

Concerning Assumption 8.4 we observe that it corresponds to property $\mathcal{S}$ found in Section 4.9 as well as to property $\mathcal{C}$ found in Section 9.6. As stated, this assumption is difficult to verify and will not hold in general. Unlike the ordinary differential equation case, in which it is shown that under suitable convexity hypotheses this condition is generic (see Section 4.9), the infinite-dimensional nature of the integrodifferential equation model precludes such a possibility here. Consequently, we content ourselves with explicit conditions which imply this assumption. To this end we introduce the following alternative assumption.

(A4′). For every $\epsilon > 0$ there exists $\delta = \delta(\epsilon) > 0$ so that if $x \in X$ satisfies $\|x - \bar{x}\| > \epsilon$, then $\mathcal{L}_0(x, u) \leq -\delta$ for all $u \in U(x)$.

Lemma 8.1 *If* (A4′) *holds, then Assumption* 8.4 *holds.*

Proof: We assume that (A4′) holds and let $x(\cdot) : \mathbb{R} \to \mathbb{R}^n$ be a trajectory (see Definition 8.1) so that (8.14) holds. We now show that $x(t) \equiv \bar{x}$ on $[0, \infty)$. Indeed, if there exists $\tau \geq 0$ for which $x(\tau) \neq \bar{x}$, the continuity of $x(\cdot)$ allows us to assert the existence of an $\epsilon_0 > 0$ and $\alpha > 0$ so that for all $t \in [\tau, \tau + \alpha)$ we have $\|x(t) - \bar{x}\| > \epsilon_0$. However, by (A4′) there exists $\delta_o > 0$ so that $\mathcal{L}_0(x(t), u) \leq -\delta_o$ for all $u \in U(x(t))$ and almost all $t \in [\tau, \tau + \alpha)$. Clearly this is a contradiction. Therefore $x(t) \equiv \bar{x}$ on $[0, \infty)$ and thus, since $x(\cdot) \in \mathcal{F}$ was arbitrary, Assumption 8.4 holds since for every $\epsilon > 0$ we can choose $t_\epsilon = 0$. ∎

The Assumption (A4′) is familiar as it appears in Chapter 4, Section 4.3 in which it is used to establish the turnpike property for the ordinary differential equation case. Its role will be the same here. The implications of (A4′) differs from the ordinary differential equation case. In particular, (A4′) implies $\mathcal{F}$ equals the singleton $\{\bar{x}\}$ in the nondelay case. In the delay case considered here it is possible (under (A4′)) that a trajectory $x(\cdot) \in \mathcal{F}$ exists which is not identically equal to the steady state $\bar{x}$ on $(-\infty, \infty)$. This is demonstrated in the following simple example. We remark however that we still essentially have $\mathcal{F}$ as the singleton $\{\bar{x}\}$, since for any $x(\cdot) \in \mathcal{F}$ with $x(t) \neq \bar{x}$ for $t < 0$ we must have (under (A4′)) that $x(t) = \bar{x}$ on $[0, \infty)$.

Example 8.1 We consider the control system

$$\begin{aligned} \dot{x}(t) &= u(t) + \int_{-\infty}^{t} 2e^{-(t-s)}x(s)ds \quad \text{a.e. on } [0,\infty), \\ x(s) &= \varphi(s) \quad \text{for all } s \leq 0, \\ x(t) &\in [-1,1] \quad \text{on } [0,\infty), \\ u(t) &\in [-1,1] \quad \text{a.e. on } [0,\infty). \end{aligned}$$

Here we assume that $\varphi : (-\infty, 0] \to [-1, 1]$ is a fixed given initial function. For the objective functional we take

$$J_T(x,u) = \int_0^T \left[9u(t) - 2x(t)\left(4x(t)^2 + 6x(t) - 9\right)\right] dt$$

for $T \geq 0$. The corresponding optimal steady-state problem becomes

$$\max\left\{9u - 2x(4x^2 + 6x - 9) : \ 0 = u + 2x, \ x, u \in [-1,1]\right\}$$

It is straightforward to show that Assumption 8.3 holds with $\bar{x} = 0$, $\bar{u} = 0$, and $\bar{p} = -9$. Thus,

$$\mathcal{L}_0(x,u) = 9u - 2x(4x^2 + 6x - 9) - 9(2x + u) = -4x^2(2x + 3)$$

and since $(x, u) \in [-1,1] \times [-1,1]$ we have

$$\mathcal{L}_0(x,u) \leq -4x^2,$$

so that if $|x - \bar{x}| = |x| > \epsilon$ we have, upon choosing $\delta = 4\epsilon^2$,

$$\mathcal{L}_0(x,u) \leq -\delta.$$

Consequently, Assumption (A4′) holds. Clearly $x(t) \equiv \bar{x} = 0$ is in $\mathcal{F}$. We now show that $\hat{x} : (-\infty, \infty) \to [-1, 1]$, defined by

$$\hat{x}(t) = \begin{cases} \frac{1}{2}(e^t - 1) & \text{for } t < 0 \\ 0 & \text{for } t \geq 0, \end{cases}$$

is also in $\mathcal{F}$. Indeed, we observe that by taking $\hat{u}(\cdot) : [0,\infty) \to [-1,1]$ to be

$$\hat{u}(t) = \frac{1}{2}e^{-t},$$

we have for all $t \geq 0$ that

$$\begin{aligned}\hat{u}(t) + \int_{-\infty}^{t} 2e^{-(t-s)}\hat{x}(s)\,ds &= \frac{1}{2}e^{-t} + \int_{-\infty}^{0} 2e^{-(t-s)}\left(\frac{1}{2}(e^{s}-1)\right) ds \\ &= \frac{1}{2}e^{-t} - \frac{1}{2}e^{-t} \\ &= 0.\end{aligned}$$

This implies that $\hat{x}(\cdot)$ is indeed a trajectory with $\mathcal{L}_0(\hat{x}(t), \hat{u}(t)) = 0$ for all $t \geq 0$ as desired.

Condition (A4′) is finite-dimensional in nature and consequently is easier to verify than Assumption 8.4. To present general conditions that are sufficient for (A4′) to hold requires stronger convexity hypotheses than those presently assumed. The following proposition provides us with sufficient conditions under which (A4′) holds.

Proposition 8.2 *Suppose that $M = \{(x,u) : \; x \in X, \; u \in U(x)\}$ is compact, convex with nonempty interior; that $f_0 : M \to \mathbb{R}$ is strictly concave, upper semicontinuous and bounded above; and that $F : M \to \mathbb{R}^n$ defined by*

$$F(x,u) = f(x,u) + \Big(\int_0^{\infty} g(s)\,ds\Big)h(x),$$

is continuous and concave (i.e., each component F_i of $F = (F_1, F_2, \ldots, F_n)$ is a concave function). Furthermore, suppose the following hold:

(i) There exists $(x,u) \in M$ so that $F_i(x,u) > 0$ for $i = 1,2,\ldots,n$

(ii) If $(x,u) \in M$ is such that $F_i(x,u) \geq 0$, $i = 1,2,\ldots,n$, then there exists $v \in U(x)$ so that $F(x,v) = 0$ and $f_0(x,v) \geq f_0(x,u)$.

Then both Assumption 8.3 and (A4′) hold.

Proof: This result is identical to that of Lemma 4.4 of Chapter 4 (with f replaced by F). Consequently we direct the reader there for its proof. ∎

Remark 8.5 *We remark that the above proposition only provides sufficient conditions for Assumptions* 8.3 *and* 8.4 *to hold. These conditions are not necessary. Indeed, the strict concavity of f_0 is not satisfied in Example* 8.1. *To see this consider the points* $(-1,0)$, $(-\frac{1}{2},0) \in M = [-1,1] \times [-1,1]$ *and observe that for* $\lambda = \frac{1}{2}$ *we have*

$$\begin{aligned} f_0\left(\lambda(-1) + (1-\lambda)\left(\frac{1}{2}\right), 0\right) &= f_0(-\frac{3}{4}, 0) = -\frac{135}{8} < -\frac{33}{2} \\ &= \lambda f_0(-1,0) + (1-\lambda) f_0(-\frac{1}{2}, 0), \end{aligned}$$

which violates the definition of concavity.

8.4.2 A Turnpike theorem

With Assumption 8.4 in hand we can now establish the following turnpike result.

Proposition 8.3 *Let $X \in \mathbb{R}^n$ be closed, let $U : X \to 2^{\mathbb{R}^m}$ be a set valued map with closed graph M, let $f_0(\cdot,\cdot) : M \to \mathbb{R}$ be upper semicontinuous, let $f(\cdot,\cdot) : M \to \mathbb{R}^n$ and $h(\cdot) : X \to \mathbb{R}^p$ both be continuous and let g be an $n \times p$ matrix function satisfying* (8.5). *Furthermore, assume that Assumptions* 8.1–8.4 *hold. If $(x(\cdot), u(\cdot)) \in A_0$ is such that*

$$\int_0^\infty \mathcal{L}_0(x(t), u(t))\, dt > -\infty, \tag{8.15}$$

then $\lim_{t\to\infty} x(t) = \bar{x}$.

Proof: We proceed by contradiction and suppose that $(x(\cdot), u(\cdot)) \in A_0$ is as above, but $\lim_{t\to\infty} x(t) \neq \bar{x}$. This implies there exists $\epsilon_0 > 0$ and times t_k, $k = 1, 2, \ldots$, increasing to positive infinity such that $\|x(t_k) - \bar{x}\| > \epsilon_0$ for all k. As a consequence of Assumption 8.4 there exists $t_0 > 0$ such that for all $t \geq t_0$ we have $\|s(t) - \bar{x}\| < \frac{\epsilon_0}{2}$ for all trajectories $s(\cdot) \in \mathcal{F}$. Define the sequences of functions $x_k(\cdot) : (-\infty, \infty) \to \mathbb{R}^n$ and $u_k(\cdot) : (-\infty, \infty) \to \mathbb{R}^m$ by the formulas

$$x_k(t) = x(t + t_k - t_0), \qquad u_k(t) = u(t + t_k - t_0).$$

Clearly, we have for all $k = 1, 2, \ldots,$ and almost all $t \geq t_0 - t_k$ that

$$\dot{x}_k(t) = f(x_k(t), u_k(t)) + \int_{-\infty}^{t} g(t-s) h(x_k(s))\, ds,$$

$$u_k(t) \in U(x_k(t)) \qquad \textit{and} \qquad x_k(t) \in X.$$

We further notice that for each $k = 1, 2, \ldots,$ x_k satisfies the "initial condition"

$$x_k(s) = x_0(s + t_k - t_0) \quad \textit{for all } x \leq t_0 - t_k.$$

As we will see, since $(t_0 - t_k, \infty)$ tends to $(-\infty, \infty)$ as $k \to \infty$, this fact is not required for our proof. Moreover, since by definition $t \to x(t)$ is bounded, the sequence $\{x_k(\cdot)\}_{k=1}^{\infty}$ is uniformly bounded and satisfies $x_k(t_0) = x(t_k)$. From the above it is evident that $x_k(\cdot)$ is a trajectory corresponding to the control $u_k(\cdot)$ (see Definition 8.1). In addition we observe that for any $T > 0$, and all k sufficiently large

$$\lim_{k\to\infty} \int_{-T}^{T} \mathcal{L}_0(x_k(t), u_k(t))\, dt = \lim_{k\to\infty} \int_{t_k - t_0 - T}^{t_k - t_0 + T} \mathcal{L}_0(x(t), u(t))\, dt$$

and thus by the growth condition Assumption 8.2 applied to f, h, and $\mathcal{L}_0$ (see Proposition 8.1) we have that the sequences of functions $z_k(t) = f(x_k(t), u_k(t))$ and $y_k(t) = h(x_k(t))$, $k = 1, 2, \ldots$, are equiabsolutely integrable on $[-T, T]$ for all $T > 0$. Thus, by a standard diagonalization process there exists locally integrable functions $z(\cdot) : (-\infty, \infty) \to \mathbb{R}^n$ and $y(\cdot) : (-\infty, \infty) \to \mathbb{R}^p$ and subsequences, say still $\{z_k(\cdot)\}_{k=1}^{\infty}$ and $\{y_k(\cdot)\}_{k=1}^{\infty}$, that converge weakly to $z(\cdot)$ and $y(\cdot)$ respectively in $L^1_{loc}((-\infty, \infty); \mathbb{R}^n)$ and $L^1_{loc}((-\infty, \infty); \mathbb{R}^p)$. In addition, as the sequence $\{x_k(\cdot)\}_{k=1}^{\infty}$ is bounded and $h : X \to \mathbb{R}^p$ is continuous, there exists $K > 0$ so that for all $t \in (-\infty, \infty)$ and all $k = 1, 2, \ldots$

$$\|y_k(t)\| = \|h(x_k(t))\| \leq K.$$

Therefore as a consequence of Theorem 8.1 (D), we can further assume (by extracting another subsequence by diagonalization) that there exists a continuous function $r(\cdot) : (-\infty, \infty) \to \mathbb{R}^n$ such that the sequence $\{(Gy_k(\cdot))(\cdot)\}_{k=1}^{\infty}$, defined for $k = 1, 2, \ldots$ by

$$(Gy_k(\cdot))(t) = \int_{-\infty}^{t} g(t - s) y_k(s)\, ds, \quad t \in (-\infty, \infty),$$

converges uniformly on compact subsets of $(-\infty, \infty)$ to $r(\cdot)$. That is,

$$r(t) = \lim_{k\to\infty} \int_{-\infty}^{t} g(t - s) y_k(s)\, ds. \tag{8.16}$$

Combining the above sequences we observe that for all $k = 1, 2, \ldots$

$$\dot{x}_k(t) = z_k(t) + (Gy_k(\cdot))(t) \quad \text{a.e. } t \geq t_0 - t_k;$$

and we conclude that the sequence $\{\dot{x}_k(\cdot)\}_{k=1}^{\infty}$ converges weakly in $L^1_{loc}((-\infty, \infty); \mathbb{R}^n)$ to the locally integrable function $t \mapsto z(t) + r(t)$. Also, since $\{x_k(t_0)\}_{k=1}^{\infty}$ is bounded, we can assume that our subsequence has been chosen so that $\lim_{t\to\infty} x_k(t_0) = \hat{x}$, where $\hat{x} \in X$. Define the function $\tilde{x}(\cdot) : (-\infty, \infty) \to \mathbb{R}^n$ by the formula

$$\tilde{x}(t) = \hat{x} + \int_{t_0}^{t} [z(s) + r(s)]\, ds$$

and observe, since for $t > t_0 - t_k$

$$x_k(t) = x_k(t_0) + \int_{t_0}^{t} [z_k(s) + Gy_k(s)]\, ds,$$

that $x_k(t) \to \tilde{x}(t)$ pointwise in $(-\infty, \infty)$. Furthermore, we observe that as X is closed we have $\tilde{x}(t) \in X$ for all $t \in (-\infty, \infty)$ and that $\tilde{x}(\cdot)$ is locally absolutely continuous. From these facts, and since $s \mapsto g(t-s)$ is Lebesgue integrable on $(-\infty, t)$ we have by the Dominated Convergence Theorem,

$$\lim_{k\to\infty}\Big(\int_{-\infty}^{t} g(t-s)h(x_k(s))\, ds\Big) = \int_{-\infty}^{t} g(t-s)h(\tilde{x}(s))\, ds.$$

This implies, for all $t \in (-\infty, \infty)$, as a consequence of (8.18), that

$$r(t) = \int_{-\infty}^{t} g(t-s)h(\tilde{x}(s))\, ds.$$

It is now easy to see that for all $t \in (-\infty, \infty)$,

$$\begin{aligned} \tilde{x}(t) &= \hat{x} + \int_{t_0}^{t}\Big(z(s) + \int_{-\infty}^{s} g(s-\tau)h(\tilde{x}(\tau))\, d\tau\Big)\, ds \\ &= \tilde{x}(0) + \int_{0}^{t}\Big(z(s) + \int_{-\infty}^{s} g(s-\tau)h(\tilde{x}(\tau))\, d\tau\Big)\, ds. \end{aligned} \tag{8.17}$$

We now wish to show that $\tilde{x}(\cdot)$ is a trajectory. To this end we appeal to the Upper Closure Theorem 8.2 using the following notation. For each integer $N = \pm1, \pm2, \ldots$, we let

(i) $\mathcal{G} = [N, N+1)$, $\mathcal{G}_k = [N, N+1)$, $k = 1, 2, \ldots$;

(ii) $\eta_k(t) = \mathcal{L}_0(x_k(t), u_k(t))$, $\xi_k(t) = z_k(t)$, $\xi(t) = z(t)$, $\lambda_k(t) \equiv 0$, $\lambda(t) \equiv 0$, $x_k(t)$ as above and $x(t) = \tilde{x}(t)$.

(iii) $\mathcal{R}(x) = \tilde{Q}\mathcal{L}_0(x)$

Observe that we also have

$$\begin{aligned} \lim_{k\to\infty} \int_{\mathcal{G}_k} \eta_k(t)\, dt &= \lim_{k\to\infty} \int_{N}^{N+1} \mathcal{L}_0(x_k(t), u_k(t))dt \\ &= \lim_{k\to\infty} \int_{N+t_k-t_0}^{N+1+t_k-t_0} \mathcal{L}_0(x(s), u(s))\, ds = 0, \end{aligned}$$

since $\mathcal{L}_0(x,u)$ is nonpositive and (8.15) holds. Thus, all the hypotheses of Theorem 8.2 are satisfied and we can conclude that there exists an integrable function $\eta_N : [N, N+1) \to \mathbb{R}$ such that for almost all $t \in [N, N+1)$

$$\hat{x}(t) \in X \quad \text{and} \quad (\eta_N(t), z(t)) \in \tilde{Q}\mathcal{L}_0(\hat{x}(t))$$

and

$$\int_N^{N+1} \eta_N(t)\, dt \geq 0.$$

By standard measurable selection theorem arguments (see Cesari[47, Theorem 11.4i], there exists a measurable function $u_N : [N, N+1) \to \mathbb{R}^m$ such that for almost all $t \in [N, N+1)$ we have $\eta_N(t) \leq \mathcal{L}_0(\tilde{x}(t), u_N(t))$, $z(t) = f(\tilde{x}(t), u_N(t))$, and $u_N(t) \in U(\tilde{x}(t))$. The desired control generating the trajectory $\tilde{x}(\cdot) : (-\infty, \infty) \to \mathbb{R}^n$ is now obtained by defining $\tilde{u}(\cdot) : (-\infty, \infty) \to \mathbb{R}^m$ by

$$\tilde{u}(t) = u_N(t) \quad \textit{for} \quad N \leq t \leq N+1,$$

$N = \pm 1, \pm 2, \ldots$. Clearly, $\tilde{u}(\cdot)$ is measurable and upon substituting into (8.17) we have for any $t \in \mathbb{R}$,

$$\tilde{x}(t) = \tilde{x}(0) + \int_0^t f(\tilde{x}(s), \tilde{u}(s))\, ds + \int_0^t \int_{-\infty}^s g(t-\tau)h(\tilde{x}(\tau))\, d\tau,$$

implying that $\tilde{x}(\cdot)$ is a trajectory. Moreover, for any integer N, we have

$$0 \geq \int_N^{N+1} \mathcal{L}_0(\tilde{x}(t), \tilde{u}(t))\, dt \geq \int_N^{N+1} \eta_N(t)\, dt \geq 0,$$

from which it follows that $\mathcal{L}_0(\tilde{x}(t), \tilde{u}(t)) = 0$ for almost all $t \in (-\infty, \infty)$ since $\mathcal{L}_0$ is nonpositive. Thus we see that $\tilde{x}(\cdot) \in \mathcal{F}$. This, however, leads to a contradiction since for all k,

$$\epsilon_0 \leq \|x_k(t_0) - \bar{x}\| \leq \|x_k(t_0) - \tilde{x}(t_0)\| + \|\tilde{x}(t_0) - \bar{x}\| < \|x_k(t_0) - \tilde{x}(t_0)\| + \frac{\epsilon_0}{2},$$

giving us a contradiction

$$0 = \lim_{k \to \infty} \|x_k(t_0) - \tilde{x}(t_0)\| \geq \frac{\epsilon_0}{2}.$$

Therefore we must have $x(t) \to \bar{x}$ as $t \to \infty$. ∎

The turnpike result given above is analogous to Lemma 4.6. As we shall see that its role in the existence theory for the present class of systems is the same as before.

The following technical lemma will be needed in our existence proof.

Lemma 8.2 *Let $X \subset \mathbb{R}^n$ be closed, let $h(\cdot) : X \to \mathbb{R}^p$ be continuous, and let $g(\cdot)$ be an $n \times p$ matrix function satisfying* (8.5). *Suppose that $(x_i(\cdot), u_i(\cdot)) \in A_0$ for $i = 1, 2$ are such that $\lim_{T\to\infty} x_i(T) = \bar{x}$, $i = 1, 2$. Then we have*

$$\lim_{T\to\infty} \int_0^T \Big(\int_{T-t}^{\infty} g(s)\, ds \Big) (h(x_1(t)) - h(x_2(t)))\, dt = 0$$

Proof: Let $\epsilon > 0$ be given. From the continuity of h and the convergence of $x_i(T)$ to $\bar{x}$ as $T \to \infty$, there exists $\tau > 0$ so that for all $T \geq \tau$,

$$\|h(x_1(T)) - h(x_2(T))\| < \epsilon.$$

For $T > \tau$ we write

$$\begin{aligned}
&\Big\| \int_0^T \Big(\int_{T-t}^{\infty} g(s)\, ds \Big) [h(x_1(t)) - h(x_2(t))]\, dt \Big\| \\
&\leq \Big\| \int_0^{\tau} \Big(\int_{T-t}^{\infty} g(s)\, ds \Big) [h(x_1(t)) - h(x_2(t))]\, dt \Big\| \\
&\quad + \Big\| \int_{\tau}^{T} \Big(\int_{T-t}^{\infty} g(s)\, ds \Big) [h(x_1(t)) - h(x_2(t))]\, dt \Big\| \\
&= I_1(T) + I_2(T),
\end{aligned}$$

and estimate $I_1(T)$ and $I_2(T)$ separately. As $x_i(\cdot)$, $i = 1, 2$, are admissible trajectories they are bounded. Thus, the continuity of h implies, there exists $H > 0$ so that for all $t \in [0, \infty)$, $\|h(x_1(t)) - h(x_2(t))\| \leq H$. This gives us

$$\begin{aligned}
I_1(T) &\leq \int_0^{\tau} \Big(\int_{T-t}^{\infty} \|g(s)\|\, ds \Big) \|h(x_1(t)) - h(x_2(t))\|\, dt \\
&\leq H \int_0^{\tau} \int_{T-t}^{\infty} \|g(s)\|\, ds\, dt \\
&= H \int_{T-\tau}^{T} \int_{t}^{\infty} \|g(s)\|\, ds\, dt \\
&= H \Big[\int_0^{T} \int_{t}^{\infty} \|g(s)\|\, ds\, dt - \int_0^{T-\tau} \int_{t}^{\infty} \|g(s)\|\, ds\, dt \Big],
\end{aligned}$$

Which tends to zero as $T \to \infty$ since

$$\begin{aligned}
\lim_{T\to\infty} \int_0^{T} \int_{t}^{\infty} \|g(s)\|\, ds\, dt &= \int_0^{\infty} \int_{t}^{\infty} \|g(s)\|\, ds\, dt \\
&= \int_0^{\infty} s\, \|g(s)\|\, ds < \infty.
\end{aligned}$$

For $I_2(T)$ we observe that

$$\begin{aligned} I_2(T) &\le \int_\tau^T \Big(\int_{T-t}^{\infty} \|g(s)\|\, ds \Big) \|h(x_1(t)) - h(x_2(t))\|\, dt \\ &< \epsilon \int_\tau^T \int_{T-t}^{\infty} \|g(s)\|\, ds\, dt \le \epsilon \int_0^{\infty} \int_t^{\infty} \|g(s)\|\, ds\, dt, \end{aligned}$$

and since $\epsilon > 0$ was arbitrary, it follows that for T sufficiently large we have that $I_2(T)$ is as small as desired. Combining these two results, we obtain

$$\lim_{T\to\infty} \int_0^T \Big(\int_{T-t}^{\infty} g(s)\, ds \Big) \left(h(x_1(t)) - h(x_2(t)) \right) dt = 0$$

■

We now present the following set of sufficient conditions for overtaking optimality.

8.4.3 A sufficient condition for overtaking optimality

As in the developments of Chapter 4 we now show that a solution to the associated problem of Lagrange is overtaking under appropriate assumptions.

Theorem 8.3 *Assume that $X \subset \mathbb{R}^n$ is closed, $U(\cdot) : X \to 2^{\mathbb{R}^m}$ is a set valued mapping with closed graph M, $f_0(\cdot,\cdot) : M \to \mathbb{R}$ is upper semicontinuous, $f(\cdot,\cdot) : M \to \mathbb{R}^n$ and $h(\cdot) : X \to \mathbb{R}^p$ are both continuous, and $g(\cdot)$ is an $n \times p$ matrix function satisfying* (8.5). *In addition, assume that Assumptions* 8.1–8.4 *hold. If $(x^*(\cdot), u^*(\cdot)) \in A_0$ is such that*

(i)

$$\int_0^{\infty} \mathcal{L}_0(x^*(t), u^*(t))\, dt > -\infty, \tag{8.18}$$

(ii)

$$\int_0^{\infty} \mathcal{L}_0(x^*(t), u^*(t))\, dt \ge \lim_{T\to\infty} \int_0^T \mathcal{L}_0(x(t), u(t))\, dt, \tag{8.19}$$

for all $(x(\cdot), u(\cdot)) \in A_0$, where the limit on the right is either finite or negative infinity, then the pair $(x^*(\cdot), u^*(\cdot))$ is overtaking optimal.

Proof: Let $(x^*(\cdot), u^*(\cdot)) \in A_0$ be as above and let $(x(\cdot), u(\cdot)) \in A_0$ be arbitrary. Then for any $T > 0$, we have

$$\begin{aligned}
&J_T(x_0(\cdot), u^*(\cdot)) - J_T(x_0(\cdot), u(\cdot)) \\
&= \int_0^T \mathcal{L}_0(x^*(t), u^*(t))\,dt - \int_0^T \mathcal{L}_0(x(t), u(t))\,dt \\
&\quad + \Big\langle \bar{p}, \int_0^T \Big[f(x(t), u(t)) + \Big(\int_0^\infty g(s)\,ds\Big) h(x(t))\Big]\,dt \Big\rangle \\
&\quad - \Big\langle \bar{p}, \int_0^T \Big[f(x^*(t), u^*(t)) + \Big(\int_0^\infty g(s)\,ds\Big) h(x^*(t))\Big]\,dt \Big\rangle \\
&= \int_0^T \mathcal{L}_0(x^*(t), u^*(t))\,dt - \int_0^T \mathcal{L}_0(x(t), u(t))\,dt \\
&\quad + \Big\langle \bar{p}, \int_0^T \Big[f(x(t), u(t)) + \Big(\int_0^{T-t} g(s)\,ds\Big) h(x(t))\Big]\,dt \Big\rangle \\
&\quad - \Big\langle \bar{p}, \int_0^T \Big[f(x^*(t), u^*(t)) + \Big(\int_0^{T-t} g(s)\,ds\Big) h(x^*(t))\Big]\,dt \Big\rangle \\
&\quad - \Big\langle \bar{p}, \int_0^T \Big(\int_{T-t}^\infty g(s)\,ds\Big) [h(x^*(t)) - h(x(t))]\,dt \Big\rangle.
\end{aligned}$$

Also for $T > 0$ and $(x(\cdot), u(\cdot)) \in A_0$, we observe that

$$\begin{aligned}
\int_0^T \Big(\int_0^{T-t} g(s)\,ds\Big) h(x(t))\,dt &= \int_0^T \Big(\int_t^T g(s-t)\Big) h(x(t))\,dt \\
&= \int_0^T \int_0^t g(t-s) h(x(s))\,ds\,dt,
\end{aligned}$$

so that

$$\begin{aligned}
J_T(x_0(\cdot), u^{*(\cdot)}) - J_T(x_0(\cdot), u(\cdot)) &= \int_0^T \mathcal{L}_0(x^*(t), u^*(t))\,dt - \int_0^T \mathcal{L}_0(x(t), u(t))\,dt \\
&\quad - \langle \bar{p}, \int_0^T (\dot{x}^*(t) - \dot{x}(t))\,dt \rangle \\
&\quad - \langle \bar{p}, \int_0^T (\int_{T-t}^\infty g(s)\,ds)[h(x^*(t)) - h(x(t))]\,dt \rangle \\
&= \int_0^T \mathcal{L}_0(x^*(t), u^*(t))\,dt - \int_0^T \mathcal{L}_0(x(t), u(t))\,dt
\end{aligned}$$

$$+\langle \bar{p}, x(T)-x^*(T)\rangle$$
$$-<\bar{p}, \int_0^T \Big(\int_{T-t}^{\infty} g(s)\,ds\Big)(h(x^*(t))-h(x(t)))\,dt> .$$

We now suppose that (8.18) holds with $(x^*(\cdot), u^*(\cdot))$ replaced by $(x(\cdot), u(\cdot))$. In this case, as a consequence of Proposition 8.2, $\lim_{T\to\infty} x(T)=\bar{x}$ and we obtain

$$\begin{aligned}
&\lim_{T\to\infty} (J_T(x_0(\cdot), u^*(\cdot)) - J_T(x_0(\cdot), u(\cdot))] \\
&= \int_0^{\infty} \mathcal{L}_0(x^*(t), u^*(t))\,dt - \int_0^{\infty} \mathcal{L}_0(x(t), u(t))\,dt \\
&\geq 0,
\end{aligned}$$

where we have used the fact that $x^*(T)\to\bar{x}$ as $T\to\infty$ and applied Lemma 8.1. Thus, it is clear that for any $\epsilon>0$ we can find $T_\epsilon = T(\epsilon, x(\cdot), u(\cdot)) \geq 0$ so that for all $T \geq T_\epsilon$

$$J_T(x_0(\cdot), u^*(\cdot)) > J_T(x_0(\cdot), u(\cdot)) - \epsilon.$$

In the case where (8.18) does not hold for $(x(\cdot), u(\cdot))$, the boundedness of both $x(\cdot)$ and $x^*(\cdot)$ imply that there exists constants Λ and Γ such that for all $T>0$,

$$|\langle \bar{p}, x^*(T)-x(T)\rangle| < \Lambda\|\bar{p}\|$$

and

$$\begin{aligned}
&\Big|\Big\langle \bar{p}, \int_0^T \Big(\int_{T-t}^{\infty} g(s)\,ds\Big)[h(x^*(t))-h(x(t))]\,dt\Big\rangle\Big| \\
&\leq \|\bar{p}\|\,\Gamma \int_0^T \Big(\int_{T-t}^{\infty} \|g(s)\|\,ds\Big)\,dt \\
&\leq \|\bar{p}\|\,\Gamma \int_0^{\infty} \Big(\int_t^{\infty} \|g(s)\|\,ds\Big)\,dt \\
&= \|\bar{p}\|\,\Gamma \int_0^{\infty} t\,\|g(t)\|\,dt.
\end{aligned}$$

Thus,

$$\begin{aligned}
&\liminf_{T\to\infty} [J_T(x_0^*(\cdot), u^*(\cdot)) - J_T(x_0(\cdot), u(\cdot))] \\
&\geq \liminf_{T\to\infty} \Big\{\int_0^T (\mathcal{L}_0(x^*(t), u^*(t)) - \mathcal{L}_0(x(t), u(t)))\,dt \\
&\qquad - \|\bar{p}\|\Lambda - \|\bar{p}\|\,\Gamma \int_0^{\infty} t\,\|g(t)\|\,dt\Big\} = \infty.
\end{aligned}$$

Therefore in each case we arrive at the desired result, and so $(x^*(\cdot), u^*(\cdot))$ is overtaking optimal. ∎

8.4.4 Existence of overtaking solutions

We are now ready to present the existence result. Clearly, the existence of an overtaking optimal solution is established if the existence of a strongly optimal solution of the associated problem consisting of maximizing

$$\tilde{J}(x_0(\cdot), u(\cdot)) = \int_0^\infty \mathcal{L}_0(x(t), u(t))\, dt \tag{8.20}$$

over all $(x(\cdot), u(\cdot)) \in A_0$. With this brief remark we give the following theorem.

Theorem 8.4 *Let $X \subset \mathbb{R}^n$ be closed, let $U(\cdot) : X \to 2^{\mathbb{R}^m}$ have a closed graph M, let $f_0(\cdot,\cdot) : M \to \mathbb{R}$ be upper semicontinuous, let $f(\cdot,\cdot) : M \to \mathbb{R}^n$ and $h(\cdot) : X \to \mathbb{R}^p$ both be continuous, and let $g(\cdot)$ be an $n \times p$ matrix valued function satisfying* (8.5). *Furthermore, suppose that Assumptions* 8.1–8.4 *hold and that there exists an admissible pair $(\hat{x}(\cdot), \hat{u}(\cdot)) \in A_0$ such that* (8.20) *is finite. Under these conditions, the optimal control problem described by* (8.1)–(8.4) *and* (8.6) *has an overtaking optimal solution.*

Proof: By hypothesis, we have

$$-\infty < \tilde{J}(x_0(\cdot), \hat{u}(\cdot)) \le \sup_{(x(\cdot), u(\cdot)) \in A_0} \tilde{J}(x_0(\cdot), u(\cdot)) \le 0$$

Thus there exists a maximizing sequence $\{(x_k(\cdot), u_k(\cdot))\}_{k=1}^\infty$ for the associated optimal control problem. As a consequence of Proposition 8.1, f, h, and $\mathcal{L}_0$ satisfy the growth condition given in Assumption 8.2 with

$$-\infty < \int_0^\infty \mathcal{L}_0(\hat{x}(t), \hat{u}(t))\, dt \le \int_0^T \mathcal{L}_0(x_k(t), u_k(t))\, dt$$

for all $T > 0$. Therefore by a standard diagonalization process, there exists a subsequence, say still $\{(x_k(\cdot), u_k(\cdot))\}_{k=1}^\infty$, and locally integrable functions $z : [0, \infty) \to \mathbb{R}^n$ and $y(\cdot) : [0, \infty) \to \mathbb{R}^p$ such that $f(x_k(\cdot), u_k(\cdot)) \mapsto z(\cdot)$ and $h(x_k(\cdot)) \mapsto y(\cdot)$ weakly in $L^1_{loc}([0, \infty); \mathbb{R}^n)$ and $L^1_{loc}([0, \infty); \mathbb{R}^p)$ respectively. This implies that for each $t \ge 0$ we have

$$\begin{aligned}
x_k(t) &= x_0(0) + \int_0^t f(x_k(s), u_k(s))\, ds + \int_0^t \Big(\int_{-\infty}^0 g(s-\tau) h(x_0(\tau))\, d\tau \Big) ds \\
&\quad + \int_0^t \Big(\int_0^s g(s-\tau) h(x_k(\tau))\, d\tau \Big) ds \\
&= x_0(0) + \int_0^t f(x_k(s), u_k(s))\, ds + \int_0^t \Big(\int_{-\infty}^0 g(s-\tau) h(x_0(\tau))\, d\tau \Big) ds \\
&\quad + \int_0^t \Big(\int_\tau^t g(s-\tau)\, ds \Big) h(x_k(\tau))\, d\tau,
\end{aligned}$$

which clearly converges pointwise to $x^*(\cdot) : [0,\infty) \to \mathbb{R}^n$ given by

$$\begin{aligned} x^*(t) &= x_0(0) + \int_0^t z(s)\,ds + \int_0^t \Big(\int_{-\infty}^0 g(s-\tau)h(x_0(\tau))\,d\tau\Big)ds \\ &\quad + \int_0^t \Big(\int_\tau^t g(s-\tau)ds\Big)y(\tau)\,d\tau \qquad (8.21)\\ &= x_0(0) + \int_0^t z(s)\,ds + \int_0^t \Big(\int_{-\infty}^0 g(s-\tau)h(x_0(\tau))\,d\tau\Big)ds \\ &\quad + \int_0^t \Big(\int_0^s g(s-\tau)y(\tau)\,d\tau\Big)\,ds. \end{aligned}$$

Moreover, this pointwise convergence gives us

$$\lim_{k\to\infty} h(x_k(t)) = h(x^*(t)),$$

so that we have $y(t) = h(x^*(t))$ for almost all $t \in [0,\infty)$. We now apply the Upper Closure Theorem, Theorem 8.2, using the following notation:

(i) $\mathcal{G} = [0,\infty)$, $\mathcal{G}_k = [0,k]$,

(ii) $\eta_k(t) = \mathcal{L}_0(x_k(t), u_k(t))$, $\xi_k(t) = f(x_k(t), u_k(t))$, $x_k(t)$ as above and $\lambda_k(t) = 0$, almost everywhere $t \in \mathcal{G}$, $k = 1, 2, \ldots$,

(iii) $\lambda(t) \equiv 0$, $\xi(t) = z(t)$, and $x(t) = x^*(t)$ almost everywhere $t \geq 0$.

(iv) $\mathcal{R}(x) = \tilde{Q}_{\mathcal{L}}(x)$.

It is easy to establish that the hypotheses of Theorem 8.2 are indeed met so that we conclude there exists $\eta(\cdot) : [0,\infty) \to \mathbb{R}$, which is integrable and satisfies

$$(\eta(t), z(t)) \in \tilde{Q}_{\mathcal{L}}(x^*(t)) \quad \text{and} \quad x^*(t) \in X \quad \text{a.e. } t \geq 0$$

and

$$-\infty < \sup_{(x(\cdot),u(\cdot))\in A_0} \tilde{J}(x_0(\cdot), u(\cdot)) \leq \limsup_{k\to\infty} \int_0^k \mathcal{L}_0(x_k(t), u_k(t))\,dt \leq \int_0^\infty \eta(t)\,dt.$$

By appealing to standard measurable selection arguments (see e.g. Cesari [47, Theorem 11.4i] there exists a measurable function $u^*(\cdot) : [0,\infty) \to \mathbb{R}^m$ such that $\eta(t) \leq \mathcal{L}_0(x^*(t), u^*(t))$, $z(t) = f(x^*(t), u^*(t))$, and $u^*(t) \in U(x^*(t))$ a.e. $0 \leq t$. Substituting this information into (8.21) we obtain

$$\begin{aligned} x^*(t) = x_0(0) &+ \int_0^t f(x^*(s), u^*(s))\,ds + \int_0^t \Big(\int_{-\infty}^0 g(s-\tau)h(x_0(\tau))\,d\tau\Big)ds \\ &+ \int_0^t \Big(\int_0^s g(s-\tau)h(x^*(\tau))\,d\tau\Big)ds, \end{aligned}$$

which clearly shows that the pair $(x^*(\cdot), u^*(\cdot))$ is an admissible pair (here we have extended x^* to $(-\infty, 0]$ by defining $x^*(s) = x_0(s)$). In addition, we further notice that

$$-\infty < \sup_{(x,u)\in A_0} \tilde{J}(x_0(\cdot), u(\cdot)) \le \int_0^\infty \eta(t)\, dt \le \int_0^\infty \mathcal{L}_0(x^*(t), u^*(t))\, dt \le 0$$

which shows that $(x^*(\cdot), u^*(\cdot)) \in A_0$ maximizes $\tilde{J}$ over A_0. The desired conclusion now follows from a direct application of Theorem 8.3. ∎

The above existence result extends the corresponding results, Theorems 4.8 and 4.10 found in Chapter 4 in two directions. The first of these is the obvious extension to a class of models exhibiting time delay in the state variable. In this regard we observe that our formulation includes as a special case a class of models with finite lag where for some fixed $r > 0$ we assume $g(s) \equiv 0$. In such a situation, the hypotheses (8.5) concerning the kernel g are valid if g is essentially bounded on $[0, r]$. The other direction in which the above result provides an extension is that both the convexity and growth hypotheses are weaker than those found in Chapter 4. Indeed, the results presented there require the set

$$\Omega = \{(x, z_0, z) : x \in X,\ z^o \le f_0(x,u),\ z = f(x,u),\ u \in U(x)\}$$

to be closed and convex. This condition is stronger than the convexity and upper semicontinuity conditions required of the sets $\tilde{Q}(x)$. On the other hand to insure that (A4′), and hence Assumption 8.4, holds we must in general assume these stronger convexity conditions. With regards to the growth condition of Assumption 8.2 we refer the reader to the discussion of the growth condition (γ) found in Chapter 7.

The most difficult hypothesis to be satisfied in the above result concerns Assumption 8.4 and the existence of an admissible pair $(\hat{x}(\cdot), \hat{u}(\cdot))$ for which $\tilde{J}$ is finite. Assumption 8.4 has been discussed thoroughly and sufficient conditions for it to hold have been provided in Proposition 8.1. For the existence of $(\hat{x}(\cdot), \hat{u}(\cdot))$ we note that in the case of finite delay, this condition can be realized by controlling from the initial function $x_0(\cdot)$ (defined on $[-r, 0]$) to the terminal function $\bar{x}(s) \equiv \bar{x}$; $-r \le s \le 0$ in some finite time. Since the terminal function is a constant function, this problem can be addressed by utilizing known null controllability results for problems with finite delay (see e.g. [3],[51], [52], and [175]). For the case of infinite delay such an approach is not applicable and the realization of this hypothesis requires further investigation.

8.5 Examples

Example 8.1′ In this example we return to Example 8.1. As we have already seen, this example satisfies both Assumptions 8.3 and 8.4. Also as $(x, u) \in [-1, 1] \times [-1, 1]$, a compact set, it is easy to see that the growth condition Assumption 8.2 is also satisfied.

Furthermore, the linearity of $f_0(x,u) = 9u - 2x(4x^2 + 6x - 9)$ and $f(x,u) = u + 2x$ with respect to u easily insures that the convexity and seminormality hypotheses of Assumption 8.1 are also met. Thus, to apply Theorem 8.4 it is sufficient to ensure that there exists an admissible pair $(\hat{x}(\cdot), \hat{u}(\cdot))$ for which

$$-\infty < \int_0^\infty \mathcal{L}_0(\hat{x}(t), \hat{u}(t))\,ds = \int_0^\infty -\left[4\hat{x}(t)^2(2\hat{x}(t) + 3)\right]\,dt \le 0.$$

In this simple example we note that by choosing the control $\hat{u}(t) \equiv 0$, we have that $\hat{x}(t) = \varphi(0)e^{-2t}$ is an admissible trajectory for every continuous function $\varphi(\cdot) : (-\infty, 0] \to [-1, 1]$. Clearly, this admissible pair satisfies the desired controllability condition. Thus there exists an overtaking optimal solution for this problem.

Example 8.2 The Ramsey model with delay. In this example we present a generalization of the classical Ramsey model of economic growth (see Section 1.4 for an introduction to this model). For this model we let $x = x(\cdot)$, $-\infty < t < \infty$, denote the stock of capital at time t and $u = u(t)$, $t \ge 0$, the consumption. To conform with the notation of this chapter we let $h(\cdot)$ denote the production function and let $f_0(\cdot)$ denote the utility function. As is standard practice we assume that $h(\cdot)$ is a smooth, increasing and strictly concave function satisfying

$$\lim_{x\to 0+} h'(x) = \infty \quad \text{and} \quad \lim_{x\to\infty} h'(x) = 0.$$

Moreover, we also assume that $f_0(\cdot)$ is smooth, decreasing, strictly concave with

$$\lim_{u\to 0+} f_0'(u) = \infty \quad \text{and} \quad \lim_{u\to\infty} f_0'(u) = 0.$$

To introduce delay into the model we let $g(\cdot) : [0,\infty) \to [0,1]$ satisfy the hypotheses outlined in (8.5) and assume

$$\int_0^\infty g(s)\,ds = 1.$$

This function is introduced to reflect a distributed delay in delivering the production of a plant or factory. Thus the term $g(t-s)h(x(s))$ represents the quantity of economic good made available at time t through the use of the stock of capital $x(s)$, at time $s < t$. With this notation we let $\lambda > 0$ be given and consider the optimal control problem

$$\max\left(\int_0^\infty f_0(u(t))\,dt\right)$$

subject to

$$\dot{x}(t) = \int_{-\infty}^t g(t-s)h(x(s))\,ds - \lambda x(t) - u(t),$$

$$x(s) = x_0(s), \quad for \quad s \leq 0,$$

$$0 \leq x(t) \leq X \quad \text{for all } t \in (-\infty, \infty),$$

$$0 \leq u(t) \leq h(x(t)) \quad a.e.\, t \geq 0.$$

The upper bound $X > 0$ given above is a standard hypothesis in the nondelay case with X chosen to be the unique positive solution to

$$h(x) = \lambda x.$$

We remark that in the nondelay case an admissible stock $x(t) > X$ is necessarily decreasing. Thus it is reasonable to ask that $x(t) \leq X$, whenever $x_0(s) \in [0, X]$, $s \leq 0$.

As a result of the boundedness of the admissible states it is an easy matter to see that the growth condition of Assumption 8.2 is satisfied. Moreover, the concavity assumption made on $f_0(\cdot)$ and the linearity of $f(x,u) = -\lambda x - u$ insure that the sets $\tilde{Q}(x)$ are convex and enjoy the upper semicontinuity condition property $(\mathcal{K})$. Therefore Assumption 8.1 is satisfied. For the optimal steady state we recall that in the nondelay case (see Section 3.4) there exists a unique pair $(\bar{x}, \bar{u})$ such that for all (x, u), $x \geq 0$, $u \geq 0$ we have

$$\mathcal{L}_0(x,u) = f_0(u) - f_0(\bar{u}) + f_0'(\bar{u})\,[h(x) - \lambda x - u] \leq 0.$$

The pair $(\bar{x}, \bar{u})$ is uniquely determined by the system of equations

$$h'(\bar{x}) = \lambda \quad \text{and} \quad \bar{u} = h(\bar{x}) - \lambda \bar{x}.$$

Moreover, the strict concavity of both $h(\cdot)$ and $f_0(\cdot)$ imply that for every $\epsilon > 0$ there exists $\delta > 0$ so that if $|x - \bar{x}| > \epsilon$, then

$$\mathcal{L}_0(x,u) < -\delta \quad \text{for all} \quad u \in [0, h(x)].$$

It is easy to see that Assumption 8.3 is satisfied with $\mathcal{L}_0$ given above. Finally, the verification of (A4′) is straightforward. Thus all the hypotheses of Theorem 8.4 hold, except for the existence of an admissible pair $(\hat{x}(\cdot), \hat{u}(\cdot)) \in A_0$ such that $\tilde{J}(x_0(\cdot), \hat{u}(\cdot)) > -\infty$, and so whenever this controllability hypothesis is met the above model has an overtaking optimal solution. As remarked earlier, in the nondelay case, or in the finite delay case, this hypothesis can be met through the application of known controllability results (in particular we refer the reader to Angell [3, Theorem 5.2]).

Example 8.3 The optimal exploitation of a renewable resource. We conclude this chapter with a modification to the classical economic growth model in which a delay is introduced through the introduction of a renewable resource. As in the previous example we let $x(t)$, $-\infty < t < \infty$, denote the stock of capital at time t and $u(t)$, $t \geq 0$, denote consumer consumption. The production of capital stock depends on a renewable resource (e.g., a forest) and we let $y(t)$, $-\infty < t < \infty$, denote the amount of the resource available at time t. This resource is harvested at a rate $v(t)$, $t \geq 0$. With these variables we introduce the following control system:

$$\dot{x}(t) = f(x(t), v(t)) - \lambda x(t)u(t) \quad \text{a.e. } t \geq 0, \tag{8.22}$$

$$\dot{y}(t) = -v(t) + \int_{-\infty}^{t} g(t-s)h(x(s), y(s))\, ds \quad \text{a.e. } t \geq 0, \tag{8.23}$$

$$\begin{pmatrix} x(s) \\ y(s) \end{pmatrix} = \begin{pmatrix} x_0(s) \\ y_0(s) \end{pmatrix} \quad \text{for all } s \in (-\infty, 0] \tag{8.24}$$

$$\begin{matrix} 0 \leq x(t) \leq X \\ 0 \leq y(t) \leq Y \end{matrix} \quad \text{for all } t \in (-\infty, \infty) \tag{8.25}$$

$$\begin{matrix} 0 \leq u(t) \leq f(x(t), v(t)) \\ 0 \leq v(t) \leq h(x(t), y(t)) \end{matrix} \quad \text{a.e. } t \geq 0. \tag{8.26}$$

Here we are letting $f(\cdot,\cdot) : [0,\infty) \times [0,\infty) \to \mathbb{R}$ denote the production function for the economic good and assume that it is a concave function that satisfies

$$\lim_{x \to 0} \frac{\partial f}{\partial x}(x, v) = \infty \quad \text{for each } v \geq 0,$$

$$\lim_{x \to \infty} \frac{\partial f}{\partial x}(x, v) = 0 \quad \text{for each } v \geq 0,$$

$$\frac{\partial f}{\partial x}(x, v) > 0 \quad \text{and} \quad \frac{\partial f}{\partial v}(x, v) > 0 \quad \text{for all } (x, v),$$

$$f(0, v) = 0 \quad \text{for each } v \geq 0.$$

The function $h(\cdot,\cdot) : [0,\infty) \times [0,\infty) \to \mathbb{R}$ denotes the rate of growth of resource y and we assume that h is concave and satisfies

$$h(x, 0) = 0 \quad \text{for all } x \geq 0, \qquad \text{and} \qquad \frac{\partial h}{\partial x}(x, y) \geq 0 \quad \text{for all } (x, y),$$

and

> for each $x \in [0,\infty)$ there exists $y_x > 0$ so that for $0 < y < y_x$ one has $h(x, y) > 0$, $h(x, y_x) = 0$ and for $y_x < x < \infty$ one has $h(x, y) < 0$.

The function $g(\cdot,\cdot) : [0,\infty) \to [0,1]$ satisfies the hypotheses outlined in (8.5) and in addition we assume

$$\int_0^\infty g(s)\, ds = 1.$$

This function is introduced to represent a distributed gestation delay in the growth of the resource. Finally, the constant $\lambda > 0$ denotes the rate of depreciation of capital and the functions are $x_0(\cdot)$ and $y_0(\cdot)$ are given initial functions. We observe that the producer influences the rate of growth of the resource through the level of captial stock as well as through the harvest rate v.

As in the previous example, our objective is to maximize the accumulated utility described by the utility function $f_0 : [0, \infty) \to \mathbb{R}$ which we assume satisfies the same hypotheses as in Example 8.2. Therefore we consider the optimal control problem

$$\max \int_0^\infty f_0(u(t))\, dt$$

over all pairs of functions $((x(\cdot), y(\cdot)), (u(\cdot), v(\cdot)))$ satisfying (8.22)–(8.26).

The associated optimal steady-state problem becomes

$$\max\ f_0(u)$$

over all $((x,y),(u,v)) \in M$ satisfying

$$\begin{pmatrix} 0 \\ 0 \end{pmatrix} = \begin{pmatrix} f(x,v) - \lambda - u \\ h(x,y) - v \end{pmatrix},$$

where

$$\begin{aligned} M = \{((x,y),(u,v)) :\ & 0 \le x \le X, \\ & 0 \le y \le Y,\ 0 \le u \le f(x,v),\ 0 \le v \le h(x,y)\}. \end{aligned}$$

It is easy to see that, under the hypotheses placed on the model under consideration, the conditions of Proposition 8.1 are satisfied. Indeed we easily see that M is compact, convex and with nonempty interior, that $F_1(x,y,u,v) = f(x,v) - \lambda x - u$ and $F_2(x,y,u,v) = h(x,y) - v$ are concave, there exists $(x,y,u,v) \in M$ so that $F_i(x,y,u,v) > 0$, $i = 1,2$, and that $f_0(\cdot)$ is strictly concave. To establish the remaining hypothesis we observe that if $(x,y,u,v) \in M$ is such that $F_i(x,y,u,v) \ge 0$ for $i = 1,2$ we may take $(\tilde{u}, \tilde{v})$ to be given by

$$\tilde{v} = h(x,y) \quad \text{and} \quad \tilde{u} = f(x,\tilde{v}) - \lambda x,$$

and observe that $\tilde{v} \ge v$ so that $f(x,\tilde{v}) \ge f(x,v)$, which implies

$$u \le f(x,v) - \lambda x \le f(x,\tilde{v}) - \lambda x = \tilde{u} \le f(x,\tilde{v}).$$

Clearly this implies that $(x,y,\tilde{u},\tilde{v}) \in M$ and that $f_0(\tilde{u}) \ge f_0(u)$ as required. Consequently Assumption 8.3 and (A4′) are satisfied. The Assumptions 8.1 and 8.2 are easily verified precisely as in Example 8.2. Therefore the only remaining hypothesis in Theorem 8.4 concerns the existence of an admissible pair $((\hat{x}(\cdot), \hat{y}(\cdot)), (\hat{u}(\cdot), \hat{v}(\cdot)))$ for which the improper integral $\tilde{J}$ given by (8.7) is finite. Here, as remarked previously, in the case of finite delay this hypothesis may be met through the application of known controllability results.

Chapter 9

Extensions to Distributed Parameter Systems

9.1 Introduction

In this chapter we extend the results of Chapter 4 so that infinite horizon optimal control problems whose dynamics are described by linear partial differential equations can be considered. That such generalizations are useful in economics is clear from the introduction of infinite dimensional models in the literature. Specifically we refer the reader to [63], [97], [104], and [107] for more precise descriptions of such models.

We begin our discussion with two examples (in Section 9.2) which illlustrate how models of this type arise in applications. Our approach; initially given in the thesis of Jabrane [108] and reported in Carlson, Haurie, and Jabrane [45]; to generalizing the theory given in Chapter 4 is to view these models as linear ordinary differential equations in a Hilbert space setting. To do this we require some basic notions concerning semigroups of operators. The necessary theory of semigroups is outlined briefly in Section 9.3. Sections 9.4 to 9.7 provide generalizations to the results of Chapter 4 to initial value problems and in Section 9.8 we extend these results further to allow for the consideration of boundary controls as well.

9.2 Examples

To motivate the theory to be presented below we introduce two examples of linear distributed parameter systems arising in economics.

Example 9.1 Regional Economic Growth This example is discussed in [107] and is concerned with a closed economy on a given bounded domain Ω, which for simplicity we take as the interval $[0, h]$. In this model we define the following variables

$$
\begin{aligned}
K(x,t) &= \text{stock of capital at time } t > 0 \text{ at the point } x \in \Omega \\
C(x,t) &= \text{consumption at time } t \geq 0 \text{ at the point } x \in \Omega \\
I(x,t) &= \text{investment at time } t \geq 0 \text{ at the point } x \in \Omega \\
U(x,t) &= \text{the flow of merchandise to the point } x \in \Omega \text{ at time } t.
\end{aligned}
$$

In addition, we suppose that the development, $D(x,t)$, of a point x is directly proportional (with proportionality constant ξ) to the stock of capital at that point. Thus the demand for development can be interpreted as an additional demand on the flow of the merchandise originating from more developed regions. We let $1/l$ be the response rate of the merchandise flow with respect to the gradient of the development. Thus we may write

$$\frac{\partial U}{\partial t}(x,t) = \frac{1}{l}\nabla_x D(x,t) = -\frac{\xi}{l}\nabla_x K(x,t). \tag{9.1}$$

Unfortunately the above relation is incomplete because there is a delay in the response to the demand from other regions. The simplest way to simulate this delay is to introduce a forcing term to the R.H.S. of (9.1) which we suppose is directly proportional to U. Thus we have

$$l\frac{\partial U}{\partial t}(x,t) = -\xi\nabla_x K(x,t) - \delta U(x,t). \tag{9.2}$$

If we let $\mu \triangleq \mu(x)$ represent the rate of depreciation of capital, the local conservation of capital permits us to write

$$\frac{\partial K}{\partial t}(x,t) = I(x,t) - \mu(x)K(x,t) - \nabla_x U(x,t). \tag{9.3}$$

Finally, the investment, I, and consumption, C, are related to a production function, $F(K)$, depending on capital stock, through the relation

$$C(x,t) + I(x,t) \leq F(K(x,t)). \tag{9.4}$$

As in Example 4.1 we interpret C and I as control variables and K and U as state variables. Combining the above results we obtain the following control system on $[0,\infty) \times \Omega$

$$
\begin{aligned}
&\frac{\partial}{\partial t}\begin{pmatrix} K \\ U \end{pmatrix} = \begin{pmatrix} -\mu & \nabla_x \\ -\frac{\xi}{l}\nabla_x & -\frac{\delta}{l} \end{pmatrix}\begin{pmatrix} K \\ U \end{pmatrix} + \begin{pmatrix} I \\ 0 \end{pmatrix} \\
&C(x,t) + I(x,t) \leq F(K(x,t))
\end{aligned}
\tag{9.5}
$$

An important special case, which we consider later, is the case when $l = 0$. In this case the relation (9.2) becomes

$$U(x,t) = -\frac{\xi}{\delta}\nabla_x K(x,t)$$

and upon substitution into (9.3) we obtain the second order partial differential equation

$$\frac{\partial K}{\partial t}(x,t) = \frac{\xi}{\delta}\frac{\partial^2}{\partial x^2}K(x,t) - \mu(x)K(x,t) + I(x,t), \tag{9.6}$$

and since the economy is closed we have the boundary conditions

$$\frac{\partial}{\partial x}K(0,t) = \frac{\partial}{\partial x}K(h,t) = 0.$$

Finally, to complete this model we suppose that we are concerned with maximizing some utility of consumption and capital denoted by

$$U(K,C) = \int_0^\infty \int_\Omega f_0(K(x,t), C(x,t))\, dx\, dt, \tag{9.7}$$

where of course, f_0 is the instantaneous utility associated with consumption and captial.

Example 9.2 The Cattle Ranching Problem This problem was originally proposed in Derzko and Sethi [63] and concerns a cattle rancher who must decide the number of cattle in different age groups to be bought and sold at each instant in order to maximize his profit. To formulate this example we proceed as follows.

Let $x(t,y)$ denote the density of cattle of age y at time t, $y \in [0,h] \subset \mathbb{R}$. We assume that all cattle are slaughtered at age h. The control variable $u(t,y)$ denotes the number of y-aged cattle bought at time t. A negative control (i.e., $u(t,y) < 0$) implies that cattle are sold. This process is governed by the system

$$\begin{aligned} \frac{\partial x}{\partial t} &= -\frac{\partial x}{\partial y} + u \quad (t,y) \in [0,\infty) \times [0,h] \\ x(0,y) &= x_0(y), \quad 0 \le y \le h \end{aligned} \tag{9.8}$$

where $x_0(\cdot) : [0,h] \to \mathbb{R}$ is the given initial state which is assumed to satisfy

$$x_0(0) = 0. \tag{9.9}$$

The birth rate of cattle at any time t is assumed to be zero. This implies the following boundary condition holds

$$x(t,0) \equiv 0 \quad \text{for } t \ge 0. \tag{9.10}$$

In addition we impose the following constraints

$$\begin{aligned} 0 &\le x(t,y) \le \alpha \quad \text{a.e. } t \ge 0,\ \forall\, y \in [0,h] \\ 0 &\le u(t,y) \qquad\quad \text{a.e. } t \ge 0,\ \forall\, y \in [0,h]. \end{aligned} \tag{9.11}$$

The first constraint represents a capacity constraint of the ranch and the second reflects the fact that no cattle are sold below the age of maturity.

The objective function for the rancher is given by

$$\begin{aligned} J_T(x_0, u(\cdot)) = \int_0^T \Big[-\int_0^h (P(y)u(t,y) + C(y)x(t,y) + \\ D[u(t,y) - d(y)]^2) dy + Qx(t,h) \Big] dt \end{aligned} \tag{9.12}$$

where $P(y)$ is the unit price of cattle of age y, $C(y)$ is the unit feeding cost of cattle of age y, Q is the selling price of mature cattle (i.e., age h), and D is a positive constant associated with a quadratic cost for deviation of the control from a desired level, $d(y)$. Therefore our goal is to determine $u : [0, \infty) \times [0, h] \to \mathbb{R}$ so that the profit functional is maximized.

The above examples are presented to motivate the theory presented here. They certainly are not all inclusive. For these and other models arising in economics we refer the reader to Isard and Liossatos [107] and to Haurie, Sethi and Hartl [104]. For examples arising in engineering we refer the reader to any book concerning the optimal control of distributed parameters (e.g., Butkovskiy [33] or Lions [128]).

9.3 Semigroups of operators and linear control systems

The control systems given in Section (9.2) can both be interpreted as linear evolution equations on a Hilbert space. The theory of such equations is well developed and utilizes semigroups of operators. In this section we present a very brief introduction to these ideas. Those interested in a more detailed treatment of these concepts are referred to any of the many texts concerning them (e.g., Dunford and Schwartz [65] or Balakrishnan [10, Chapter 4]).

We begin with the following definition.

Definition 9.1 *Let E be a separable Hilbert space (i.e., a separable, complete inner product space) with inner product $\ll \cdot, \cdot \gg$, and let $\{S(t) : t \geq 0\}$ be a family of bounded linear operators from E into itself. We say this family is a semigroup of operators if the following two properties hold:*

$$S(t+s) = S(t)S(s) \quad \text{for all} \quad s, t \geq 0 \tag{9.13}$$

$$S(0) = I, \qquad \text{the identity operator.} \tag{9.14}$$

In addition we say the semigroup is strongly continuous if for each $x \in E$ we have

$$\lim_{t \to 0^+} \|S(t)x - x\| = 0. \tag{9.15}$$

where $\|\cdot\|$ denotes the usual norm on E generated by the inner product $\ll \cdot, \cdot \gg$.

Remark 9.1 *For our discussion we will always assume that our semigroups are strongly continuous.*

Given a strongly continuous semigroup it is possible to define an operator $A : \mathcal{D}(A) \subset E \rightarrow E$ which is called the infinitesimal generator and defined as follows.

Definition 9.2 *Let $\{S(t) : t \geq 0\}$ be a strongly continuous semigroup and let*

$$\mathcal{D}(A) \triangleq \left\{ x \in E : \lim_{h \to 0^+} \frac{S(h)x - x}{h} \quad exists \right\}. \tag{9.16}$$

The infinitesimal generator of the semigroup $\{S(t) : t \geq 0\}$, $A : \mathcal{D}(A) \rightarrow E$, is defined by

$$Ax = \lim_{h \to 0^+} \frac{S(h)x - x}{h} \tag{9.17}$$

Concerning the infinitesimal generator of a semigroup $\{S(t) : t \geq 0\}$ we have the following result.

Lemma 9.1 *Let $A : \mathcal{D}(A) \rightarrow E$ be the infinitesimal generator of a strongly continuous semigroup $\{S(t) : t \geq 0\}$. Then the following properties hold:*

(i) *The set $\mathcal{D}(A)$ is a dense linear manifold in E.*

(ii) *A is a closed linear map on $\mathcal{D}(A)$ (i.e., if $x_n \rightarrow x$ and $Ax_n \rightarrow y$ as $n \rightarrow \infty$, then $Ax = y$).*

(iii) *If $x \in \mathcal{D}(A)$, then $S(t)x \in \mathcal{D}(A)$ for all $t \geq 0$ and moreover*

$$\frac{d}{dt} S(t)x = AS(t)x = S(t)Ax.$$

Proof: Dunford and Schwartz [65, Chapter VIII.1.5, Lemmas 7 and 8]. ∎

From the above it is clear that if we are given a semigroup $\{S(t) : t \geq 0\}$, then the infinitesimal generator A of $\{S(t) : t \geq 0\}$ is well defined on $\mathcal{D}(A)$. In applications, however, it happens that we are given a linear operator A, which is closed and densely defined on E, which may or may not be the infinitesimal generator of a strongly continuous semigroup. Consequently, we are really interested in knowing if such an operator is the infinitesimal generator of a strongly continuous semigroup of operators. Regarding this question we have the following results.

Theorem 9.1 (Hille-Yosida-Phillips). *A necessary and sufficient condition that a closed, densely defined, linear operator A be the infinitesimal generator of a strongly continuous semigroup is that there exists real numbers M and ω such that for all $\lambda > \omega$ for which $(\lambda I - A)^{-1}$ exists as a bounded linear operator on E, it happens that for each positive integer n,*

$$\|(\lambda I - A \;)^{-n}\| \triangleq \sup_{\|x\|=1} \|(\lambda I - A)^{-n} x\| \leq \frac{M}{(\lambda - \omega)^n} \tag{9.18}$$

Proof: Dunford and Schwartz [65, Theorem 13, p. 624] ∎

Corollary 9.1 *A necessary and sufficient condition that a closed densely defined operator A with dense domain $\mathcal{D}(A)$ generates a strongly continuous semigroup $\{S(t) : t \geq 0\}$ such that $\|S(t)\| \leq e^{\omega t}$ for some real number ω is that*

$$\|(\lambda I - A)^{-1}\| \leq \frac{1}{\lambda - \omega} \qquad \textit{for} \quad \lambda > \omega. \tag{9.19}$$

Proof: Dunford and Schwartz [65, Corollary 14, p. 626]. ■

Corollary 9.2 *A necessary and sufficient condition that a closed densely defined operator A on E generates a strongly continuous semigroup is that there exists a strongly continuous family $\{S(t) : t \geq 0\}$ of bounded operators satisfying $S(0) = I$, $\|S(t)\| \leq Me^{\omega t}$ for real numbers M and ω, and such that*

$$(\lambda I - A)^{-1}x = \int_0^\infty e^{-\lambda t} S(t)x\, dt \quad \textit{for} \quad \lambda > \omega. \tag{9.20}$$

In this case $\{S(t) : t \geq 0\}$ is the semigroup generated by A.

Proof: Dunford and Schwartz [65, Corollary 16, p. 627]. ■

Example (9.2′) In the cattle ranching problem we are concerned with the partial differential equation

$$\begin{aligned} \frac{\partial}{\partial t}x(t,y) &= -\frac{\partial}{\partial y}x(t,y) + u(t,y) \quad t \geq 0, \quad y \in [0,h] \\ x(t,0) &= 0 \qquad\qquad \text{for} \qquad t > 0 \\ x(0,y) &= x_0(y) \qquad \text{for} \qquad y \in [0,h]. \end{aligned}$$

To view this problem in a Hilbert space setting, we let $E = L^2([0,h]; \mathbb{R})$ and view a solution as a map $x(\cdot) : [0,\infty) \to E$ (i.e., $x(t)(y) = x(t,y)$ for all $t \geq 0$ and $y \in [0,h]$). In this way we can write the above differential equation as an ordinary differential equation on the space E by

$$\frac{d}{dt}x(t) = Ax(t) + u(t)$$

where $u(\cdot) : [0,\infty) \to E$ is known and the operator A is defined as follows

$$\begin{aligned} \mathcal{D}(A) &= \{x \in E : \frac{\partial x}{\partial y} \in E \text{ and } x(0) = 0\} \\ (Ax)(y) &= \frac{-\partial}{\partial y}x(y). \end{aligned}$$

We observe that A is a closed, densely defined linear operator. We now show, by applying Corollary 9.2, that A generates a strongly continuous semigroup. To this end consider the equation, for $\lambda > 0$ and $x \in \mathcal{D}(A)$, and $u \in E$

$$(\lambda I - A)x(y) = u(y) \quad \text{for} \quad 0 \leq y \leq h.$$

This is simply the ordinary differential equation

$$\begin{aligned} \lambda x(y) \; + \; x'(y) &= u(y) \\ x(0) \; &= \; 0 \end{aligned}$$

which has the unique solution

$$\begin{aligned} x(y) \; &= \; \int_0^y e^{-\lambda(y-s)} u(s)\, ds \\ &= \int_0^y e^{-\lambda s} u(y-s)\, ds. \end{aligned}$$

Thus, for $u \in E$ we have

$$(\lambda I - A)^{-1} u(y) = \int_0^y e^{-\lambda s} u(y-s)\, ds.$$

Define the family of operators $S(t) : E \to E$ by

$$(S(t)u)(y) = \begin{cases} u(y-t) & \text{if } 0 \leq t \leq y \\ 0 & \text{otherwise} \end{cases}$$

and observe we can write

$$(\lambda I - A)^{-1} u(y) = \int_0^\infty e^{-\lambda st} (S(t)u)(y)\, dst.$$

In addition we also have

(i) $(S(0)u)(y) = u(y) \qquad \text{for} \quad 0 \leq y \leq h$

(ii) if $0 \leq t \leq h$ we have

$$\begin{aligned} \|S(t)u\|_{L^2} \; &= \; \left(\int_0^h |S(t)u(y)|^2\, dy \right)^{\frac{1}{2}} \\ &= \left(\int_t^h |u(y-t)|^2\, dy \right)^{\frac{1}{2}} \\ &= \left(\int_0^{h-t} |u(y)|^2\, dy \right)^{\frac{1}{2}} \\ &\leq \|u\|_{L^2} \end{aligned}$$

and if $t \geq h$ we have

$$\|S(t)u\|_{L^2} = \left(\int_0^h |S(t)u(y)|^2\, dy\right)^{\frac{1}{2}} = 0$$

and thus for all $t \geq 0$

$$\|S(t)\| \triangleq \sup_{\|u\|=1} \|S(t)u\|_{L^2} \leq 1$$

and finally for $0 \leq t < h$ and $u \in E$ we have

(iii)

$$\lim_{t\to 0^+} \|S(t)u\|_{L^2} = \lim_{t\to 0^+} \left(\int_0^{h-t} |u(y)|^2\, dy\right)^{\frac{1}{2}} = \|u\|_{L^2}$$

from which it follows that $\|S(t)u\|$ is strongly continuous. Therefore from Corollary 9.2 we see that A generates the semigroup $\{S(t) : t \geq 0\}$.

Example (9.1′) : In our regional economic model, with $l = 0$, we are concerned with the partial differential equation (9.6)

$$\frac{\partial}{\partial x}K(t,y) = \alpha\frac{\partial^2}{\partial y^2}K(t,y) - \mu(y)K(t,y) + I(t,y),$$

on $[0,\infty) \times [0,h]$ where $\alpha \triangleq \xi/\delta$ is a constant. In this case we see that, as a result of (9.2) with $l = 0$, the function $\frac{\partial K}{\partial y}$ can be interpreted as a measure of the flow of goods to $y \in [0,h]$ at time t. As we are dealing with a closed economy this imposes the boundary conditions

$$\frac{\partial}{\partial y}K(t,0) = 0 = \frac{\partial}{\partial y}K(t,h) \text{ for all } t \geq 0.$$

In addition we assume we have the following initial condition

$$K(0,y) = K_0(y) \text{ for all } y \in [0,h].$$

To place this system in a Hilbert space setting once again we let $E = L^2([0,h];\mathbb{R}^n)$ and observe that the above system takes the form

$$\dot{K}(t) = AK(t) + I(t) \text{ for } t \geq 0$$

with $K(0) = K_0 \in E$, where I is known and A is the operator

$$\begin{aligned}(AK)(y) &= A_1K(y) - \mu(y)K(y),\ 0 \leq y \leq h\\ &= \frac{\partial^2}{\partial y^2}K(y) - \mu(y)K(y)\end{aligned}$$

with domain

$$\mathcal{D}(A) = \mathcal{D}(A_1) = \{K \in E : \; K', K'' \in E \text{ and } K'(0) = K'(h) = 0\}.$$

To see that A generates a semigroup we observe first that for $\lambda > 0$ and $\mathcal{I} \in E$, the two-point boundary value problem

$$\begin{aligned} (\lambda I - A_1)K &= \mathcal{I}(y) \, , \, 0 \leq y \leq h \\ K'(0) &= 0 \\ K'(h) &= 0 \end{aligned}$$

has a unique solution. Thus $(\lambda I - A_1)^{-1} : E \to E$ is well defined for each $\lambda > 0$. Moreover we also have, for $\mathcal{J} = (\lambda I - A_1)K$,

$$\begin{aligned} \|\mathcal{J}\|_{L^2}^2 &= \int_0^h [(\lambda I - A_1)K(y)]^2 \, dy \\ &= \int_0^h [\lambda K(y) - K''(y)]^2 \, dy \\ &= \lambda^2 \int_0^h K(y)^2 \, dy - 2\lambda \int_0^h K''(y)K(y) \, dy + \int_0^h K''(y)^2 \, dy \\ &= \lambda^2 \int_0^h K(y)^2 \, dy + 2\lambda \int_0^h K'(y)^2 \, dy + \int_0^h K''(y)^2 \, dy \\ &\geq \lambda^2 \int_0^h K(y)^2 \, dy \\ &= \lambda^2 \|K\|_{L^2}^2 \end{aligned}$$

Thus, for all $\mathcal{I} \in E$,

$$\|(\lambda I - A_1)^{-1}\mathcal{I}\|_{L^2}^2 \leq \frac{1}{\lambda^2} \|\mathcal{I}\|_{L^2}^2$$

which implies

$$\|(\lambda I - A_1)^{-1}\| \leq \frac{1}{\lambda}.$$

Hence by Corollary 9.1 we have that A_1 is the generator of a strongly continuous semigroup on E, say $\{S_1(t) : t \geq 0\}$. To see that the operator A generates a strongly continuous semigroup we define the strongly continuous semigroup $\{S(t) : t \geq 0\}$ by the formula

$$(S(t)K)(y) = e^{-\mu(y)t}(S_1(t)x)(y).$$

Since $\mu(\cdot) : [0, h] \to \mathbb{R}$ is continuous it is easy to see that $\{S(t) : t \geq 0\}$ is a strongly continuous semigroup. Moreover if $K \in \mathcal{D}(A)$ we have for each $y \in [0, h]$

$$\lim_{t \to 0+} \left\{ \frac{(S(t)K)(y) - K(y)}{t} \right\} = \lim_{t \to 0+} \left\{ \frac{e^{-\mu(y)t}(S_1(t)K)(y) - K(y)}{t} \right\}$$

$$
\begin{aligned}
&= \lim_{t \to 0+} \left\{ e^{-\mu(y)t} \frac{[(S_1(t)K)(y) - K(y)]}{t} \right. \\
&\qquad \left. + \frac{e^{-\mu(y)t} - 1}{t} K(y) \right\} \\
&= (A_1 K)(y) - \frac{d}{dt} e^{-\mu(y)t} \bigg|_{t=0} K(y) \\
&= (A_1 K)(y) - \mu(y) K(y).
\end{aligned}
$$

From this it clearly follows that A is the infinitesimal generator of $\{S(t) : t \geq 0\}$, as desired. In addition we observe that this semigroup is a contraction semigroup. That is, for all $t \geq 0$,

$$\|S(t)\| = \|e^{-\mu(\cdot)t} S_1(t)\| \leq \|S_1(t)\| \leq 1,$$

in which the last inequality follows from Corollary 9.1.

Both of the examples given in Section 9.2, when viewed in a Hilbert space setting, give rise to an abstract Cauchy problem of the form

$$
\begin{aligned}
\dot{x}(t) &= Ax(t) + u(t) \ , \ t \geq 0 \\
x(0) &= x_0
\end{aligned}
\tag{9.21}
$$

in which $A : \mathcal{D}(A) \subset E \to E$ is the generator of a strongly continuious semigroup, $x_0 \in E$ is given and $u \in L^2_{loc}([0,\infty); E)$, the space of locally square integrable functions from $[0,\infty)$ into E. To proceed further in our discussion, it is necessary to present some results concerning the existence of solutions to this problem. We begin with the following result.

Theorem 9.2 *If in* (9.21) *we have* $u(t) \equiv 0$ *and* $x_0 \in \mathcal{D}(A)$, *then there exists a unique solution* $x : [0,\infty) \to E$ *such that*

$$
\begin{aligned}
&\text{(i)} \quad x(t) \in \mathcal{D}(A) \\
&\text{(ii)} \quad x \text{ is locally absolutely continuous on } [0,\infty) \\
&\text{(iii)} \quad \lim_{t \to 0+} \|x(t) - x_0\| = 0.
\end{aligned}
\tag{9.22}
$$

Moreover, this solution is given by

$$x(t) = S(t)x_0 \tag{9.23}$$

where $\{S(t) : t \geq 0\}$ *is the strongly continuous semigroup generated by* A.

Proof: Balakrishnan [10, Theorem 4.8.1]. ∎

Theorem 9.3 *Suppose $x_0 \in \mathcal{D}(A)$ and $u : [0,\infty) \to E$ is strongly continuously differentiable in $(0,\infty)$ with continuous derivative in $[0,\infty)$. Then* (9.21) *has a unique solution satisfying* (9.22) *which is given by the formula*

$$x(t) = S(t)x_0 + \int_0^t S(t-s)u(s)\,ds. \tag{9.24}$$

Proof: Balakrishnan [10, Theorem 4.8.2]. ∎

The above results show that if $u(t) \in E$ and $x_0 \in E$ are sufficiently smooth, then the abstract Cauchy problem (9.21) has a unique solution. Unfortunately, in optimal control theory it is unreasonable to expect the control $u(t)$ to be this smooth. On the other hand we observe that if $u \in L^2_{loc}([0,\infty); E)$, then the function $x : [0,\infty) \to E$ defined by (9.24) is continuous on $[0,\infty)$. With this we give the following definition.

Definition 9.3 *A function $x(\cdot) : [0,\infty) \to E$ is called a mild solution of the abstract Cauchy problem* (9.21) *if $x(\cdot)$ is given by* (9.24).

With regards to this definition we have the following.

Theorem 9.4 *Suppose that in* (9.21) *we have $u \in L^2_{loc}([0,\infty); E)$ and that $x_0 \in E$. Then there exists one and only one function $x : [0,\infty) \to E$ satisfying for every $p \in \mathcal{D}(A^*)$, the domain of A^* (the adjoint operator associated with A),*

(i) *the map $t \mapsto \ll x(t), p \gg$ is absolutely continuous;*

(ii) $\dfrac{d}{dt} \ll x(t), p \gg = \ll x(t), A^* y \gg + \ll u(t), p \gg$ *a.e. on $[0,\infty)$; and*

(iii) $\lim_{t \to 0^+} \ll x(t) - x_0, p \gg = 0.$

Moreover, this solution is given by (9.24). *That is, x is a mild solution of* (9.21).

Proof: Balakrishnan [10, Theorem 4.8.3]. ∎

9.4 The optimal control problem

Motivated by the examples of Section 9.2 we consider a control system described by the input-output system

$$x(t) = S(t)x_0 + \int_0^t S(t-s)Bu(s)\,ds, \qquad t \geq 0, \tag{9.25}$$

where we let E and F denote two separable Hilbert spaces, $x_0 \in E$, $\{S(t) : t \geq 0\}$ is a strongly continuous semigroup on E with infinitesimal generator A, $u \in L^2_{loc}([0,\infty); F)$, and $B : F \to E$ is a bounded linear operator.

Thus, for a given control function u, $x(\cdot\,; u)$ given

by (9.25) is the mild solution of the abstract Cauchy problem

$$\begin{aligned} \dot{x}(t) &= Ax(t) + Bu(t);\ t \geq 0 \\ x(0) &= x_0 \end{aligned} \tag{9.26}$$

As a result of Theorem 9.4 we know that for every $p \in \mathcal{D}(A^*)$, where A^* is the adjoint operator associated with A, we have

$$\begin{aligned} \frac{d}{dt} \ll x(t), p \gg &= \ll x(t), A^* \gg + \ll Bu(t), p \gg \ \text{a.e. on } [0,\infty) \\ \lim_{t\to 0^+} \ll x(t), p \gg &= \ll x_0, p \gg, \end{aligned} \tag{9.27}$$

where $\ll \cdot, \cdot \gg$ denotes the inner product associated with E.

We further impose the following additional constraints on the state and control variables

$$x(t) \in X \qquad \forall\, t \geq 0 \tag{9.28}$$

where X is a closed convex subset of E and

$$u(t) \in U(x(t)),\ \text{a.e. } t \geq 0, \tag{9.29}$$

where $U(\cdot) : X \to 2^F$ is a biconvex set-valued mapping (see Definition 4.1) with weakly closed graph (i.e., if $u_n \overset{w}{\to} u$ and $x_n \overset{w}{\to} x$ as $n \to \infty$, with $u_n \in U(x_n)$, then $u \in U(x)$ where $\overset{w}{\to}$ stands for weak convergence in E or F). In particular, we note that this implies that $U(x)$ is a weakly closed subset of F for each $x \in X$.

As in our previous problems we associate to the control system (9.25), (9.28), (9.29) a performance criterion given by a cost functional

$$J_T(x_0, u(\cdot)) \triangleq \int_0^T f_0(x(s), u(s))\, ds \tag{9.30}$$

defined for each $T > 0$, in which $f_0 : E \times F \to \mathbb{R}$ is a concave functional which is upper semicontinuous on $E \times F$ and satisfies the following growth condition:

There exists $K_1 > 0$ and $K > 0$ such that

$$\|x\|^2 + \|u\|^2 > K_1 \Longrightarrow f_0(x,u) \leq -K\left(\|x\|^2 + \|u\|^2\right). \tag{9.31}$$

With the above notation, we conclude our formulation with the following definition of an admissible pair.

Definition 9.4 *A pair of functions* $(x(\cdot), u(\cdot)) : [0,\infty) \to E \times F$ *is called an* admissible pair *if the following conditions are met.*

(i) $u(\cdot) \in L^2_{loc}([0,\infty); F)$.

(ii) $x(\cdot) : [0,\infty) \to E$ *is continuous and satisfies* (9.25).

(iii) $x(t) \in X$ *and* $u(t) \in U(x(t))$ *for all* $t \geq 0$.

(iv) *the map* $t \to f_0(x(t), u(t))$ *is locally Lebesgue integrable on* $[0,\infty)$.

As usual, we refer to x as an *admissible trajectory* and u as an *admissible control.* For brevity we let $\mathcal{S}(x_0)$ denote the set of all admissible pairs.

9.5 The turnpike properties

In this section we extend the results of Section 4.3 to the infinite dimensional systems considered here. We begin our discussion with the following assumption concerning the optimal steady state problem (OSSP).

Assumption 9.1 *The OSSP consisting of*

maximize

$$f_0(x,u)$$

over all $(x,u) \in E \times F$ *satisfying*

$$\begin{aligned} 0 &= \ll x, A^*p \gg + \ll Bu, p \gg \quad \forall\, p \in \mathcal{D}(A^*) \\ x &\in X \text{ and } u \in U(x) \end{aligned}$$

has a solution, denoted by $(\bar{x}, \bar{u})$, *with* $\bar{x}$ *uniquely determined.*

As a consequence of the hypotheses on f_0, X, and $U(\cdot)$, the OSSP is a concave programming problem in $E \times F$. Hence, there exists a vector $\bar{p} \in \mathcal{D}(A^*)$ (see Ekeland and Temam [67]) such that for all $x \in X$ and $u \in U(x)$, we have

$$f_0(\bar{x}, \bar{u}) \geq f_0(x,u) + \ll x, A^*\bar{p} \gg + \ll Bu, \bar{p} \gg . \tag{9.32}$$

As in Section 4.8 we introduce the nonpositive function $\mathcal{L}_0(\cdot,\cdot) : E \times F \to [0, -\infty)$ by the formula

$$\mathcal{L}_0(x,u) \triangleq \begin{cases} f_0(x,u) - f_0(\bar{x},\bar{u}) + \ll x, A^*\bar{p} \gg + \ll Bu, \bar{p} \gg & \text{if } x \in X \text{ and } u \in U(x) \\ -\infty & \text{otherwise} \end{cases} \tag{9.33}$$

We remark, that since $\mathcal{L}_0(\cdot,\cdot)$ differs from $f_0(\cdot,\cdot)$ by an affine functional of (x,u) it still satisfies the same hypotheses (particularly the growth condition (9.31)).

As a consequence of these hypotheses we have the following technical lemma.

Lemma 9.2 *If* $(\tilde{x}(\cdot), \tilde{u}(\cdot)) \in \mathcal{S}(x_0)$ *is such that*

$$\int_0^\infty \mathcal{L}_0(\tilde{x}(s), \tilde{u}(s))\, ds > -\infty, \tag{9.34}$$

then for every $T > 0$ *there exists a constant* $C_T > 0$ *such that for all* $t \geq 0$

$$\int_t^{t+T} \|\tilde{u}(s)\|^2\, ds \leq C_T. \tag{9.35}$$

Proof: Let $(\tilde{x}(\cdot), \tilde{u}(\cdot))$ be as given above. To show (9.35) we proceed by contradiction and suppose there exists $T > 0$ and $t_k \to \infty$ as $k \to \infty$ such that

$$\lim_{k\to\infty} \int_{t_k}^{t_k+T} \|\tilde{u}(s)\|^2\, ds = \infty$$

For each index k, let $\Omega_k = \{t \in [t_k, t_k+T] : \|\tilde{u}(t)\|^2 \geq K_1\}$, where K_1 is as in (9.31) with f_0 replaced by $\mathcal{L}_0$, and we let Ω'_k denote its complement in $[t_k, t_k+T]$. For each index k we observe,

$$\int_{\Omega'_k} \|\tilde{u}(s)\|^2\, ds \leq K_1 r(\Omega'_k) \leq K_1 T,$$

and thus

$$\lim_{k\to\infty} \int_{\Omega_k} \|\tilde{u}(s)\|^2\, ds = \infty$$

On the other hand, as a result of the growth condition (9.31) and (9.34)

$$\begin{aligned} -\infty &< \int_0^\infty \mathcal{L}_0(\tilde{x}(s), \tilde{u}(s))\, ds \\ &\leq \int_{\Omega_k} \mathcal{L}_0(\tilde{x}(s), \tilde{u}(s))\, ds \\ &\leq -K \int_{\Omega_k} \|\tilde{u}(s)\|^2\, ds, \end{aligned}$$

for each index k. However, letting $k \to \infty$ in this last expression clearly implies

$$\int_0^\infty \mathcal{L}_0(\tilde{x}(s), \tilde{u}(s))\, ds = -\infty,$$

which contradicts (9.34). Hence there must exist $C_T > 0$ such that for all $t \geq 0$ (9.35) holds. ■

With this technical lemma, we return to our control problem and establish the following *Weak Turnpike Theorem.*

Theorem 9.5 *Under Assumption* 9.1, *if* $(\tilde{x}(\cdot), \tilde{u}(\cdot)) \in \mathcal{S}(x_0)$ *is such that*

$$\liminf_{T\to\infty} \left[\int_0^T (f_0(\tilde{x}(t), \tilde{u}(t)) - f_0(\bar{x}, \bar{u}))\, dt \right] > -\infty, \tag{9.36}$$

then

$$\frac{1}{T} \int_0^T \tilde{x}(t)\, dt \xrightarrow{w} \bar{x} \text{ as } T \to \infty. \tag{9.37}$$

Proof: We begin by showing that, under Assumption 9.1, the set

$$\Lambda = \left\{ \left(\frac{1}{T} \int_0^T \tilde{x}(s)\, ds, \frac{1}{T} \int_0^T \tilde{u}(s)\, ds \right) : T > 0 \right\} \tag{9.38}$$

is norm bounded in $E \times F$. This of course implies that this set is relatively weakly compact in $E \times F$. To see that this is the case, we assume first that

$$\|\frac{1}{T} \int_0^T \tilde{x}(s)\, ds\|$$

is not bounded for $T > 0$. Hence there exists a sequence $T_k \to \infty$ such that

$$\lim_{k\to\infty} \|\frac{1}{T_k} \int_0^{T_k} \tilde{x}(s)\, ds\| = \infty, \tag{9.39}$$

and so for all k sufficiently large we have

$$\|\frac{1}{T_k} \int_0^{T_k} \tilde{x}(s)\, ds\| > K_1,$$

where K_1 is as in the growth condition (9.31). Appealing to Jensen's Inequality and to the growth condition (9.31) we see that for all k sufficiently large,

$$\begin{aligned} &\frac{1}{T_k} \int_0^{T_k} [f_0(\tilde{x}(s), \tilde{u}(s)) - f_0(\bar{x}, \bar{u})]\, ds \\ &\leq f_0\left(\frac{1}{T_k} \int_0^{T_k} \tilde{x}(s)\, ds, \frac{1}{T_k} \int_0^{T_k} \tilde{u}(s)\, ds \right) - f_0(\bar{x}, \bar{u}) \\ &\leq -K\left(\|\frac{1}{T_k} \int_0^{T_k} \tilde{x}(s)\, ds\|^2 + \|\frac{1}{T_k} \int_0^{T_k} \tilde{u}(s)\, ds\|^2 \right) - f_0(\bar{x}, \bar{u}). \end{aligned} \tag{9.40}$$

From the hypothesis (9.36) it is clear that the left hand side of (9.40) remains bounded as $k \to \infty$, while the right hand side tends to negative infinity as $k \to \infty$. Clearly this is a contradiction. An analogous argument shows that

$$\|\frac{1}{T}\int_0^T \tilde{u}(s)\,ds\|,$$

is bounded as well. Consequently we see that Λ is a bounded subset of $E \times F$ and hence has weak limit points.

We now show every weak limit point of Λ, say $(\hat{x}, \hat{u})$, satisfies $\hat{x} = \bar{x}$. To see this we let $(\hat{x}, \hat{u})$ be any weak limit point of Λ and observe from (9.36) and (9.40) there exists $\alpha > 0$ such that for all $k = 1, 2, \ldots$

$$-\frac{\alpha}{T_k} + f_0(\bar{x}, \bar{u}) \leq f_0\Big(\frac{1}{T_k}\int_0^{T_k} \tilde{x}(s)\,ds, \frac{1}{T_k}\int_0^{T_k} \tilde{u}(s)\,ds\Big), \tag{9.41}$$

where $\left(\frac{1}{T_k}\int_0^{T_k} \tilde{x}(s)\,ds, \frac{1}{T_k}\int_0^{T_k} \tilde{u}(s)\,ds\right) \xrightarrow{w} (\hat{x}, \hat{u})$ as $k \to \infty$. Letting $k \to \infty$ we obtain, from the upper semicontinuity of f_0,

$$f_0(\bar{x}, \bar{u}) \leq f_0(\hat{x}, \hat{u}). \tag{9.42}$$

Moreover, since $(\tilde{x}(\cdot), \tilde{u}(\cdot)) \in \mathcal{S}(x_0)$ we have for all $p \in \mathcal{D}(A^*)$,

$$\begin{aligned}\frac{1}{T_k}[\ll \tilde{x}(T_k), p \gg \; - \; \ll x_0, p \gg] \\ = \frac{1}{T_k}\int_0^{T_k} [\ll \tilde{x}(s), A^*p \gg + \ll B\tilde{u}(s), p \gg]\,ds.\end{aligned}$$

Since $t \to \tilde{x}(t)$ is bounded it clear that the left hand side tends to zero as $k \to \infty$. Hence,

$$\begin{aligned} 0 &= \lim_{k\to\infty} \frac{1}{T_k}\int_0^{T_k} [\ll \tilde{x}(s), A^*p \gg + \ll B\tilde{u}(s), p \gg]\,ds \\ &= \lim_{k\to\infty}\Big[\ll \frac{1}{T_k}\int_0^{T_k} \tilde{x}(s)\,ds, A^*p \gg + \ll \frac{1}{T_k}\int_0^{T_k} B\tilde{u}(s)\,ds, p \gg\Big] \\ &= \ll \hat{x}, A^*p \gg + \ll B\hat{u}, p \gg . \end{aligned}$$

Further since $X \subset E$ is weakly closed and $x \to U(x)$ has a weakly closed graph it follows that $(\hat{x}, \hat{u})$ is a feasible point of OSSP and as a result of Assumption 9.1 and relation (9.42) we must have $\hat{x} = \bar{x}$, as desired. ∎

To obtain a stronger turnpike theorem we follow Section 4.9 and define the set

$$\mathcal{G} = \{x \in E : \mathcal{L}_0(x, u) = 0 \text{ for some } u \in F\} \tag{9.43}$$

and give the following definition.

Definition 9.5 *Let $\mathcal{F}$ be the family of all trajectories $x \in \mathcal{S}(X) = \bigcup\{\mathcal{S}(x_0) : x_0 \in X\}$ such that*

$$x(t) \in \mathcal{G}, \text{ a.e. on } [0, \infty). \tag{9.44}$$

We say $\mathcal{G}$ has the convergence property $\mathcal{C}$ if $x(t) \xrightarrow{w} \bar{x}$ as $t \to \infty$ uniformly in $\mathcal{F}$.

Under the hypotheses of the convergence property $\mathcal{C}$ we give the following turnpike result.

Theorem 9.6 *Under Assumption 9.1, if the set $\mathcal{G}$ has property $\mathcal{C}$ and if there exists a feasible pair $(\tilde{x}(\cdot), \tilde{u}(\cdot)) \in \mathcal{S}(x_0)$ such that*

$$\int_0^\infty \mathcal{L}_0(\tilde{x}(s), \tilde{u}(s))\, ds > -\infty, \tag{9.45}$$

then $\tilde{x}(t) \xrightarrow{w} \bar{x}$ as $t \to \infty$.

Proof: We proceed by contradiction and suppose that there exists $\bar{y} \in E$, $\bar{y} \neq 0$, times $t_k \to \infty$ and $\epsilon > 0$ such that for $k = 1, 2, \ldots,$

$$| \ll \tilde{x}(t_k) - \bar{x}, \bar{y} \gg | \geq \epsilon \tag{9.46}$$

Let $t_0 > 0$ be chosen such that for all $x \in \mathcal{F}$ and $t \geq t_0$,

$$| \ll x(t) - \bar{x}, \bar{y} \gg | < \frac{\epsilon}{2} \tag{9.47}$$

Define the sequence of functions $\{(x_k(\cdot), u_k(\cdot))\}_{k=1}^{\infty}$ from $[0, \infty)$ into $E \times F$ by the formulas,

$$(x_k(t), u_k(t)) \triangleq (\tilde{x}(t_k - t_0 + t), \tilde{u}(t_k - t_0 + t)),$$

for $k = 1, 2, \ldots$. A simple calculation shows that for each k, $(x_k(\cdot), u_k(\cdot)) \in \mathcal{S}(\tilde{x}(t_k - t_0))$. Further, Lemma 9.2 shows that for each $T > 0$, there exists $C_T > 0$ such that

$$\begin{aligned} \int_0^T \|u_k(s)\|^2\, ds &= \int_0^T \|\tilde{u}(t_k - t_0 + s)\|^2\, ds \\ &= \int_{t_k - t_0}^{t_k - t_0 + T} \|\tilde{u}(s)\|^2\, ds \\ &\leq C_T. \end{aligned}$$

Hence $\{u_k(\cdot)\}_{k=1}^{\infty}$ is bounded in $L^2([0, T]; F)$ for each $T > 0$. Further, since $t \to \tilde{x}(t)$ is bounded in E we easily see that $\{x_k(0)\}_{k=1}^{\infty}$ is bounded in E. These facts allow us to assert the existence of $x_0^* \in E$ and $u^*(\cdot) \in L^2_{loc}([0,\infty);F)$ such that (by extracting appropriate sequences when necessary),

$$u_k(\cdot) \xrightarrow{w} u^*(\cdot) \text{ in } L^2_{loc}([0,\infty);F)$$

and

$$x_k(0) \xrightarrow{w} x_0^* \text{ in } E.$$

Define $x^*(\cdot) : [0,\infty) \to E$ by the formula,

$$x^*(t) = S(t)x_0^* + \int_0^t S(t-s)Bu^*(s)\,ds \tag{9.48}$$

and observe that the pair $(x^*(\cdot), u^*(\cdot))$ satisfies (9.27). In addition, as a result of the convergence properties above, for each $t \geq 0$, and all $p \in E$ we have

$$\begin{aligned}\lim_{k\to\infty} \ll x_k(t) - x^*(t), p \gg &= \lim_{k\to\infty}\Big(\ll S(t)(x_k(0) - x^{0*}), p \gg \\ &\quad + \int_0^T \ll S(t-s)B(u_k(s) - u^*(s)), p \gg \, ds\Big) \\ &= 0,\end{aligned}$$

and so $x_k(t) \xrightarrow{w} x^*(t)$ pointwise in E. Further, from the hypotheses placed on X and the map $x \to U(x)$, it follows that $x^*(t) \in X$ for $t \geq 0$ and $u^*(t) \in U(x^*(t))$ a.e. on $[0,\infty)$. Hence $(x^*(\cdot), u^*(\cdot)) \in \mathcal{S}(x_0)$. We further observe that since $\mathcal{L}_0 \leq 0$ and the fact (9.45) holds we have for each $T > 0$

$$\lim_{k\to\infty} \int_0^T \mathcal{L}_0(x_k(s), u_k(s))\,ds = \lim_{k\to\infty} \int_{t_k-t_0}^{T+t_k-t_0} \mathcal{L}_0(\tilde{x}(s), \tilde{u}(s))\,ds = 0 \tag{9.49}$$

On the other hand, f_0 being upper semicontinuous implies the functional $\varphi : L^2([0,T];E) \times L^2([0,T];F) \to \mathbb{R} \cup \{-\infty\}$ defined by

$$\varphi(x(\cdot), u(\cdot)) = \int_0^T \mathcal{L}_0(x(s), u(s))\,ds$$

is concave and upper semicontinuous. In addition, since $\mathcal{L}_0$ is concave and upper semicontinuous, it follows that for every $K \in \mathbb{R}$, the set

$$\Phi_K = \Big\{(x(\cdot), u(\cdot)) \in L^2([0,T];E) \times L^2([0,T];F) : \varphi(x(\cdot), u(\cdot)) \geq K\Big\}$$

is convex and closed. It is a well known fact (see e.g., [67]) that a convex set is closed if and only if it is weakly closed. Hence Φ_K is weakly closed implying that φ is weakly upper semicontinuous. Thus from (9.49) we obtain

$$0 = \limsup_{k\to\infty} \int_0^T \mathcal{L}_0(x_k(s), u_k(s))\,ds \leq \int_0^T \mathcal{L}_0(x^*(s), u^*(s))\,ds \leq 0,$$

so that for all $T \geq 0$

$$\int_0^T \mathcal{L}_0(x^*(s), u^*(s))\, ds = 0.$$

This of course implies $\mathcal{L}_0(x^*(s), u^*(s)) = 0$ a.e. on $[0, \infty)$ and so, $x^*(\cdot) \in \mathcal{F}$. As a result of (9.47) we have that

$$\ll x^*(t) - \bar{x}, \bar{y} \gg < \frac{\epsilon}{2} \text{ on } [t_0, \infty),$$

which implies, when combined with (9.46), for all k sufficiently large that

$$\begin{aligned} \epsilon &\leq | \ll \tilde{x}(t_k) - \bar{x}, \bar{y} \gg | = | \ll x_k(t_0) - \bar{x}, \bar{y} \gg | \\ &\leq | \ll x_k(t_0) - x^*(t_0), \bar{y} \gg | + | \ll x^*(t_0) - \bar{x}, \bar{y} \gg | < \epsilon \end{aligned}$$

an obvious contradiction of (9.46). ∎

The utility of the turnpike results given above is well established in Chapter 4, in a finite dimensional setting. Their role for the infinite dimensional setting is precisely the same. In particular, we present the following result.

Corollary 9.3 *If the hypotheses of Theorem* 9.6 *hold and if there exists an overtaking optimal solution, say* $(\hat{x}(\cdot), \hat{u}(\cdot)) \in \mathcal{S}(x_0)$ *over the class of all bounded trajectories, then*

$$\hat{x}(t) \xrightarrow{w} \bar{x} \text{ as } t \to \infty .$$

Proof: Since the hypotheses of Theorem 9.6 hold, there exists $(\tilde{x}(\cdot), \tilde{u}(\cdot)) \in \mathcal{S}(x_0)$ such that

$$\int_0^\infty \mathcal{L}_0(\tilde{x}(s), \tilde{u}(s))\, ds > -\infty$$

and hence, as a result of Lemma 9.2, $\tilde{x}(\cdot)$ is bounded. The optimality of $(\hat{x}(\cdot), \hat{u}(\cdot))$ therefore implies that for any $\epsilon > 0$ there exists $T(\epsilon) > 0$ such that for all $T \geq T(\epsilon)$ we have

$$\int_0^T f_0(\hat{x}(s), \hat{u}(s))\, ds > \int_0^T f_0(\tilde{x}(s), \tilde{u}(s))\, ds - \epsilon.$$

Hence from the definition of $\mathcal{L}_0$ we can write,

$$\begin{aligned} &\int_0^T [\mathcal{L}_0(\hat{x}(s), \hat{u}(s) - \frac{d}{ds} \ll \hat{x}(s), \bar{p} \gg]\, ds \\ &\quad \geq \int_0^T [\mathcal{L}_0(\tilde{x}(s), \tilde{u}(s)) - \frac{d}{ds} \ll \hat{x}(s), \bar{p} \gg]\, ds - \epsilon. \end{aligned}$$

Thus, for all $T \geq T(\epsilon)$,

$$\begin{aligned} 0 &\geq \int_0^T \mathcal{L}_0(\hat{x}(s), \hat{u}(s)\, ds \\ &\geq \int_0^T \mathcal{L}_0(\tilde{x}(s), \tilde{u}(s))\, ds + \ll \hat{x}(T) - \tilde{x}(T), \bar{p} \gg -\epsilon, \end{aligned}$$

which implies, since both $\hat{x}(\cdot)$ and $\tilde{x}(\cdot)$ are bounded that,

$$0 \geq \int_0^\infty \mathcal{L}_0(\hat{x}(s), \hat{u}(s))\, ds > -K,$$

for some constant $K > 0$. Hence, by Theorem 9.6 it follows that $\hat{x}(t) \xrightarrow{w} \bar{x}$ as $t \to \infty$. ∎

9.6 Existence of overtaking optimal solutions

With the requisite turnpike theorems presented in Section 9.5 we are now able to discuss the existence of optimal solutions. These results parallel the discussions presented in Sections 4.8 and 4.9. We begin our treatment by defining the corresponding associated problem of Lagrange (APL) as consisting of maximizing the integral functional

$$J(x_0, u(\cdot)) = \int_0^\infty \mathcal{L}_0(x(s), u(s))\, ds \tag{9.50}$$

over all pairs $(x(\cdot), u(\cdot)) \in \mathcal{S}(x_0)$.

Assumption 9.2 *There exists an admissible pair $(\tilde{x}(\cdot), \tilde{u}(\cdot)) \in \mathcal{S}(x_0)$ such that*

$$J(\tilde{x}_0, \tilde{u}(\cdot)) > -\infty.$$

The next result shows that under Assumptions 9.1 and 9.2 the APL has a strongly optimal solution.

Lemma 9.3 *Under Assumptions* 9.1 *and* 9.2, *there exists $(\hat{x}(\cdot), \hat{u}(\cdot)) \in \mathcal{S}(x_0)$ such that for all $(x(\cdot), u(\cdot)) \in \mathcal{S}(x_0)$,*

$$\int_0^\infty \mathcal{L}_0(\hat{x}(s), \hat{u}(s))\, ds \geq \lim_{T \to \infty} \int_0^T \mathcal{L}_0(x(s), u(s))\, ds. \tag{9.51}$$

Proof: We begin by observing that the nonpositiveness of $\mathcal{L}_0$ and Assumption 9.2 imply that the functional $J(x_0, u(\cdot)\,)$ has a finite supremum which we denote by σ. This permits us to assert the existence of a sequence $\{(x_n(\cdot)\,, u_n(\cdot)\,)\}_{n=1}^{\infty}$ in $\mathcal{S}(x_0)$ such that

$$\lim_{n\to\infty} J(x_0, u_n(\cdot)\,) = \sigma \tag{9.52}$$

We claim that this sequence is bounded in $L^2([0,T];E) \times L^2([0,T];F)$ for each $T > 0$. To see this we assume the contrary. That is, we suppose for some fixed $T > 0$ there exists a subsequence, say $\{(x_{n_k}(\cdot)\,, u_{n_k}(\cdot)\,)\}_{n=1}^{\infty}$, which is divergent. For each integer k, we define the sets Ω_k by

$$\Omega_k = \left\{t \in [0,T] : \|x_{n_k}(t)\|^2 + \|u_{n_k}(t)\|^2 > K_1\right\}.$$

Clearly we have,

$$\int_{\Omega_k} [\|x_{n_k}(t)\|^2 + \|u_{n_k}(t)\|^2\,]dt \;\to \infty \text{ as } k \to \infty. \tag{9.53}$$

On the other hand, the growth condition (9.31) implies (with f_0 replaced by $\mathcal{L}_0$)

$$\begin{aligned}\int_0^{\infty} \mathcal{L}_0(x_{n_k}(s), u_{n_k}(s))\,ds &\le \int_{\Omega_k} \mathcal{L}_0(x_{n_k}(s), u_{n_k}(s))\,ds \\ &\le -K \int_{\Omega_k} [\|x_{n_k}(s)\|^2 + \|u_{n_k}(s)\|^2]\,ds\end{aligned}$$

Letting k tend to infinity leads to a contradiction of (9.52) so that our claim is true. Therefore, by diagonalizing if necessary, there exist a pair $(x^*(\cdot)\,, u^*(\cdot)\,)$ in $L^2([0,\infty);E) \times L^2([0,\infty);F)$ such that for each $T > 0$,

$$u_n(\cdot) \xrightarrow{w} u^*(\cdot) \text{ in } L^2([0,T];F)$$

and

$$x_n(\cdot) \xrightarrow{w} x^*(\cdot) \text{ in } L^2([0,T];E)$$

as $n \to \infty$. Moreover, from our hypotheses on X and $U(\cdot)$ it follows (see the proof of Theorem 9.6) that $(x^*(\cdot)\,, u^*(\cdot)\,) \in \mathcal{S}(x_0)$. These relations, combined with the weak upper semicontinuity of the functionals $\varphi_T : L^2([0,T];E) \times L^2([0,T];F) \to \mathbb{R} \cup \{-\infty\}$ given by

$$\varphi_T(x(\cdot)\,, u(\cdot)\,) = \int_0^T \mathcal{L}_0(x(s), u(s))\,ds$$

imply,

$$\limsup_{n\to\infty} \int_0^T \mathcal{L}_0(x_n(s), u_n(s))\,ds \le \int_0^T \mathcal{L}_0(x^*(s), u^*(s)\,)ds \le 0$$

for all $T > 0$. Hence we obtain, upon letting $T \to \infty$,

$$\sigma = \limsup_{n\to\infty} \int_0^{\infty} \mathcal{L}_0(x_n(s), u_n(s))\,ds \le \int_0^{\infty} \mathcal{L}_0(x^*(s), u^*(s))ds \le \sigma,$$

our desired conclusion. ∎

The next two results are analogous to Theorems 4.9 and 4.10 respectively. In fact, as their proofs are immediate adaptations of these results, we present them without proof.

Theorem 9.7 *Under Assumptions* 9.1 *and* 9.2 *the optimal control problem* (9.27) *to* (9.30) *has a weakly overtaking optimal solution over* $\mathcal{S}(x_0)$.

Theorem 9.8 *If, in addition to Assumptions* 9.1 *and* 9.2 *one assumes that the set* $\mathcal{G}$, *given* (9.43), *enjoys the convergence property* $\mathcal{C}$, *then the optimal control problem* (9.27) *to* (9.30) *has an overtaking optimal solution over* $\mathcal{S}_B(x_0)$ *the set of all pairs* $(x(\cdot), u(\cdot)) \in \mathcal{S}(x_0)$ *having the additional property that the state* $x(\cdot)$ *is bounded in* E *on* $[0,\infty)$.

Remark 9.2 *The convergence property* $\mathcal{C}$ *is analogous to Property* $\mathcal{S}$ *of Section* 4.9. *Unlike property* $\mathcal{S}$, *there are at present no known conditions under which property* $\mathcal{C}$ *holds (except, of course, when* $\mathcal{G} = \{\bar{x}\}$ *(i.e.,* $\mathcal{G}$ *is a singleton set)).*

9.7 More on the examples

In this section we apply the above theory to our motivating examples 9.1 and 9.2. We begin with the regional economic growth model.

Example (9.1″) We recall that the regional economic growth model introduced in Section 9.2 is briefly described as follows

$$\text{Maximize} \int_0^{\infty}\int_0^{h} f_0(K(t,y), C(t,y))\, dt$$

over all triples $(K(\cdot,\cdot), I(\cdot,\cdot), C(\cdot,\cdot)) : [0,\infty) \times [0,h] \to \mathbb{R}^3$ satisfying for $(t,y) \in [0,\infty) \times [0,h]$

$$\begin{aligned}
\frac{\partial}{\partial t}K(t,y) &= I(t,y) + \alpha\frac{\partial^2}{\partial y^2}K(t,y) - \mu(y)K(t,y), \\
I(t,y) &+ C(t,y) \le F(K(t,y)), \\
I(t,y) &\ge 0, \text{ and } C(t,y) \ge 0, \\
0 &\le K(t,y) \le \tilde{K} \\
K(0,y) &= K_0(y), \text{ for } y \in [0,h], \\
\frac{\partial}{\partial y}K(t,0) &= \frac{\partial}{\partial y}K(t,h) = 0, \text{ for } t \ge 0.
\end{aligned}$$

For this model we assume

(i) $F : [0, \tilde{K}] \to \mathbb{R}$ is concave and Lipschitz continuous;

(ii) $\mu(\cdot) : [0, h] \to \mathbb{R}$ is continuous and nonnegative;

(iii) $f_0(\cdot, \cdot) : \mathbb{R}^2 \to \mathbb{R}$ is strictly concave in (K, C) and decreasing in K.

(iv) $K_0(\cdot) : [0, h] \to \mathbb{R}$ and $\tilde{K} \in \mathbb{R}$ are given and fixed.

To formulate this problem in the setting of the previous sections we let

$$\begin{aligned} E &\triangleq L^2([0,h]; \mathbb{R}) \text{ and } F = E \times E; \\ X &= \left\{x \in E : \ 0 \le x(y) \le \tilde{K} \text{ a.e.}\right\}; \\ U(x) &= \{u = (u_1, u_2) \in F : \ u_i(y) \ge 0 \text{ a.e. for } i = 1, 2, \\ &\qquad \text{and } u_1(y) + u_2(y) \le F(x(y)) \text{ a.e. on } [0,h]\}; \\ \mathcal{D}(A) &= \{x \in E : \ x', x'' \in E, \text{ and } x'(0) = x'(h) = 0\}. \end{aligned}$$

where $A : \mathcal{D}(A) \to E$ is given by

$$(Ax)(y) = \alpha x''(y) - \mu(y)x(y),$$

in which x' and x'' denote the first and second distributional derivatives of x. The function $F_0 : E \times F \to \mathbb{R}$ is defined by

$$F_0(x, u) = \int_0^h f_0(x(y), u_2(y))\, dy,$$

and $B : F \to E$ is given as

$$Bu = \begin{pmatrix} I & 0 \\ 0 & 0 \end{pmatrix} \begin{pmatrix} u_1 \\ u_2 \end{pmatrix} = u_1.$$

With this notation, we obtain the following optimal control problem

$$\text{Maximize} \int_0^\infty F_0(x(t), u(t))\, dt$$

over all pairs of functions $(x(\cdot), u(\cdot)) :\to [0, \infty) \to E \times F$ satisfying

$$\begin{aligned} \dot{x}(t) &= Ax(t) + Bu(t) \\ x(0) &= K_0 \\ x(t) &\in X \\ u(t) &\in U(x(t)). \end{aligned}$$

Now that we have given the correct formulation (recall we have already shown that A generates a strongly continuous semigroup) we will show that there exists an overtaking (weakly overtaking) optimal solution for any initial condition K_0 which is *strictly sustainable* in the sense defined below.

Definition 9.6 *An initial state $K_0 \in \mathcal{D}(A)$ will be called* strictly sustainable *if there exists $u_0 = (u_{01}, u_{02}) \in U(K_0)$ such that*

$$K_0''(y) - \mu(y)K_0(y) + u_0(y) = 0, \text{ for } 0 < y < h$$

and such that the inequalities in the definition of $U(\cdot)$ being strictly satisfied.

We now establish the following properties:

(a) The OSSP has a unique solution $(\bar{x}, \bar{u}) \in E \times F$,

(b) The steady state $\bar{x}$ is asymptotically reachable from any strictly sustainable state,

(c) The set $\mathcal{G}$, given by (9.43) has the convergence property $\mathcal{C}$.

(a) The OSSP for this problem consists of maximizing

$$F_0(x, u) = \int_0^h f_0(x(y), u_2(y))\, dt$$

over all triples $(x, u_1, u_2) : [0, h] \to \mathbb{R}^3$ satisfying

$$\begin{aligned} x''(y) - \mu(y)x(y) + u_1(y) &= 0, \text{ a.e. on } [0, h]; \\ 0 \le x(y) &\le \tilde{K} \text{ a.e. on } [0, h]; \\ u_1(y) + u_2(y) &\le F(x(y)) \text{ a.e. on } [0, h]; \\ x'(0) = x'(h) &= 0. \end{aligned}$$

This problem is a standard Lagrange-type problem in optimal control and it is well known (see e.g., Lee and Markus [115]) that an optimal solution exists. Moreover, since f_0 is strictly concave with respect to u_1 it follows that $\bar{u}_1$ is uniquely determined. Further, since f_0 is decreasing in x, the output constraint is active at the optimum. Hence, $\bar{x}$ satisfies

$$\bar{x}''(y) - \mu(y)\bar{x}(y) + F(\bar{x}(y)) - \bar{u}_1(y) = 0$$

so that the uniqueness of $\bar{u}_1$ and the Lipschitz hypothesis on F imply $\bar{x}$ is uniquely determined. Thus, for this problem Assumption 9.1 is satisfied and the multiplier $\bar{p}$ is defined as the adjoint variable in the Pontryagin Maximum Principle applied to the steady state problem.

(b) To establish the asymptotic reachability of $\bar{x}$ from K_0, consider the function $x_\alpha : [0,\infty) \times [0,h] \to \mathbb{R}$ given by

$$x_\alpha(t,y) = e^{-\alpha t} K_0(y) + (1 - e^{-\alpha t})\bar{x}(y),$$

where α is a fixed parameter. Further define

$$\begin{aligned} u_1(t,y) &= e^{-\alpha t}\left[u_{01}(y) - \alpha(K_0(y) - \bar{x}(y))\right] + (1 - e^{-\alpha t})\bar{u}_1(y) \\ u_2(t,y) &= e^{-\alpha t}\left[u_{02}(y) - \alpha(K_0(y) - \bar{x}(y))\right] + (1 - e^{-\alpha t})\bar{u}_2(y). \end{aligned}$$

We observe that for α sufficiently small, the strict sustainability of K_0 implies that $u_1(\cdot)$ and $u_2(\cdot)$ are nonnegative on $[0,\infty) \times [0,h]$. Moreover, the concavity of F insures that

$$\begin{aligned} u_1(t,y) + u_2(t,y) &= e^{-\alpha t}\left(u_{01}(y) + u_{02}(y)\right) + \left(1 - e^{-\alpha t}\right)\left(\bar{u}_1(y) + \bar{u}_2(y)\right) \\ &\leq e^{-\alpha t} F(K_0(y)) + \left(1 - e^{-\alpha t}\right) F(\bar{x}(y)) \\ &\leq F\left(e^{-\alpha t} K_0(y) + \left(1 - e^{-\alpha t}\right)\bar{x}(y)\right) \\ &= F\left(x_\alpha(t,y)\right). \end{aligned}$$

Finally, it is easy to see that the triple $(x_\alpha(\cdot,\cdot), u_1(\cdot,\cdot), u_2(\cdot,\cdot))$ as defined above is an admissible pair and that $x_\alpha(t,y) \mapsto \bar{x}(y)$, as $\alpha \mapsto 0$, for each $y \in [0,h]$, and in addition,

$$\begin{aligned} &\int_0^T \left[F_0(x_\alpha(t,\cdot), u(t,\cdot)) - F_0(\bar{x},\bar{u})\right] dt \\ &= \int_0^T \left[\int_0^h f_0\left(x_\alpha(t,y), u_2(t,y)\right) - f_0\left(\bar{x}(y), \bar{u}_2(y)\right)\right] dt \\ &= \int_0^T \left[\int_0^h \left\{ f_0(e^{-\alpha t} K_0(y) + (1 - e^{-\alpha t})\bar{x}(y), e^{-\alpha t} u_2(y) + (1 - e^{-\alpha t})\bar{u}(y)\right.\right. \\ &\qquad \left.\left. + e^{-\alpha t}\alpha\left(K_0(y) - \bar{x}(y)\right)\right) - f_0\left(\bar{x}(y), \bar{u}_2(y)\right)\right\} dy\Big] dt \\ &\geq \int_0^T \Big[\int_0^h \left\{ e^{-\alpha t} f_0\left(K_0(y), u_{02}(y) + \alpha(K_0(y) - \bar{x}(y))\right)\right. \\ &\qquad \left. + (1 - e^{-\alpha t}) f_0\left(\bar{x}(y), \bar{u}_2(y)\right) - f_0\left(\bar{x}(y), \bar{u}_2(y)\right)\right\} dy\Big] dt \\ &= \int_0^T e^{-\alpha t}\left[\int_0^h \left\{ f_0\left(K_0(y), u_{02}(y) + \alpha(K_0(y) - \bar{x}(y))\right)\right.\right. \\ &\qquad \left.\left. - f_0\left(\bar{x}(y), \bar{u}(y)\right)\right\} dy\right] dt, \end{aligned}$$

which clearly remains bounded below as $T \to \infty$. From this it is easy to see that Assumption 9.2 holds, and therefore Theorem 9.7 is applicable. Thus, we can conclude that there exists a weakly overtaking optimal solution.

(c) Finally, since $f_0(\cdot,\cdot)$ is strictly concave it follows that the optimal steady state $\bar{x}$ is unique so that the set $\mathcal{G}$ reduces to a singleton, and so the convergence property $\mathcal{C}$ holds trivially.

Hence, Theorem 9.9 is applicable and we further conclude that this problem has an overtaking optimal solution.

Example (9.2″) Recall that the cattle ranching model consists of maximizing

$$\int_0^\infty \left\{ \int_0^h - \left[P(y)u(t,y) + C(y)x(t,y) + D\left[u(t,y) - d(y)\right]^2 \right] dy + Qx(t,h) \right\} dt$$

over all admissible pairs $(x(\cdot), u(\cdot)) : [0,\infty) \times [0,h] \to \mathbb{R}^2$ satisfying for $(t,y) \in [0,\infty) \times [0,h]$:

$$\begin{aligned} \frac{\partial}{\partial t}x(t,y) &= -\frac{\partial}{\partial y}x(t,y) + u(t,y) \\ x(0,y) &= x_0(y) \\ x(t,0) &= 0 \\ 0 &\le x(t,y) \le \alpha \\ 0 &\le u(t,y) \end{aligned}$$

in which $x_0(\cdot) : [0,h] \to \mathbb{R}$ is a given function satisfying $x_0(0) = 0$. To place this problem in our abstract setting we set

$$\begin{aligned} E &= L^2([0,h],\mathbb{R}) = F \\ \mathcal{D}(A) &= \{x \in E : x' \in E,\ x(0) = 0\} \\ (Ax)(y) &= -x'(y) \\ X &= \{x(\cdot) \in E : x(0) = 0,\ 0 \le x(y) \le \alpha \text{ a.e. on } [0,h]\} \\ U(x(\cdot)) &= U = \{u(\cdot) \in F : 0 \le u(y) \text{ a.e. on } [0,h]\} \end{aligned}$$

$f_0(\cdot,\cdot) : E \times F \to \mathbb{R}$ is defined by

$$f_0(x(\cdot), u(\cdot)) = Qx(h) - \int_0^h \left[P(y)u(y) + C(y)x(y) + D[u(y) - d(y)]^2 \right] dy.$$

Here we assume $P(\cdot)$, $C(\cdot)$, and $d(\cdot)$ are continuous on $[0,h]$ and that both Q and D are positive constants. With this notation, the cattle ranching problem has the desired form. However as $E = L^2([0,h],\mathbb{R})$ the function f_0 is not well defined since $x(h)$ need not exist. Nevertheless, we shall see that the above theory can still be applied with relatively minor adjustments. In particular, we shall show that

(a) The OSSP is well defined;

(b) There exists a functional $\mathcal{L}_0'$ slightly different from $\mathcal{L}_0$ which will permit us to arrive at our results; and

(c) The reachability of the optimal steady state and the convergence property $\mathcal{C}$ are guaranteed.

(a) For this example the OSSP is a Bolza problem of optimal control and is briefly written

maximize

$$\left[Qx(h) - \int_0^h \left[C(\tau)x(\tau) + P(\tau)u(\tau) + D\left[u(\tau) - d(\tau)\right]^2\right] d\tau\right]$$

over all pairs of functions $(x(\cdot), u(\cdot)) : [0,h] \to \mathbb{R}^2$ satisfying

$$\begin{aligned} \frac{d}{d\tau}x(\tau) &= u(\tau) \\ x(0) &= 0 \\ 0 &\leq x(\tau) \leq \alpha \\ 0 &\leq u(\tau). \end{aligned}$$

It is well known that his problem admits a unique solution $(\bar{x}(\cdot), \bar{u}(\cdot))$, with $\bar{x}(\cdot)$ continuous, satisfying the Pontryagin Maximum Principle (see Chapter 2). Thus there exists $\bar{p}(\cdot) : [0,h] \to \mathbb{R}$ such that

$$\begin{aligned} \frac{d}{dy}\bar{p}(y) &= -C(y) \text{ a.e. on } [0,h] \\ \bar{p}(h) &= -Q \end{aligned}$$

and such that the optimal control maximizes the function

$$\begin{aligned} -[C(y)\bar{x}(y) &+ P(y)u + D\,(u - d(y))^2\,] + \bar{p}(y)u \\ &= -C(y)\bar{x}(y) - P(y)u - D\left(u^2 - 2d(y)u + d(y)^2\right) + \bar{p}(y)u \\ &= -C(y)\bar{x}(y) - Dd(y)^2 + u\,(\bar{p}(y) - P(y) + 2Dd(y) - Du) \end{aligned}$$

as a function of $u \geq 0$ for almost all $y \in [0,h]$. Therefore the optimal control must be defined as follows

$$\bar{u}(y) = \begin{cases} 0 & \text{if } \bar{p}(y) - P(y) + 2Dd(y) < 0 \\ \dfrac{1}{2D}\left[\bar{p}(y) - P(y) + 2Dd(y)\right] & \text{otherwise} \end{cases}$$

(b) We now define the function $\mathcal{L}_0'(\cdot,\cdot) : E \times F \to (-\infty, 0]$ by the formula,

$$\mathcal{L}_0'(x,u) = \begin{cases} -\int_0^h [P(y)u(y) + C(y)x(y) + D\,(u(y) - d(y))^2\,]\,dy \\ \qquad + \;\; Q\bar{x}(h) + \int_0^h [P(y)\bar{u}(y) + C(y)\bar{x}(y) + D\,(\bar{u}(y) - d(y))^2\,]\,dy \\ \qquad + \;\; \int_0^h \bar{p}(y)u(y) + \left(\frac{d}{dy}\bar{p}(y)\right)x(y)\,dy \text{ if } (x,u) \in X \times U \\ -\infty \qquad \text{otherwise} \end{cases}$$

We further observe that, as a result of the conditions satisfied by $\bar{p}(\cdot)$ and $\bar{x}(\cdot)$, we have

$$Q\bar{x}(h) = -\int_0^h \frac{d}{dy}\left(\bar{p}(y)\bar{x}(y)\right)\,dy,$$

and this fact in conjunction with our convexity hypotheses shows that $\bar{x}(y)$ is a "supported" trajectory on $[0,h]$. This clearly implies that $\mathcal{L}_0'$ is a nonpositive concave functional. Further, it is also easy to deduce that it satisfies the growth condition (9.31) and that it is weakly upper semicontinuous on $E \times F$. As a consequence of these facts the existence theory given in the previous sections can be easily be developed for this problem with $\mathcal{L}_0'$ replacing $\mathcal{L}_0$.

(c) Define the control $\tilde{u}(\cdot,\cdot) : [0,\infty) \times [0,h] \to [0,\infty)$ by

$$\tilde{u}(t,y) = \begin{cases} 0 & \text{if } t < h \\ \bar{u}(y) & \text{if } t \geq h. \end{cases}$$

It is easy to deduce that $\tilde{u}$ generates an admissible control $\tilde{x}(\cdot)$ with the property that $\tilde{x}(t,y) = \bar{x}(y)$ for all $t > h$. Consequently,

$$\int_0^\infty \mathcal{L}_0'(\tilde{x}(t), \tilde{u}(t))\,dt = \int_0^h \mathcal{L}_0'(\tilde{x}(t), \tilde{u}(t))\,dt > -\infty.$$

Moreover, the convergence property $\mathcal{C}$ is easily satisfied because of the uniqueness of $(\bar{x}(\cdot), \bar{u}(\cdot))$. Hence, there exists an overtaking optimal solution for this problem for any initial condition x_0.

9.8 The extension to systems with distributed parameters and boundary controls

The examples, and the resulting theory, considered in the previous sections is concerned only with the case where the control is a distributed control (e.g., $u(t,\cdot)$ is defined on all of $[0,h]$). In certain engineering applications, it is not possible or perhaps not desirable to use such controls. Instead one applies a control only to the boundary (e.g., $u(t,0)$ and $u(t,h)$ are defined but $u(t,\cdot)$ is not). In this section, we extend the results of the previous sections to this case. These results are obtained by appealing to the boundary control theory of Fattorini [73] and Barbu [13]. In this way we transform our system into a linear evolution equation on an appropriate Hilbert space. This clearly permits us to exploit the results presented previously.

We consider the following control system defined for $t \geq 0$

$$\begin{aligned} \dot{x}(t) &= \sigma x(t) + B_1 u_1(t) \\ \gamma x(t) &= B_2 u_2(t) \in E \\ x(0) &= x_0 \\ x(t) &\in X \subset E_1 \\ u_i(t) &\in U_i(x(t)) \subset F_i \text{ for } i = 1,2 \end{aligned} \tag{9.54}$$

in which E_1, E_2, F_1, and F_2 are separable Hilbert spaces, $x_0 \in E_1$ is a fixed initial condition, σ is a closed, linear, densely defined operator on E_1, γ is a linear operator (*the boundary operator*) with domain in E_1 and range in E_2, and $B_i : F_i \to E_i$ for $i = 1,2$, are continuous linear operators.

We will assume that the sets X and $U_i(x)$, $i = 1,2$ and $x \in X$, satisfy the hypotheses given in Section 9.4. That is, we assume that X is a closed, convex subset of E_1 and that for $i = 1,2$ $U_i : X \to 2^{F_i}$ is biconvex with a weakly closed graph.

A solution to the control system (9.54) satisfies the input-output relation

$$\begin{aligned} x(t) = S(t)x_0 \; + & \int_0^t S(t-s)\,[B_1 u_1(s) + \sigma B u_2(s)]\; ds \\ - & \; A\int_0^t S(t-s) B u_2(s)\, ds\;, \end{aligned} \tag{9.55}$$

in which the operators A, B, and $S(\cdot)$ are defined in the following assumptions.

Assumption 9.3 *We assume that $\mathcal{D}(\sigma) \subset \mathcal{D}(\gamma)$ (here $\mathcal{D}(\cdot)$ denotes the domain), and that the restriction of γ to $\mathcal{D}(\sigma)$ is continuous with respect to the graph norm of $\mathcal{D}(\sigma)$.*

Assumption 9.4 *The operator $A : \mathcal{D}(A) \subset E_1 \to E_1$ is defined by*

$$\begin{aligned} \mathcal{D}(A) &\triangleq \{y \in \mathcal{D}(\sigma) : \; \gamma y = 0\} \\ Ay &= \sigma y \; \textit{for } y \in \mathcal{D}(A). \end{aligned} \tag{9.56}$$

We further assume that A generates a strongly continuous semigroup $\{S(t) : \; t \geq 0\}$ on E_1.

Assumption 9.5 *There exists a bounded linear operator $B : F_2 \to E_1$ such that*

$$\begin{aligned} &\text{(i)} \quad \sigma B \text{ is a bounded linear operator from } F_2 \text{ to } E_1. \\ &\text{(ii)} \quad \gamma(Bv) = B_2 v \text{ for all } v \in F_2. \\ &\text{(iii)} \quad \|Bv\|_{E_1} \le c\|B_2 v\|_{E_2} \text{ for all } v \in F_2. \end{aligned} \tag{9.57}$$

Assumption 9.6 *For each $t \ge 0$ and $v \in L^2_{loc}([0,\infty); F_2)$ we have,*

$$\int_0^t S(t-s)Bv(s)\,ds \in \mathcal{D}(A) \tag{9.58}$$

and there exists a $\delta \in L^2_{loc}([0,\infty); \mathbb{R}^+)$ and a constant $\mathcal{K}$ such that

$$\begin{aligned} \|AS(t)B\| &\le \delta(t) \ \text{a.e. on } [0,\infty) \\ \int_t^{t+h} \delta(s)\,ds &\le \mathcal{K} \text{ for each } t \ge 0 \text{ and } h \in [0,1]. \end{aligned} \tag{9.59}$$

With these hypotheses, the right hand side of (9.55) is well defined whenever $u_i \in L^2_{loc}([0,\infty); F_i)$, for $i = 1, 2$. Further, we remark that our formulation agrees with Barbu [13].

The performance of the system (9.54) on $[0,t]$, for any $t > 0$, will (as before) be evaluated by

$$J_T(x_0, u(\cdot)) = \int_0^T f_0(x(s), u_1(s), u_2(s))\,ds, \tag{9.60}$$

where $f_0 : E_1 \times F_1 \times F_2 \to \mathbb{R}$ is a concave, upper semicontinuous functional which satisfies the following coercivity conditions:

There exists $K_1 > 0$ and $K > 0$ so that if $(x, u_1, u_2) \in E_1 \times F_1 \times F_2$ satisfies

$$\|x\|^2_{E_1} + \|u_1\|^2_{F_1} + \|u_2\|^2_{F_2} \ge K_1,$$

then

$$f_0(x, u_1, u_2) \le -K\left(\|x\|^2_{E_1} + \|u_1\|^2_{F_1} + \|u_2\|^2_{F_2}\right). \tag{9.61}$$

With this notation we now define an admissible pair as the triple of functions $(x(\cdot), u_1(\cdot), u_2(\cdot)) : [0,\infty) \to E_1 \times F_1 \times F_2$; with x continuous, $u_i \in L^2_{loc}([0,\infty); F_i)$, $i = 1, 2$; the relation (9.55) and the constraints given in (9.54) hold; the map $t \mapsto \|x(t)\|_{E_1}$ is bounded on $[0,\infty)$ and the map $t \mapsto f_0(x(t), u_1(t), u_2(t))$ is locally integrable on $[0,\infty)$.

Proceeding as in the preceding sections we describe the optimal steady state problem (OSSP) as maximize

$$f_0(x, u_1, u_2)$$

subject to

$$\begin{aligned} 0 = \ll x - Bu_2, A^*z \gg + \ll B_1u_1 + \sigma Bu_2, z \gg, &\quad \text{for all } z \in \mathcal{D}(A^*) \\ x \in X \text{ and } u_i \in U_i(x) &\quad \text{for } i = 1,2 \\ (x, u_1, u_2) \in E_1 \times F_1 \times F_2. & \end{aligned} \tag{9.62}$$

With regards to this problem we make the following two assumptions.

Assumption 9.7 *The OSSP given above has a unique solution denoted by* $(\bar{x}, \bar{u}_1, \bar{u}_2)$.

Assumption 9.8 *There exists* $\bar{p} \in \mathcal{D}(A^*)$ *such that the functional* $\mathcal{L}_0(\cdot,\cdot) : E_1 \times F_1 \times F_2 \to \mathbb{R} \cup \{-\infty\}$ *defined by*

$$\mathcal{L}_0(x, u_1, u_2) = \begin{cases} f_0(x, u_1, u_2) - f_0(\bar{x}, \bar{u}_1, \bar{u}_2) - \ll x - Bu_2, A^*\bar{p} \gg \\ \qquad - \ll B_1u_1 + \sigma Bu_2, \bar{p} \gg, \\ \qquad \text{if } x \in X,\ u_i \in U_i(x),\ i = 1,2 \\ -\infty \qquad \text{otherwise} \end{cases} \tag{9.63}$$

is nonpositive.

With these hypotheses we have the following results which are analogous to those of the previous sections.

Theorem 9.9 *Under the assumptions made above, if* $(x(\cdot), u_1(\cdot), u_2(\cdot))$ *is an admissible pair satisfying*

$$\liminf_{T \to \infty} \int_0^T [f_0(x(t), u_1(t), u_2(t)) - f_0(\bar{x}, \bar{u}_1, \bar{u}_2)]\, dt > -\infty \tag{9.64}$$

then

$$\frac{1}{T} \int_0^T x(t)\, dt \xrightarrow{w} \bar{x} \text{ as } T \to \infty. \tag{9.65}$$

Proof: Preceding exactly as in Theorem 9.3 we see that the set

$$\Lambda = \left\{ \left(\frac{1}{T} \int_0^T x(t)\, dt, \frac{1}{T} \int_0^T u_1(t)\, dt, \frac{1}{T} \int_0^T u_2(t)\, dt \right) : T > 0 \right\}$$

is bounded in $E_1 \times F_1 \times F_2$, and thus is relatively weakly compact. Moreover it is easy to see that if $(\hat{x}, \hat{u}_1, \hat{u}_2)$ is a weak cluster point of Λ, then (9.64) permits us to conclude (see the proof of Theorem 9.5) that

$$f_0(\bar{x}, \bar{u}_1, \bar{u}_2) \leq f_0(\hat{x}, \hat{u}_1, \hat{u}_2).$$

Thus, the proof is complete if we can show that $(\hat{x}, \hat{u}_1, \hat{u}_2)$ is feasible of OSSP. We clearly have that $\hat{x} \in X$ and $\hat{u}_i \in U_i(\hat{x})$ for $i = 1,2$ both hold. Thus it remains to show that $(\hat{x}, \hat{u}_1, \hat{u}_2)$ satisfies the equality constraint in (9.62). To to this we observe from (9.55) we have

$$x(t) = \alpha(t) + \beta(t) - A\eta(t),$$

where

$$\begin{aligned} \alpha(t) &= S(t)x_0 + \int_0^t S(t-s)B_1u_1(s)\,ds, \\ \beta(t) &= \int_0^t S(t-s)\sigma Bu_2(s)\,ds, \end{aligned} \tag{9.66}$$

and

$$\eta(t) = \int_0^t S(t-s)Bu_2(s)\,ds.$$

The functions $\alpha(\cdot)$ and $\beta(\cdot)$ are mild solutons of the systems $\dot{\alpha}(t) = A\alpha(t) + B_1u_1(t)$, $\alpha(0) = x_0$, and $\dot{\beta}(t) = A\beta(t) + \sigma Bu_2(t)$, $\beta(0) = 0$, respectively. Hence we know that for almost all $t \geq 0$ and all $z \in \mathcal{D}(A^*)$,

$$\begin{aligned} &\ll \alpha(t), A^*z \gg = \frac{d}{dt} \ll \alpha(t), z \gg - \ll B_1u_1(t), z \gg \\ \text{and} \\ &\ll \beta(t), A^*z \gg = \frac{d}{dt} \ll \beta(t), z \gg - \ll \sigma Bu_2(t), z \gg . \end{aligned} \tag{9.67}$$

Further, since $\eta(\cdot) \in \mathcal{D}(A)$ (Assumption 9.6) it follows that $\eta(\cdot)$ is the weak solution of $\dot{\eta}(t) = A\eta(t) + Bu_2(t)$, $\eta(0) = 0$. This, of course, means that for all $y \in E_1$,

$$\ll A\eta(t), y \gg = \frac{d}{dt} \ll \eta(t), y \gg - \ll Bu_2(t), y \gg . \tag{9.68}$$

Combining (9.66) and (9.67), with $y = A^*z$, we obtain

$$\begin{aligned} \ll x(t), A^*z \gg &= \frac{d}{dt}[\ll \alpha(t) + \beta(t), z \gg - \ll \eta(t), A^*z \gg] \\ &\quad - \ll B_1u_1(t), z \gg - \ll \sigma Bu_2(t), z \gg + \ll Bu_2(t), A^*z \gg \\ &= \frac{d}{dt} \ll x(t), z \gg - \ll B_1u_1(t), z \gg - \ll \sigma Bu_2(t), z \gg \\ &\quad + \ll Bu_2(t), A^*z \gg \end{aligned} \tag{9.69}$$

for all $z \in \mathcal{D}(A^*)$. Integrating (9.69) and using (9.66) gives

$$\frac{1}{T}\int_0^T \ll x(t), A^*z \gg \, dt$$

$$= \frac{1}{T} \ll x(T) - x_0, z \gg - \frac{1}{T} \int_0^T \ll B_1 u_1(t) + \sigma B u_2(t), z \gg dt$$

$$- \int_0^T \ll B u_2(t), A^* z \gg dt$$

for all $z \in \mathcal{D}(A^*)$. This implies, upon letting $T \to \infty$, that

$$\ll \hat{x}, A^* z \gg = - \ll B_1 \hat{u}_1 + \sigma B \hat{u}_2, z \gg + \ll B \hat{u}_2, A^* z \gg,$$

where we have used the boundedness of $t \to \|x(t)\|$ on $[0, \infty)$. This of course imples $(\hat{x}, \hat{u}_1, \hat{u}_2)$ is feasible for OSSP and hence, because of Assumption 9.7, we must have $(\hat{x}, \hat{u}_1, \hat{u}_2) = (\bar{x}, \bar{u}_1, \bar{u}_2)$.

We now state the analogues of Theorems 9.6, 9.7 and 9.8 respectively, remarking that their proofs are identical to the distributed control case, with appropriate modifications to account for the boundary control u_2.

Theorem 9.10 *Under Assumptions* 9.3 *to* 9.8, *if the set*

$$\mathcal{G} = \{x \in E_1 : \mathcal{L}_0(x, u_1, u_2) = 0 \text{ for some } (u_1, u_2) \in F_1 \times F_2\} \tag{9.70}$$

has the convergence property C, *then every admissible pair* $(x(\cdot), u_1(\cdot), u_2(\cdot))$ *which satisfies*

$$\int_0^\infty \mathcal{L}_0(x(t), u_1(t), u_2(t))\, dt > -\infty \tag{9.71}$$

has the property that

$$x(t) \xrightarrow{w} \bar{x} \text{ as } t \to \infty \tag{9.72}$$

Theorem 9.11 *Under Assumptions* 9.3 *to* 9.8, *if there exists* $(\tilde{x}(\cdot), \tilde{u}_1(\cdot), \tilde{u}_2(\cdot))$ *admissible such that* (9.71) *holds, then ther exists a weakly overtaking optimal solution to the infinite horizon distributed and boundary optimal control problem.*

Theorem 9.12 *If in addition to the hypotheses of Theorem* 9.11, *the set* $\mathcal{G}$ *given by* (9.70) *enjoys the convergence property* C, *then there exists an overtaking optimal solution to the infinite horizon distributed and boundary control problem.*

We conclude our treatment of distributed parameter systems by presenting the following example.

Example 9.3 This example is the infinite horizon analogue of the *Mixed Dirichlet Problem* considered in Barbu [13]. Specifically, we let $\Omega \subset \mathbb{R}^n$ be a bounded open set with smooth boundary Γ, and we let X and U_2 denote fixed bounded closed intervals of $\mathbb{R}$. The boundary control system we consider is described as follows.

$$\left.\begin{aligned}\frac{\partial}{\partial t}x(t,\tau) &= \Delta_\tau x(t,\tau) \text{ for } (t,\tau) \in [0,\infty) \times \Omega \\ x|_\Gamma &= u_2 \text{ for } t \in [0,\infty) \\ x(0,\tau) &= x_0(\tau) \text{ for } \tau \in \Omega \\ x(t,\tau) &\in X \text{ for } (t,\tau) \in [0,\infty) \times \Omega \\ u(t,\tau) &\in U_2 \text{ for } (t,\tau) \in [0,\infty) \times \Omega.\end{aligned}\right\} \quad (9.73)$$

Following Barbu [13] we let $E_1 = F_1 = L^2(\Omega; \mathbb{R})$, $E_2 = H^{-\frac{1}{2}}(\Gamma; \mathbb{R})$, and $F_2 = L^2(\Gamma; \mathbb{R})$. We further define $B_i : F_i \to E_i$, $i = 1,2$ by $B_1 \triangleq 0$ and $B_2 = I$ (the identity), $\sigma : \mathcal{D}(\sigma) \to E_1$ as $\sigma = \Delta_\tau$ with $\mathcal{D}(\sigma) = \{x \in L^2(\Omega; \mathbb{R}) : \Delta x \in L^2(\Omega; \mathbb{R})\}$ and $\gamma : E_1 \to E_2$ as the trace operator which is well defined and belongs to $H^{-\frac{1}{2}}(\Gamma : \mathbb{R})$ for each $y \in \mathcal{D}(\sigma)$. Finally we take $A : \mathcal{D}(A) \to E_1$ to be $A = \Delta$, with $\mathcal{D}(A) = H_0^1(\Omega; \mathbb{R}) \cap H^2(\Omega; \mathbb{R})$, and we define the linear map $B : F_2 \to F_1$ by $Bu = \omega_u$ in which $\omega_u \in L^2(\Omega; \mathbb{R})$ is the unique (generalized) solution to the Dirichlet Boundary-value problem

$$\begin{aligned}\Delta\omega_u &= 0 \\ \omega_u|_\Gamma &= u.\end{aligned}$$

Under these hypotheses it follows, as in [13], that Assumptions 9.3 to 9.6 are satisfied.

To describe the performance of the system we let $g(\cdot,\cdot) : \Omega \times \mathbb{R} \to \mathbb{R}$ and $h(\cdot,\cdot) : U_2 \to \mathbb{R}$ be two given functions satisfying the following conditions:

(i) $g(\cdot,\cdot)$ is continuous, concave in its second argument, measurable in its first argument and satisfies

$$|g(\tau,x)| \geq -c|x|^2 + \xi(t) \text{ a.e. in } \Omega \times \mathbb{R},$$

where $c > 0$ is a constant and $\xi \in L^2(\Omega; \mathbb{R})$.

(ii) h is concave and upper semicontinuous.

With these functions we define $f_0 : E_1 \times F_2 \to \mathbb{R}$ by

$$f_0(x,u_2) = \int_\Omega g(\tau, x(\tau))\, d\tau + \int_\Gamma h(u_2(\sigma))\, d\sigma.$$

From [13] it is shown that f_0 is concave and upper semicontinuous.

With the above formulation the optimal control problem we consider consists of "maximizing"

$$\int_0^\infty f_0(x(t), u_2(t))\, dt$$

over all pairs of functions $(x(\cdot)\,,u_2(\cdot)\,) : [0,\infty) \to E_1 \times F_2$ satisfying $x(t)(\cdot) \in L^2(\Omega; X)$ a.e. on $[0,\infty)$, $u_2(t)(\cdot) \in L^2(\Omega; U_2)$ a.e. on $[0,\infty)$ and the input-output relation (9.55), where the operators $\{S(t) : t \geq 0\}$; A; B_i, $i = 1,2$; and B are as above. The corresponding OSSP consists of maximizing

$$f_0(x(\cdot)\,,u_2(\cdot)\,)$$

over all pairs $(x(\cdot)\,,u_2(\cdot)\,) \in L^2(\Omega;\mathbb{R}) \times L^2(\Gamma;\mathbb{R})$ satisfying

$$\begin{aligned} \Delta_\tau x(\tau) &= 0 \text{ in } \Omega \\ x|_\Gamma &= u_2 \\ x(\tau) \in X \qquad & \text{and } u_2(\tau) \in U_2 \text{ a.e.} \end{aligned}$$

for $u_2(\cdot) \in L^2(\Gamma;\mathbb{R})$ the above Dirichlet problem has a unique generalized solution $x(\cdot) \in L^2(\Omega;\mathbb{R})$. Consequently if X is taken to be a sufficiently large interval, the OSSP will have feasible pairs $(x(\cdot)\,,u_2(\cdot)\,)$ and consequently OSSP will have solutions. The uniqueness assumption will be assured if we require f_0 to be strictly concave.

From the above it is clear that Theorems 9.11 and 9.12 can be applied if one can show that there exists an admissible pair $(\tilde{x}(\cdot)\,,\tilde{u}_2(\cdot)\,)$ such that

$$\int_0^\infty \mathcal{L}_0(\tilde{x}(t),\tilde{u}_2(t))\,dt > -\infty,$$

where $\mathcal{L}_0$ is as given in (9.63). Of course to verify condition C we will require additional information on the function f_0.

Chapter 10

Stochastic Control with the Overtaking Criterion

10.1 Introduction

In this chapter we consider infinite-horizon optimal control of stochastic systems. The methods which we present may be applied to a variety of systems; nevertheless we will mainly concentrate on controlled diffusion processes in $\mathbb{R}^n$. Other systems which may be studied similarly are controlled Markov chains with finite (or denumerable) number of states, and random evolution piecewise deterministic controlled systems, which will be further considered in Chapter 11.

In parallel to the situation with deterministic systems, an important step toward establishing overtaking optimality is to achieve a reduction to finite costs. This can be established once there exists a solution of a certain type to the infinite-horizon Bellman equation. If $\mathcal{L}^u$ is the generator of the process while employing the conrol u, and $k(x,u)$ is the running cost, then the infinite-horizon Bellman equation is

$$\min_u \{\mathcal{L}^u \phi(x) + k(x,u)\} = \mu \tag{10.1}$$

which is an *additive eigenvalue* equation both for the function ϕ and the constant μ.

We will describe some consequences of the existence of a solution to the Bellman equation (10.1). In particular, under certain conditions, the cost process may be represented as

$$C_T(u) = \mu T + \phi(x_0) - \phi(x_T) + M_T \tag{10.2}$$

where μ and $\phi(\cdot)$ are as in (10.1), and $\{M_T\}_{T\geq 0}$ is a submartingale which turns out to be a martingale whenever $u(\cdot)$ is optimal. There is much similarity between (10.2) in the stochastic case and the equation (5.1) in the deterministic framework. Instead of a submartingale $\{M_T\}_{T\geq 0}$ we have in (5.1) a nondecreasing function, which is identically zero for good controls.

We will discuss the important Linear-Quadratic-Gaussian system which tracks a periodic trajectory with incomplete information. It turns out that there exists a unique overtaking solution given by a feedback law which generalizes the deterministic result according to the separation and the certainty-equivalence principles.

Another implication of the existence of certain solutions to (10.1) is the possibility of almost-sure optimization with the overtaking criterion. A discussion of this issue will conclude the chapter.

10.2 The reduction to finite costs and the infinite-horizon Bellman equation

We begin by describing Problem I, which is the infinite-horizon optimal control of diffusion processes in $\mathbb{R}^n$. Later we will describe two other problems to which similar overtaking optimality considerations may be applied.

Problem I. The dynamics of the system is given by

$$dx_t = b(x_t, u_t) + \sigma(x_t)d\beta_t, \; x \in \mathbb{R}^n, \; u \in \mathbb{R}^m, \; x_0 \text{ is given.} \tag{10.3}$$

In (10.3) $\{\beta_t\}_{t\geq 0}$ is a Brownian motion on a probability space $(\Omega, \mathcal{F}, P)$. Let $\{\mathcal{F}_t\}_{t\geq 0}$ be the filtration generated by $\{\beta_t\}_{t\geq 0}$. Then an admissible control $\{u_t\}_{t\geq 0}$ is an $\{\mathcal{F}_t\}_{t\geq 0}$ adapted process in $\mathbb{R}^n$ which satisfies the consraints

$$u_t \in U(x_t), \; t \geq 0 \tag{10.4}$$

where $x \to U(x)$ is a measurable set-valued mapping from $\mathbb{R}^n$ into $\mathbb{R}^m$. We call the solution of (10.3) which corresponds to the admissible control $\{u_t\}_{t\geq 0}$ *the response to* $\{u_t\}_{t\geq 0}$.

With a control $\mathbf{u} = \{u_t\}_{t\geq 0}$ and its response $\mathbf{x} = \{x_t\}_{t\geq 0}$ there is associated a cost-flow process

$$c_T(\mathbf{u}) = \int_0^T k(x_t, u_t)\, dt, \tag{10.5}$$

where $k(\cdot,\cdot)$ is a continuous function. Additional assumptions will be imposed on $k(\cdot,\cdot)$ in (10.10) and (10.11) below. The optimal control problem is to minimize $E_{x_0}[c_T(u)]$ as $T \to \infty$, where E_{x_0} denotes the expectation conditioned on $x(0) = x_0$. Thus one may try to minimize the functional

$$J(\mathbf{u}) = \limsup_{T\to\infty} E_{x_0}\left[\frac{c_T(\mathbf{u})}{T}\right]. \tag{10.6}$$

Alternatively, one may look for a control $\mathbf{u}^*$ which will be optimal in the overtaking, or weakly overtaking sense.

The infinite-horizon Bellman equation (10.1) for this problem is

$$\frac{1}{2}\sum_{i,j=1}^{N} a_{ij}(x)\frac{\partial^2\phi}{\partial x_i \partial x_j} + \min_{u\in U(x)}\left\{k(x,u) + b(x,u)'\frac{\partial\phi}{\partial x}\right\} = \mu \tag{10.7}$$

which is an equation for the constant μ and the unknown function $\phi(\cdot) : \mathbb{R}^n \to \mathbb{R}$. The following result guarantees a solution (ϕ, μ) to (10.7).

Theorem 10.1 (Borkar [21], Theorem 3.1.) *Assume that $b(\cdot,\cdot)$ and $\sigma(\cdot)$ satisfy the following conditions:*

(i) $b(\cdot,\cdot) : \mathbb{R}^n \times U \to \mathbb{R}^n$ *is bounded, continuous, and Lipschitz continuous in x for every $u \in U$, where $U \subset \mathbb{R}^m$ is a compact set and in* (10.4) $U(x) = U$ *holds for every $x \in \mathbb{R}^n$.*

(ii) $\sigma(\cdot) : \mathbb{R}^n \to \mathbb{R}^{n\times n}$ *is bounded and Lipschitz continuous, and* $\|\sigma^T(x)z\|^2 \geq \lambda_0|z|^2$ *for every x, $z \in \mathbb{R}^n$, for some $\lambda_0 > 0$.*

(iii) *Let $J(\cdot)$ be as in* (10.6) *and let*

$$\mu_0 = \inf\{J(\mathbf{u}) : \mathbf{u} \textit{ is admissible }\}.$$

Assume that the running cost $k(\cdot,\cdot)$ satisfies

$$\liminf_{|x|\to\infty}\left\{\inf_{u\in U} k(x,u)\right\} > \mu_0.$$

Then there exists a function $\phi(\cdot) : \mathbb{R}^n \to \mathbb{R}$ which satisfies (10.7) *with $\mu = \mu_0$. Moreover, $\phi(\cdot)$ is bounded below and $\phi \in C^{2,\beta}$ for every* $0 < \beta < 1$.

As indicated in the theorem the constant μ in (10.7) is the minimal cost-growth rate. The function $\phi(\cdot)$ can be given a simple interpretation if we can achieve *a reduction to finite costs with the constant* μ, which we define as follows.

Definition 10.1 *We say that there exists a reduction to finite costs with the constant μ if for every bounded set $Q \subset \mathbb{R}^n$ there exists a constant $M > 0$ such that*

$$\limsup_{T\to\infty} E_{x_0}[c_T(\mathbf{u}) - \mu T] > -M \tag{10.8}$$

for every $x_0 \in Q$ and every admissible control $\mathbf{u}$. Moreover, there is a feedback control $x \to u^*(x)$ such that

$$\limsup_{T\to\infty} E_{x_0}[c_T(\mathbf{u}^*) - \mu T] < M \tag{10.9}$$

holds for every $x_0 \in Q$.

Let $\mathcal{A}$ be the following class of functions

$$\mathcal{A} = \left\{ \begin{array}{r} \phi(\cdot) \; : \phi(\cdot) \in C^2, \; \phi(\cdot) \text{ is bounded below, and } \phi(x) \leq c_1 h(x) + c_2 \\ \text{for some } c_1, \; c_2 > 0 \end{array} \right\} \tag{10.10}$$

where we denote

$$h(x) = \inf\{k(x,u) : u \in \mathbb{R}^n\}. \tag{10.11}$$

Theorem 10.2 *If there exists a solution* $(\phi(\cdot), \mu)$ *to* (10.7) *with* $\phi(\cdot) \in \mathcal{A}$, *then there exists a reduction to finite costs with* μ.

Proof: It follows from (10.7), by applying the Ito's Lemma to the process $\{x_t\}_{t\geq 0}$ that for every admissible control $\{u_t\}_{t\geq 0}$

$$E_{x_0}\Big[c_T(\mathbf{u}) \geq \mu T + \phi(x_0)\Big] - E_{x_0}\Big[\phi(x_T)\Big] \tag{10.12}$$

for every $T > 0$. If

$$\liminf_{T\to\infty} E_{x_0}[\phi(x_T) > c_1\mu + c_2],$$

where c_1 and c_2 are as in (10.10), then, by (10.11) we have

$$\liminf_{T\to\infty} \frac{1}{T} E\Big[\int_0^T k(x_t, u_t)\, dt > \mu\Big],$$

which clearly implies (10.8). If, however,

$$\liminf_{T\to\infty} E_{x_0}[\phi(x_T) \leq c_1\mu + c_2]$$

then clearly (10.12) implies (10.8) with the choice

$$M = |\phi(x_0)| + c_1|\mu| + c_2.$$

To establish (10.9) for some $\mathbf{u}$ and a given x_0 we define the set-valued function

$$F(x) = \{u \in U : k(x,u) + \mathcal{L}^u\phi(x) = \mu\}$$

where we denote by $\mathcal{L}^u$ the following operator

$$\mathcal{L}^u\psi(x) = \frac{1}{2}\sum_{i,j=1}^{n} a_{ij}(x)\psi_{x_i,x_j} + b(x,u)'\frac{\partial\psi}{\partial x}$$

for C^2 functions $\psi(\cdot)$. By assumption, $F(x)$ is nonempty for every $x \in \mathbb{R}^n$ and it follows that it is a measurable set-valued function. Hence there is a measurable selection $u_0(\cdot) : \mathbb{R}^n \to \mathbb{R}^m$ such that $u_0(x) \in F(x)$ for every $x \in \mathbb{R}^n$. We consider the response

$\{x_t\}_{t\geq 0}$ to the feedback control $u_t = u_0(x_t)$, $t \geq 0$, and apply Ito's Lemma to the process $\{\phi(x_t)\}_{t\geq 0}$. We then obtain, analogous to (10.12), that

$$E_{x_0}[c_T(\mathbf{u}_0)] = \mu T + \phi(x_0) - E_{x_0}[\phi(x_T)],$$

which implies (10.9) since $\phi(\cdot)$ is bounded below on $\mathbb{R}^n$, and is bounded on the bounded set Q. This completes the proof of the Theorem. ∎

In Section 10.4 below we will describe a situation where there exists a solution $(\phi(\cdot)\,,\mu)$ to (10.7) with $\phi \in \mathcal{A}$.

We demonstrate now how the converse implication applies, namely how the existence of a reduction to finite costs may imply the existence of a solution to (10.7). We consider (10.3) with a feedback control $u(\cdot)$ such that

$$\|b(x, u_0(x))\| \leq a_1\|x\| + a_2 \text{ for every } x \in \mathbb{R}^n, \tag{10.13}$$

for some a_1, $a_2 > 0$. We assume that $\sigma(\cdot)$ is bounded and Lipschitz continuous and satisfies

$$\|\sigma(x)^T y\|^2 \geq \lambda_0 \|y\|^2 \text{ for every } x,\ y \in \mathbb{R}^n \tag{10.14}$$

for some $\lambda_0 > 0$.

Theorem 10.3 *Assume that there exists a reduction to finite costs with the constant μ and the feedback control $u_0(\cdot)$. Assume that* (10.13) *and* (10.14) *hold. Then there exists a solution $(\phi(\cdot)\,,\mu)$ for* (10.7) *where $\phi(\cdot)$ is a C^2 function.*

Sketch of the proof: Let S be the unit sphere in $\mathbb{R}^n$. Consider the positively recurrent diffusion $\{x_t\}_{t\geq 0}$ which is the response to $u_0(\cdot)$. Let τ_{x_0} be the first hitting time of $\{x_t\}_{t\geq 0}$ at S and define

$$\psi(x_0) = E_{x_0}\left[\int_0^{\tau_{x_0}} [k(x_t, u_0(x_t)) - \mu]\,dt + \phi(x_{\tau_{x_0}})\right] \tag{10.15}$$

where $\phi(\cdot) : S \to \mathbb{R}$ is some continuous function on S. It follows that $\psi(\cdot)$ satisfies the equation

$$\begin{cases} L^{u_0(x)}\psi(x) = -[k(x, u_0(x)) - \mu] \\ \psi(x) = \phi(x) \text{ on } S \end{cases}$$

hence it satisfies

$$\frac{1}{2}\sum_{i,j=1}^{n} a_{ij}(x)\psi_{x_i,x_j}(x) + b(x, u_0(x))'\nabla\psi(x) + k(x, u_0(x)) = \mu. \tag{10.16}$$

If we choose in (10.15) the function ϕ to be the following

$$\phi(x_0) = \inf_{\mathbf{u}}\left\{\liminf_{T\to\infty} E_{x_0}[C_T(\mathbf{u}) - \mu T]\right\}$$

then $\phi(\cdot)$ is finite, bounded below and continuous. It follows then that also $\psi(\cdot)$ in (10.15) is bounded below. Since it satisfies (10.16) it follows that

$$\frac{1}{2}\sum_{i,j=1}^{n} a_{ij}(x)\psi_{x_i,x_j}(x) + \min_{u\in U(x)} \{b(x,u)'\nabla\psi(x) + k(x,u)\} \leq \mu. \tag{10.17}$$

If in (10.17) there occures strict inequality on a set of positive Lebesgue measure in $\mathbb{R}^n$, then we get a contradiction to the fact that μ is the minimal cost-growth rate. (See below the proof of Theorem 10.9 in Section 10.4.) Hence the equality in (10.17) holds for almost every x in $\mathbb{R}^n$, and since ψ is C^2 it holds everywhere in $\mathbb{R}^n$. ∎

The existence of a solution $(\phi(\cdot), \mu)$ to (10.17) implies that the cost-process $\{c_T(u)\}_{T\geq 0}$ in (10.5) has a special structure. This is the content of the following result.

Theorem 10.4 *Suppose that $(\phi(\cdot), \mu)$ is a solution to* (10.7). *Then the cost-process $\{c_T(u)\}_{T\geq 0}$ can be represented as follows:*

$$c_T(\mathbf{u}) = \mu T + \phi(x_0) - \phi(x_T) + N_T \tag{10.18}$$

where $\{N_T\}_{T\geq 0}$ is a submartingale. Moreover, there exists a feedback control $u_0(\cdot)$ such that

$$c_T(\mathbf{u}_0) = \mu T + \phi(x_0) - \phi(y_T) + M_T \tag{10.19}$$

where $\{y_t\}_{t\geq 0}$ is the response to u_0, and $\{M_T\}_{T\geq 0}$ is a martingale.

Proof: Let $\mathbf{u} = \{u_t\}_{t\geq 0}$ be an admissible control with a response $\{x_t\}_{t\geq 0}$ which corresponds to the initial value x_0. Equation (10.12) states that

$$E_{x_0}[c_T(\mathbf{u}) - \mu T] \geq \phi(x_0) - E_{x_0}\phi(x_T),$$

which implies that

$$E\Big[\int_{T_1}^{T_2} [k(x_t, u_t) - \mu]\, dt | \mathcal{F}_{T_1}\Big] \geq \phi(x_{T_1}) - E_{x_{T_1}}\phi(x_{T_2}). \tag{10.20}$$

But in view of (10.18), we have that (10.20) is equivalent to

$$E[N_{T_2}|\mathcal{F}_{T_1}] \geq N_{T_1}$$

for every $T_2 \geq T_1 \geq 0$, proving that $\{N_T\}_{T\geq 0}$ is a submartingale.

If $u_0(\cdot)$ is chosen so that $u_0(x)$ is a minimizer of $u \to k(x,u) + b(x,u)'\nabla\phi(x)$, then $\{N_T\}_{T\geq 0}$ in (10.18) turns out to be a martingale, and we thus obtain (10.19). ∎

Remark 10.1 *In certain linear quadratic systems it follows that all the relevant controls have a common value for* $\lim_{T\to\infty} \phi(x_T)$. *In this situation Theorem* 10.4 *implies that* $u_0(\cdot)$ *is an overtaking optimal control.* (*See Section* 10.3 *for the discussion of the infinite-horizon LQG problem.*)

We conclude this section with a brief description of two other stochastic problems to which similar methods may be applied.

Problem II. *Optimal control of Markov chains with finite number of states.*
We consider the optimal control of a discrete time stochastic process $\{j_k\}_{k\geq 0}$ which is defined on a probability space $(\Omega, \mathcal{F}, P)$. The state space of the controlled process is the finite set $\{1, \ldots, n\}$. We denote the discrete time variable by $k = 0, 1, 2, \ldots$. If we apply a control u at time k while the system occupies the state $1 \leq i \leq n$, then as a result the system will occupy the state j at time $k+1$ with probability $p_{ij}(u)$, which does not depend on k.

The control variable u takes values in some complete metric space U. A control process $\{u_k\}_{k=0}^{\infty}$ is an admissible control if it is nonaticipative, namely u_k is a functional of $\{j_0, \ldots, j_k\}$ for every $k \geq 0$. If $\phi(j, u)$ is the cost incurred while applying the control u when the system is at state j, then the cost of applying the admissible control on the $[0, N]$ time interval is

$$C_N(u) = \sum_{k=0}^{N-1} \phi(j_k, u_k).$$

The typical situation is that the expected cost $E_{j_0} C_N(u)$ will diverge to infinity as $T \to \infty$. The goal is to minimize the expected cost-flow $\{E_{j_0} C_N(u)\}_{N=1}^{\infty}$ in the overtaking, or weakly overtaking sense.

Problem III. *Optimal control of random evolution systems.*
We consider the control system

$$\dot{x} = f_i(x, u), \; x \in \mathbb{R}^n, \; u \in \mathbb{R}^m$$

where $f_i(\cdot, \cdot)$ is a smooth function on $\mathbb{R}^n \times \mathbb{R}^m$ for every $1 \leq i \leq N$. The system's dynamics is given by the state equation

$$\dot{x}(t) = f_{j_t}(x(t), u(t)), \; x(0) = x_0$$

for a prescribed $x_0 \in \mathbb{R}^n$, where $\{j_t\}_{t\geq 0}$ is a finite state Markov chain such that $j_t \in \{1, \ldots, N\}$ for every $t \geq 0$. The transition probabilities of $\{j_t\}_{t\geq 0}$ are known. Thus the system follows a deterministic differential equation for a random time interval at the end of which it starts following another differential equation, etc. Thus there are N different possible deterministic differential equations and the system shifts from one to another at random times.

A control function $t \to u(t)$ is admissible if it is nonanticipative in the sense that $u(t)$ is a functional of x_0 and of $\{j_s, x_s\}_{s=0}^{t}$. With an admissible control $\mathbf{u}$ there is associated a cost-process

$$C_T(\mathbf{u}) = \int_0^T f_0(x_t, u_t)\, dt$$

and the goal is to minimize the expected cost-flow $\{E_{x_0,j_0}[C_T(\mathbf{u})]\}_{T\geq 0}$ as $T \to \infty$. Since typically the expected cost-flow diverges to infinity, this minimization is in the overtaking, or weakly overtaking sense.

10.3 Infinite-horizon stochastic tracking

We apply the ideas described in Section (10.2) to study the standard LQG problem with partial information and with the overtaking criterion. We establish the existence of a reduction to finite costs, from which we conclude the existence of a unique overtaking optimal control. We will show that the optimal control is obtained while using the well-known affine feedback control, according to the separation and certainty-equivalence principles. We will discuss some asymptotic properties and the limiting behaviour as time grows to infinity of the controlled process.

We consider the stochastic analog of the tracking problem which was discussed in Section 5.5. Let $(\Omega, \mathcal{F}, P)$ be a fixed probability space, and the system under consideration is

$$\begin{aligned} dx(t) &= (Ax(t) + Bu(t))\, dt + Gd\beta(t),\ x(0) = x_0 \\ dy(t) &= Dx(t)\, dt + Hd\tilde{\beta}(t), \end{aligned} \tag{10.21}$$

$x(\cdot)$ and $y(\cdot)$ being stochastic processes in $\mathbb{R}^n$ and $\mathbb{R}^k$ respectively. In (10.21) $\beta(\cdot)$ and $\tilde{\beta}(\cdot)$ are p and $\tilde{p}$ dimensional Brownian motions respectively. Let $\mathcal{Y}_t$ be the σ-algebra generated by $\{y(s)\}_{0\leq s\leq t}$. For the special case where we choose $u(\cdot)$ to be identically zero, then we denote the corresponding increasing family of σ-algbras by $\{\mathcal{Y}_t^0\}_{t\geq 0}$.

Definition 10.2 *The* **admissible controls** *are stochastic processes in* $\mathbb{R}^m$ *such that*

(i) $u(t)$ *is* $\mathcal{Y}_t$*-measurable for every* $t \geq 0$.

(ii) *There is a constant* α *(dependent on* $u(\cdot)$ *) such that the solution* $x(\cdot)$ *of* (10.21) *satisfies*

$$E\left[\|x(t)\|^2\right] < \alpha \text{ for every } t \geq 0.$$

(iii) *The* σ*-algebra* $\mathcal{Y}_t$*, which generally contains* $\mathcal{Y}_t^0$*, is in fact equal to* $\mathcal{Y}_t^0$.

Remark 10.2 *It will be clear from the structure of the cost expression* (10.23) *below that we can restrict our attention to the class* U_0 *of controls* $u(\cdot)$ *which satisfy* $E\Big[\int_0^T \|u(t)\|^2 dt\Big] < \infty$ *for all* $T > 0$.

Remark 10.3 *The claim in* (iii) *that* $\mathcal{Y}_t^0 \subset \mathcal{Y}_t$ *for every* $t \geq 0$ *can be easily established (see Lemma* 10.2 *below).*

Considering the assumption in (iii) the reader is referred to Balakrishnan[11], Chapter 7. In particular, it is proved there (Theorem 7.1, page 69) that controls which are linear transformations of the observed process $y(\cdot)$ are such that $\mathcal{Y}_t^0 = \mathcal{Y}_t$. Also see Willems [181] for a discussion concerning the admissible controls. It turns out that our class of admissible controls contains the locally Lipschitz continuous transformations used, e.g., by Wonham [182] for the Linear-Quadratic problem.

The matrices A, B, C, D and H are constant and of appropriate dimensions.

Along with (10.21) a trajectory

$$\Gamma(\cdot) : [0, \infty) \to \mathbb{R}^n \tag{10.22}$$

is given and assumed to be periodic with period T_0, i.e.

$$\Gamma(t + T_0) = \Gamma(t) \text{ for all } t \geq 0 \,.$$

We assume that $\Gamma(\cdot)$ is piecewise continuous.

To every admissible control there corresponds a unique solution of (10.21) which we call *the response to* $u(\cdot)$ and denote by $x(\cdot, u(\cdot), x_0)$ (see Friedman [80] Theorem 5.1, page 218). The dependence on $u(\cdot)$ and x_0 will be suppressed if no confusion may arise.

The cost of using the control $u(\cdot)$ with its respnse $x(\cdot)$ over the [0, T] time interval is

$$c_T(u(\cdot)) = E\Big[\int_0^T \left\{\|x(t) - \Gamma(t)\|_Q^2 + \|u(t)\|_R^2\right\} dt\Big] \tag{10.23}$$

where R is a given positive-definite symmetric matrix and Q is a positive-semidefinite symmetric matrix.

The following is assumed throughout:

A STANDING HYPOTHESIS. *The pair $(A,\ B)$ is controllable and the pair $(A,\ Q)$ is observable. The matrix A is stable.*

We associate with (10.21) its deterministic analog, which was studied in Section 5.5

$$\dot{X}(t) = AX(t) + BU(t),\ X(0) = x_0,\ X \in \mathbb{R}^n,\ U \in \mathbb{R}^n. \tag{10.24}$$

We will use capital letters for the corresponding variables of the deterministic system. *Admissible deterministic controls* $U(\cdot)$ are measurable functions such that $\int_0^T \|U(t)\|^2\,dt < \infty$ for every $T > 0$. If $u(\cdot)$ is an admissible stochastic control and $x(\cdot)$ is its response then $Eu(\cdot)$ is an admissible deterministic control with the response $Ex(\cdot)$, and the expectations $Eu(\cdot)$ and $Ex(\cdot)$ satisfy equation (10.24). We will use the cost expression

$$C_T(U(\cdot)) = \int_0^T \left\{\|X(t) - \Gamma(t)\|_Q^2 + \|U(t)\|_R^2\right\}\, dt \tag{10.25}$$

for a deterministic control $U(\cdot)$ with the response $X(\cdot)$.

Even if $\Gamma(t) \equiv 0$ for every $t \geq 0$, then $c_T(\mathbf{u}) \to \infty$ *as* $T \to \infty$ for every choice of an admissible control. This is in contrast to the situation for deterministic system where there are controls $U(\cdot)$ for which the cost $C_\infty(U)$ of regulating on $[0, \infty)$ is finite. It thus follows that for stochastic systems the introduction of the overtaking criterion to handle infinite-horizon optimization problems is even more natural then in the deterministic context.

The main result is the following Theorem:

Theorem 10.5 *For every initial condition x_0 there exists a unique overtaking optimal control $u^*(\cdot)$ given by the feedback control law*

$$u^*(t) = -R^{-1}B'[K\hat{x}(t) - g(t)]. \tag{10.26}$$

Here $\hat{x}(t)$ is the best estimate of $x(t)$ given $\{y(s),\ 0 \leq s \leq t\}$, K is the positive-definite symmetric solution of the algebraic Riccati equation

$$KA + AK' - KBR^{-1}B'K + Q = 0 \tag{10.27}$$

and $g(\cdot)$ is the periodic function

$$g(t) = \int_t^\infty e^{(s-t)F'}Q\Gamma(s)ds\ , \tag{10.28}$$

denoting $F' = A' - KBR^{-1}B'$.

The following stronger version of overtaking optimality holds: If $u(\cdot)$ is any other admissible control then there is a time t_0 and an $\epsilon \geq 0$ (depending on t_0 and $u(\cdot)$) such that

$$c_t(u^*(\cdot)) + \epsilon < c_t(u(\cdot)) \textit{ for all } t \geq t_0.$$

Remark 10.4 *The feedback control law* (10.26) *is a straightforward generalization of the result for deterministic systems obtained in Section* 5.5. *The only difference is the replacement of* $x(\cdot)$ *by* $\hat{x}(\cdot)$ *since there is partial information and* $x(t)$ *cannot be measured at time* t. *In case that there is complete information then the feedback law* (10.26) *coincides with the one for deterministic systems.*

The following result serves as an auxilliary tool in the proof of Theorem 10.5.

Lemma 10.1 (Åstrom [8] Lemma 7.1, page 287). *Let* K *be the positive definite solution of the Riccati equation* (10.27) *and let* $u(\cdot)$ *be an admissible control with a response* $x(\cdot)$ *on* $[0,\infty)$. *Then the following identity holds*

$$\begin{aligned}\int_0^T [x(t)'Qx(t) + u(t)'Ru(t)]\,dt &= x(0)'Kx(0) - x(T)'Kx(T)\\ &\quad + \int_0^T \|u + R^{-1}B'Kx(t)\|_R^2\,dt + \int_0^T trG'KG\,dt\\ &\quad + \int_0^T d\beta(t)'G'Kx(t)\,dt + \int_0^T x(t)KGd\beta(t).\end{aligned} \tag{10.29}$$

We use (10.29) to express the cost $c_T(u)$ as follows:

$$\begin{aligned}c_T(u) &= E_{x_0}\Big[\int_0^T [x'Qx + u'Ru]\,dt\Big] - 2E_{x_0}\Big[\int_0^T \Gamma(t)'Qx(t)\,dt + \int_0^T \Gamma'Q\Gamma\,dt\Big]\\ &= E_{x_0}\Big[\int_0^T \|u + R^{-1}B'Kx(t)\|_R^2\,dt\Big] - 2E_{x_0}\Big[\int_0^T \Gamma(t)'Qx(t)\,dt + \int_0^T [\mathrm{tr}\,G'KG + \Gamma'Q\Gamma]\,dt\Big]\\ &\quad + E_{x_0}\Big[\int_0^T d\beta(t)'G'Kx(t) + \int_0^T x(t)'KGd\beta(t) + x_0'Kx_0 - x(T)'Kx(T)\Big].\end{aligned}$$

The expectations of the stochastic integrals vanish by (ii) of Definition 10.2, and we obtain

$$\begin{aligned}c_T(u) &= E_{x_0}\Big[\int_0^T \|u + R^{-1}B'Kx(t)\|_R^2\,dt\Big] - 2E_{x_0}\Big[\int_0^T \Gamma(t)'Qx(t)\,dt - x(T)'Kx(T)\Big]\\ &\quad + \Big\{\int_0^T [\mathrm{tr}G'KG + \Gamma'Q\Gamma]\,dt + x_0'Kx_0\Big\}.\end{aligned} \tag{10.30}$$

The second term is common to all the admissible controls. We will see below that $\lim_{T\to\infty} E_{x_0}\big[x(T)'Kx(T)\big]$ exists and has the same value for all the "relevant" admissible controls (which will be precisely formulated below). The essential contribution

$$E_{x_0}\left[\int_0^T \|u + R^{-1}B'Kx(t)\|_R^2\, dt\right] - 2E_{x_0}\left[\int_0^T \Gamma(t)'Qx(t)\, dt\right]$$

will be estimated by employing the deterministic result obtained in Section 5.5. The following facts will be needed in the sequel:

Lemma 10.2 *Let $u(\cdot)$ be an admissible control, $x(\cdot)$ and $y(\cdot)$ the corresponding response and observed process in* (10.21), *and $\mathcal{Y}_t$ be the σ-algebra generated by $\{y(s),\ 0 \le s \le t\}$. Let $\mathcal{Y}_t^0$ be this σ-algebra in the case where $u(t) \equiv 0$. Then*

(i) $\mathcal{Y}_t^0 \subseteq \mathcal{Y}_t$.

(ii) *The process $\{x(t) - \hat{x}(t)\}_{t\ge 0}$, where $\hat{x}(t) = E[x(t)|\mathcal{Y}_t^0]$ does not depend on the choice of the control $u(\cdot)$.*

(iii) *If $u(t)$ is a linear transformation of the observed process up to time t, namely $u(t)$ is a linear functional of the values of $y(\cdot)$ in the $[0,t]$ time interval, then $\mathcal{Y}_t^0 = \mathcal{Y}_t$ and* (iii) *of Definition* 10.2 *holds.*

Proof: Let $x_0(\cdot)$ and $y_0(\cdot)$ be the solution and observed process of (10.21) corresponding to the control $u_0(t) \equiv 0$. Denoting

$$x_u(t) = \int_0^T e^{A(t-s)}Bu(s)ds$$

we get $x(t) = x_0(t) + x_u(t)$, and since $x_u(t)$ is $\mathcal{Y}_t$-measurable, it follows from $y(t) = y_0(t) + D\int_0^t x_u(s)ds$ that $y_0(t)$ is $\mathcal{Y}_t$ -measurable, hence $\mathcal{Y}_t^0 \subseteq \mathcal{Y}_t$ for every $t \ge 0$, proving (i).

Since $u(\cdot)$ is an admissible control then we have $\mathcal{Y}_t^0 = \mathcal{Y}_t$, hence $\hat{x}(t) = \hat{x}_0(t) + x_u(t)$, which together with $x(t) = x_0(t) + x_u(t)$ yields (ii). The claim in (iii) is asserted and proved in Balakrishnan [11] (Theorem 7.1, page 169). ∎

With an admissible control $u(\cdot)$ we associate the following process

$$v(t) = u(t) + R^{-1}B'[K\hat{x}(t) - g(t)] \tag{10.31}$$

and it follows that $v(t)$ is $\mathcal{Y}_t^0$-measurable. The process $v(\cdot)$ measures how far is the control $u(\cdot)$ from the feedback law (10.26), which is our candidate to be optimal. In terms of $v(\cdot)$ the cost $c_T(u)$ in (10.30) can be written as

$$c_T(u(\cdot)) = E_{x_0}\left[\int_0^T \|v(t) + R^{-1}B'g(t) + R^{-1}B'K[x(t) - \hat{x}(t)]\|_R^2 - x(T)'Kx(T) - \right.$$

$$\left. -2\int_0^T \Gamma(t)'Qx(t)\, dt\right] + \int_0^T [\operatorname{tr} G'KG + \Gamma(t)'Q\Gamma(t)]\, dt + x_0'Kx_0.$$

By Lemma 10.2 $[x(t)-\hat{x}(t)]$ does not depend on $u(\cdot)$ and is independent of $\mathcal{Y}_t^0$, and we obtain

$$\begin{aligned} c_T(u(\cdot)) &= E_{x_0}\Big[\int_0^T \|v(t)\|_R^2\,dt + 2\int_0^T g(t)'Bv(t)\,dt - 2\int_0^T \Gamma(t)'Qx(t)\,dt - x(T)'Kx(T)\Big] \\ &\quad +\Big\{x_0'Kx_0 + (\operatorname{tr} G'KG)T + \int_0^T \Gamma(t)'Q\Gamma(t)\,dt \\ &\quad + E_{x_0}\Big[\int_0^T [x(t)-\hat{x}(t)]'KBR^{-1}B'[x(t)-\hat{x}(t)]\,dt\Big] + \int_0^T \|B'g(t)\|_{R^{-1}}^2\,dt\Big\}. \end{aligned} \tag{10.32}$$

The second term in (10.32) is common to all the admissible controls, thus as long as overtaking optimality is concerned we may consider the cost functional

$$\begin{aligned} m_T(u(\cdot)) &= E_{x_0}\Big[\int_0^T \|v(t)\|_R^2\,dt + 2\int_0^T g(t)'Bv(t)\,dt \\ &\quad -2\int_0^T \Gamma(t)'Qx(t)\,dt - x(T)'Kx(T)\Big] \end{aligned} \tag{10.33}$$

and a control $u(\cdot)$ is overtaking optimal with respect to the cost-flow $\{c_T(u)\}_{T\geq 0}$ if and only if it is overtaking optimal with respect to the cost-flow $\{m_T(u)\}_{T\geq 0}$ for $m_T(\cdot)$ in (10.33). Note that (10.32) yields an expression for the minimal cost-growth rate

$$\begin{aligned} \mu = \operatorname{tr} G'KG + \lim_{T\to\infty}\Big\{\frac{1}{T}E_{x_0}\Big[\int_0^T [x(t)-\hat{x}(t)]'KBR^{-1}B'K[x(t)-\hat{x}(t)]\,dt\Big]\Big\} \\ + \frac{1}{T_0}\int_0^{T_0}\Big\{\Gamma(t)'Q\Gamma(t) + \|B'g(t)\|_{R^{-1}}^2 - 2\Gamma(t)'Q\gamma^*(t)\Big\}\,dt \end{aligned} \tag{10.34}$$

when this limit exists. In (10.34) $\gamma^*(\cdot)$ is the limitting trajectory of the deterministic system discussed in Chapter 5. (In Theorem 10.6 below there are presented conditions which guarantee the existense of the limit in (10.34).) It will follow from the proof of Theorem 10.5 that then there exists a reduction to finite costs, namely the limit $\lim_{T\to\infty}[c_T(\mathbf{u}) - \mu T]$ exists (either finite or plus infinity), and there are admissible controls for which it is finite.

Proof of Theorem 10.5: As discussed above a control $u^*(\cdot)$ is overtaking optimal if and only if it minimizes the functional $u(\cdot) \to m_\infty(u(\cdot))$ for $m_T(\cdot)$ in (10.33). For every control $u(\cdot)$ we define the corresponding expected values

$$U(t) = E_{x_0}[u(t)],\ V(t) = E_{x_0}[v(t)],\ X(t) = E_{x_0}[x(t)]. \tag{10.35}$$

Notice that the cost $C_T(U(\cdot))$ of the deterministic control $U(\cdot)$ over the $[0,\infty)$ is similar to the right-hand-side of (10.33), and in fact

$$M_T(U(\cdot)) = \int_0^T \|V(t)\|_R^2\,dt + 2\int_0^T g(t)'B'V(t)\,dt - 2\int_0^T \Gamma(t)'QX(t)\,dt - X(T)'KX(T). \tag{10.36}$$

In particular, for the control $u^*(\cdot)$ in (10.26) we have $v^*(t) \equiv 0$ hence we obtain

$$m_T(u^*) = -2\int_0^T \Gamma(t)'QX^*(t)\,dt - E_{x_0}\Big[x^*(T)'Kx^*(T)\Big] \tag{10.37}$$

and the corresponding deterministic control $U^*(\cdot)$, which is known to be the unique optimal control, satisfies

$$M_T(U^*) = -2\int_0^T \Gamma(t)'QX^*(t)\,dt - X^*(T)'KX^*(T). \tag{10.38}$$

We distinguish between two cases:

Case a: $E_{x_0}\Big[\int_0^\infty \|v(t)\|_R^2\,dt\Big] = \infty.$

We will show that in this case

$$\lim_{T\to\infty}[m_T(u(\cdot)) - m_T(u^*(\cdot))] = \infty, \tag{10.39}$$

which implies

$$\lim_{T\to\infty}[c_T(u(\cdot)) - c_T(u^*(\cdot))] = \infty$$

so that $u^*(\cdot)$ overtakes $u(\cdot)$. To show the validity of (10.39) we will have to show that $T \to E\Big[\|x^*(T)\|^2\Big]$ is a bounded function, which is proved below. We next split the discussion into two possibilities:

(i) If $\int_0^\infty \|V(t)\|_R^2 dt < \infty$, then (10.39) follows immediately from (10.33), (10.36), (10.37), (10.38), the boundedness of $T \to E\Big[\|x(T)\|^2\Big]$ and the overtaking optimality of $U^*(\cdot)$ (for the associated deterministic problem).

(ii) If $\int_0^\infty \|V(t)\|_R^2\,dt = \infty$, then it is easy to show (see [120] Lemma 3.5) that the deterministic pair $(U(\cdot), X(\cdot))$ is such that

$$\lim_{T\to\infty}[M_T(U) - M_T(U^*)] = \infty.$$

It follows from (10.33), (10.36), (10.37) and (10.38) that

$$m_T(u) - m_T(u^*) \geq M_T(U) - M_T(U^*) + [X(T)'KX(T) - Ex(T)'Kx(T)$$
$$-X^*(T)'KX^*(T) + Ex^*(T)'Kx^*(T)]$$

and since $T \to E\left[\|x^*(T)\|^2\right], T \to E\left[\|x(T)\|^2\right]$ are bounded, it follows that $m_T(u) - m_T(u^*) \to \infty$ as $T \to \infty$, which is the claim in (10.39). This concludes the discussion in this case.

Case b: $E_{x_0}\left[\int_0^\infty \|v(t)\|_R^2\, dt\right] < \infty$.

From (10.33), (10.36), (10.37) and (10.38) we have the relations

$$m_T(u) \geq M_T(U) + X(T)'KX(T) - E_{x_0}[x(T)'Kx(T)] \tag{10.40}$$

and

$$m_T(u^*) = M_T(U^*) + X^*(T)'KX^*(T) - E_{x_0}\left[x^*(T)'Kx^*(T)\right]. \tag{10.41}$$

By the results of Chapter 5 we know that $X^*(T) - \gamma^*(T) \to 0$ as $T \to \infty$. Once we have shown that $T \to E\left[\|x^*(T)\|^2\right]$ is a bounded function on $[0,\infty)$ and $E\left[\|x(T) - x^*(T)\|^2\right] \to 0$ as $T \to \infty$, then it will follow that $T \to E\left[\|x(T)\|^2\right]$ is a bounded function on $[0,\infty)$ and, moreover, that $X(T) - X^*(T) \to 0$ and $E[x(T)'Kx(T)] - E[x^*(T)'Kx^*(T)] \to 0$ as $T \to \infty$. This will prove that $u^*(\cdot)$ overtakes $u(\cdot)$. By the equality $x'Kx = \mathrm{tr}Kxx'$ it is thus sufficient to prove that

$$\lim_{T\to\infty} E_{x_0}[x(T)x(T)'] - E_{x_0}[x^*(T)x^*(T)'] = 0. \tag{10.42}$$

The equations satisfied by $x(\cdot)$ and $x^*(\cdot)$ are

$$dx^*(t) = [Fx^*(t) + BR^{-1}B'g(t)]\, dt + Gd\beta + BR^{-1}B'K(x^* - \hat{x}^*)(t)\, dt$$

and

$$dx(t) = [Fx(t) + BR^{-1}B'g(t)]\, dt + Bv(t)\, dt + Gd\beta_t + BR^{-1}B'K(x - \hat{x})(t)\, dt$$

respectively where F is a stable matrix which satiafies $\|e^{Ft}\| \leq \gamma e^{-\alpha t}$ for some constants γ, $\alpha > 0$ and all $t \geq 0$, and $(x - \hat{x})(\cdot)$ does not depend on the choice of $u(\cdot)$. It will be convenient to denote

$$\Delta(t) = BR^{-1}B'K(x - \hat{x})(t).$$

The stability of A in (10.21) implies that there is a bound b such that

$$E\left[\|\Delta(t)\|^2\right] \leq b \text{ for all } t \geq 0.$$

We will denote

$$\phi(t) = BR^{-1}B'g(t)$$

and $\phi(\cdot)$ is a periodic function of period T_0. It follows from the above differential equations for $x(\cdot)$ and $x^*(\cdot)$ that

$$x^*(t) = e^{Ft}x_0 + \int_0^t e^{F(t-s)}\phi(s)ds + \int_0^t e^{F(t-s)}Gd\beta(s) + \int_0^t e^{F(t-s)}\Delta(s)ds \tag{10.43}$$

and

$$x(t) = x^*(t) + \int_0^t e^{F(t-s)}Bv(s)ds. \tag{10.44}$$

We first consider $x^*(\cdot)$ and prove that $t \to E\Big[\|x^*(t)\|^2\Big]$ is a bounded function on $[0,\infty)$. To this end it is enough to show that each term in (10.43) has a bounded variance. Now $\int_0^T e^{F(t-s)}\phi(s)ds$ is bounded on $[0,\infty)$ due to the stability of F and the periodicity of $\phi(\cdot)$. We have

$$E\Big[\Big(\int_0^t e^{F(t-s)}Gd\beta(s)\Big)\Big(\int_0^t e^{F(t-s)}Gd\beta(s)\Big)'\Big] = \int_0^t e^{F(t-s)}GG'e^{F'(t-s)}ds$$

which is bounded since F is stable.

It follows from the Cauchy-Schwarz inequality that there is a constant κ such that any two random variables Z_1 and Z_2 satisfy

$$E\|Z_1Z_2'\| \le \kappa(E\|Z_1\|^{\frac{1}{2}})(E\|Z_2\|^{\frac{1}{2}}). \tag{10.45}$$

For the last term in (10.43) we thus have the following estimate:

$$\Big\|E\Big[\Big(\int_0^t e^{F(t-s)}\Delta(s)ds\Big)\Big(\int_0^t e^{F(t-s)}\Delta(s)ds\Big)'\Big]\Big\|$$

$$= \Big\|\int_0^t\int_0^t e^{F(t-s_1)}E[\Delta(s_1)\Delta(s_2)']e^{F(t-s_2)}ds_1ds_2\Big\|$$

$$\le \gamma^2\kappa b\int_0^t\int_0^t e^{-\alpha(t-s_1)}e^{-\alpha(t-s_2)}ds_1ds_2$$

using the stability of F, the inequality (10.45) and the uniform bound $E\Big[\|\Delta(t)\|^2\Big] \le b$. Hence this term has a bounded variance too and the boundedness of $\Big[\|x^*(t)\|^2\Big]$ follows.

From (10.44) we have

$$E[x(t)x(t)'] = E[x^*(t)x^*(t)'] + E\Big[[x^*(t)]\Big[\int_0^t e^{F(t-s)}Bv(s)ds\Big]'\Big]$$

$$+E\Big[\Big[\int_0^t e^{F(t-s)}Bv(s)ds\Big]x^*(t)'\Big] + E\Big[\Big[\int_0^t e^{F(t-s)}Bv(s)ds\Big]\Big[\int_0^t e^{F(t-s)}Bv(s)ds\Big]'\Big].$$

By (10.45) it will be sufficient to prove that

$$E\Big[\Big\|\int_0^t e^{F(t-s)}Bv(s)ds\Big\|^2\Big] \to 0, \text{ as } t \to \infty,$$

in order to establish $E[x(t)x(t)' - x^*(t)x^*(t)'] \to 0$. Let $0 < t_0 < t$ and compute

$$\Big\|\int_0^t e^{F(t-s)}Bv(s)ds\Big\|^2 \le 2\Big\|\int_0^{t_0} e^{F(t-s)}Bv(s)ds\Big\|^2 + 2\Big\|\int_{t_0}^t e^{F(t-s)}Bv(s)ds\Big\|^2$$

$$\le 2\Big[\int_0^{t_0} \|e^{F(t-s)}\|\, |Bv(s)|ds\Big]^2 + 2\|\int_{t_0}^t e^{F(t-s)}Bv(s)ds\|^2.$$

The matrix F is stable so that $\|e^{Ft}\| \le \gamma e^{-\alpha t}$ for some γ, $\alpha > 0$ and all $t > 0$, therefore

$$E_{x_0}\Big[\Big\|\int_0^t e^{F(t-s)}Bv(s)ds\Big\|^2\Big] \le 2\gamma^2 e^{-2\alpha(t-t_0)}E_{x_0}\Big[\int_0^{t_0} \|Bv(s)\|ds\Big]^2$$

$$+2E_{x_0}\Big[\Big\|\int_{t_0}^t e^{F(t-s)}Bv(s)ds\Big\|^2\Big].$$

Once t_0 is fixed the first term decreases to zero when t grows to infinity. For the second term we have by the Cauchy-Schwarz inequality, that for some constant κ_0

$$\Big\|\int_{t_0}^t e^{F(t-s)}Bv(s)ds\Big\|^2 \le \kappa_0 \int_{t_0}^t \|e^{F(t-s)}\|^2 ds \int_{t_0}^t \|Bv(s)\|^2 ds$$

$$\le [\kappa_0\gamma^2 \cdot \|B\|^2 \cdot \int_{t_0}^t e^{-2\alpha(t-s)}ds] \cdot \int_{t_0}^\infty \|v(s)\|^2 ds.$$

Thus we obtain the estimate

$$E_{x_0}\Big[\Big\|\int_{t_0}^t e^{F(t-s)}Bv(s)ds\Big\|^2\Big] \le \frac{\gamma^2\kappa_0\|B\|^2}{2\alpha}E_{x_0}\Big[\int_{t_0}^\infty \|v(s)\|^2 ds\Big]$$

and the last expression is arbitrarily small if t_0 is sufficiently large. This proves (10.42) and concludes the proof of the overtaking optimality of the control $u^*(\cdot)$ in (10.26).

The stronger version of overtaking optimality is obtained as follows. If $u(\cdot)$ is such that $V(\cdot)$ is not identically zero, then the claim follows from (10.42) and the strong overtaking optimality for deterministic systems, as asserted in theorem 5.5. If, however, $V(\cdot)$ is identically zero, then $U(\cdot)$ and $X(\cdot)$ coincide with $U^*(\cdot)$ and $X^*(\cdot)$, but in order to have an equality in (10.40) we must have that

$$E\big[v(t)'Rv(t)\big] = \big(E[v(t)]\big)'R\big(E[v(t)]\big) = 0$$

for almost every $t \geq 0$, hence $v(t) = 0$ for almost every $t \geq 0$, and in fact $u(\cdot)$ coincides with $u^*(\cdot)$. This concludes the proof. ■

We now assume, in addition to the standing hypothesis, the following:

Assumption 10.1 *The pair (A, G) in (10.18) is controllable, the pair (A, D) is observable and the matrix HH' is nonsingular.*

The following result is standard, see e.g. Bucy and Joseph [32], Theorem 4.5, p.53 and Theorem 5.4, p.73.

Theorem 10.6 *Let Assumption 10.1 hold and denote*

$$P(t) = E\Big[(x(t) - \hat{x}(t))(x(t) - \hat{x}(t))'\Big].$$

Then $P(\cdot)$ satisfies

$$\begin{aligned} P'(t) &= AP(t) + P(t)A' - PD'(HH')^{-1}DP + GG', \\ P(0) &= 0. \end{aligned}$$

Let P_0 be the positive-definite solution of

$$AP_0 + P_0A' - P_0D'(HH')^{-1}DP_0 + GG' = 0$$

then $P(t) \to P_0$ exponentially as $t \to \infty$. The optimal filter equation is

$$dx(t) = [A - P(t)D'(HH')^{-1}D]x(t)\,dt + P(t)D'(HH')^{-1}dy(t)$$

and the unforced filter equation

$$d\xi(t) = [A - P(t)D'(HH')^{-1}D]\xi(t)\,dt$$

is uniformly asymptotically stable

We shall consider now the minimal rate of cost-growth and derive an explicit expression for it.

Theorem 10.7 *The minimal growth-rate of cost of solutions of (10.21) is given by*

$$\begin{aligned} \frac{1}{T_0}\Big\{\int_0^{T_0} [g(t)'BR^{-1}B'g(t) + \Gamma(t)Q\Gamma(t) - 2\Gamma(t)'Q\gamma^*(t)]\,dt\Big\} \\ + \mathrm{tr}G'KG + \mathrm{tr}KBR^{-1}B'KP_0 \end{aligned} \tag{10.46}$$

and is independent of the initial value.

Remark 10.5 *Since the minimal growth-rate is independent of the initial value the rate in* (10.46) *will also be the minimal one when the initial value is a bounded random variable rather than fixed.*

Remark 10.6 *In the minimal growth expression* (10.46) *we identify the first term as the minimal deterministic growth-rate, while the second*

$$\mathrm{tr}[G'KG + KBR^{-1}B'KP_0]$$

results from the stochastic terms in (10.21) *and is quadratic in* G. *It is strictly positive whenever* G *is nonzero. Thus it follows that even if* $\Gamma(\cdot) \equiv 0$ *then, in the stochastic case, the cost grows to infinity for every control.*

Proof of Theorem 10.7: For every initial value the minimal growth-rate is obtained by

$$\lim_{T\to\infty} \frac{1}{T} Ec_T(u^*(\cdot))$$

where $u^*(\cdot)$ is the overtaking optimal control for this initial value. Substituting in (10.32) $u^*(t) = -R^{-1}B'K\hat{x}(t) + R^{-1}B'g(t)$, then dividing by T and letting it grow to infinity we obtain

$$\lim_{T\to\infty} \frac{1}{T} E[c_T(u^*(t))] = \frac{1}{T_0} \int_0^{T_0} [g(t)'BR^{-1}B'g(t) + \Gamma(t)'Q\Gamma(t)]\, dt + \mathrm{tr}G'KG$$

$$-2 \lim_{T\to\infty} \frac{1}{T} \int_0^T E[\Gamma(t)'Qx^*(t)]\, dt + \lim_{T\to\infty} E\Big[\int_0^T \|B'K(x^*(t) - \hat{x}^*(t))\|^2_{R^{-1}}\, dt\Big].$$

We observe that since $\|(x^* - \gamma^*)(t)\| \to 0$ as $t \to \infty$ we have

$$\lim_{T\to\infty} \frac{1}{T} E\Big[\int_0^T \Gamma(t)'Qx^*(t)\, dt\Big] = \frac{1}{T_0} \int_0^{T_0} \Gamma(t)'Q\gamma^*(t)\, dt.$$

It follows from Theorem 10.6 that

$$\lim_{T\to\infty} E\Big[\int_0^T \|B'K(x^* - \hat{x}^*)(t)\|^2_{R^{-1}}\, dt\Big] = \mathrm{tr}KBR^{-1}B'KP_0$$

which concludes the proof of the theorem. ∎

10.4 Optimal control of nonlinear diffusions in $\mathbb{R}^n$

We consider the framework which was described in Problem I in Section 10.2. Let $J(\cdot)$ be the following functional on admissible controls

$$J(u(\cdot)) = \limsup_{T\to\infty} \frac{1}{T} E[c_T(u(\cdot))]$$

and let μ be defined by

$$\mu = \inf\{J(u(\cdot)) : u(\cdot) \text{ is admissible}\}. \tag{10.47}$$

Let the operator $\mathcal{L}$ be such that $\mathcal{L}\phi$ is the left-hand side of (10.7) for a C^2 function $\phi(\cdot)$. We have the following result (see [118] Theorem 3.2):

Theorem 10.8 *Let μ be as in* (10.47)*, the infimal cost-growth rate. Then*

$$\sup_{\phi\in\mathcal{A}} \inf_{x\in\mathbb{R}^n} \mathcal{L}\phi(x) \le \mu \le \inf_{\phi\in\mathcal{A}} \sup_{x\in\mathbb{R}^n} \mathcal{L}\phi(x) \tag{10.48}$$

(*where $\mathcal{A}$ is as in* (10.10)).

It is clear that if equation (10.7) has a solution $(\phi(\cdot), \mu_0)$ with $\phi(\cdot) \in \mathcal{A}$ then there hold equalities in (10.47) and $\mu = \mu_0$. Conditions for this to happen for certain systems will be given in Theorem 10.11 below. In this situation (namely when a solution $(\phi(\cdot), \mu_0)$ to (10.7) exists), we define the set-valued function $x \to F(x)$, $x \in \mathbb{R}^n$, by

$$F(x) = \left\{u \in U(x) : \frac{1}{2}\sum_{i,j=1}^{n} a_{ij}\frac{\partial^2\phi}{\partial x_i \partial x_j} + \sum_{i=1}^{n} b_i^u(x)\frac{\partial\phi}{\partial x_i} + k(x,u) = \mu\right\}. \tag{10.49}$$

We assume that for every x, $\eta \in \mathbb{R}^n$ the function $u \to k(x,u) + \eta' b(x,u)$ attains its minimal value in $U(x)$. Then $F(x)$ is nonempty for every $x \in \mathbb{R}^n$. It is easy to see that there exists a measurable function $u_0(\cdot) : \mathbb{R}^n \to \mathbb{R}^m$ such that $u_0(x) \in F(x)$ for a.e. $x \in \mathbb{R}^n$. We denote by $\mathcal{U}_0$ the set of all such functions, namely

$$\mathcal{U}_0 = \left\{ \begin{array}{l} u(\cdot) : \mathbb{R}^n \to \mathbb{R}^m : \ u(\cdot) \text{ is measurable and} \\ \qquad\qquad\qquad\quad u(x) \in F(x) \text{ for a.e. } x \in \mathbb{R}^n \end{array} \right\}. \tag{10.50}$$

A control $u(\cdot)$ is called a *Markov control* if there exists a function $g : \mathbb{R}^n \to \mathbb{R}^m$ which is measurable and satisfies $g(x) \in U(x)$ for every $x \in \mathbb{R}^n$, and such that $u(t) = g(x_t)$. We have the following result:

Theorem 10.9 *Assume that* (10.7) *has a solution* $(\phi(\cdot), \mu_0)$ *such that* $\phi(\cdot) \in \mathcal{A}$. *Assume that the function* $h(\cdot)$ *defined by*

$$h(x) = \min\{k(x,u) : u \in U(x)\}$$

satisfies $h(x) \to \infty$ *as* $|x| \to \infty$. *Then* $J(u(\cdot)) \geq \mu_0$ *for every admissible control. Moreover, if* $u(\cdot)$ *is a Markov control then* $J(u(\cdot)) = \mu$ *if and only if* $u(\cdot) \in \mathcal{U}_0$, *(*$\mu$ *is as in* (10.47)*). It thus follows that there exists a Markov control* $u_0(\cdot)$ *such that* $J(u_0(\cdot)) \leq J(u(\cdot))$ *for every admissible control* $u(\cdot)$.

To prove Theorem 10.9 we will need the following auxilliary result:

Lemma 10.3 *Let* $u(\cdot)$ *be a control such that the associated diffusion in* $\mathbb{R}^n$ *is positively recurrent and let* $\nu^u(dy)$ *be the corresponding equilibrium measure. Let* $m(dx)$ *denote the Lebesgue measure on* $\mathbb{R}^n$. *Then* $P^u(t,x,dy)$ *and* $m(dy)$ *are mutually absolutely continuous, and if* K *is a Borel set in* $\mathbb{R}^n$ *such that* $m(K) > 0$ *then* $\nu^u(K) > 0$.

Proof of Theorem 10.9: Since $\phi(\cdot) \in \mathcal{A}$ it follows that the extreme sides of (10.48) are equal to μ_0, hence, by Theorem 10.8, μ_0 is indeed the infimal long-run average cost, and $\mu_0 = \mu$. As in the proof of Theorem 10.2 there is a control $u_0(\cdot) \in \mathcal{U}_0$ such that $J(u_0(\cdot)) = \mu$. It remains to prove that $J(u(\cdot)) > \mu$ if $u(\cdot) \notin \mathcal{U}_0$.

If $u(\cdot)$ is such that the corresponding diffusion in $\mathbb{R}^n$ is either transient or null-recurrent, then it follows from the property $h(x) \to \infty$ as $|x| \to \infty$ that

$$\lim_{T\to\infty} E_x^u h(x_T) = \infty,$$

which implies that $\lim_{T\to\infty} E_x^u k(x_T, u(x_T)) = \infty$, hence $J(u(\cdot)) = \infty$. We therefore consider a control $u(\cdot) \notin \mathcal{U}_0$ such that the corresponding diffusion $\{x_t\}_{t\geq 0}$ in $\mathbb{R}^n$ is positively recurrent with an equilibrium measure $\nu^u(dy)$.

Denote by $\mathcal{L}^u$ the operator

$$\mathcal{L}^u\phi(x) = \frac{1}{2}\sum_{i,j}^{n} a_{ij}\frac{\partial^2\phi}{\partial x_i \partial x_j} + \sum_{i=1}^{n} b_i(x,u(x))\frac{\partial \phi}{\partial x_i}$$

and for the control $u(\cdot)$ we define the function $\Delta_{u(\cdot)} : \mathbb{R}^n \to \mathbb{R}$

$$\Delta_{u(\cdot)}(x) = k(x,u(x)) + \mathcal{L}^{u(x)}\phi(x) - \mu$$

(where $(\phi(\cdot), \mu)$ is the solution to (10.7)). Then $\Delta_{u(\cdot)}$ is a measurablle, nonnegative function. It follows from the assumption $u(\cdot) \notin \mathcal{U}_0$ that there is a set $K \subset \mathbb{R}^n$ with $m(K) > 0$ such that $\Delta_{u(\cdot)}(x) > 0$ for $x \in K$ ($m(dy)$ is the Lebesgue measure on $\mathbb{R}^n$). It follows from Lemma 10.3 that

$$\int_{\mathbb{R}^n} \Delta_{u(\cdot)}(x)\nu^u(dx) = \epsilon > 0.$$

In fact, if $l > 0$ is a constant and we set

$$\Delta_{u(\cdot)}^l(x) = \min\{\Delta_{u(\cdot)}, l\},$$

then for a certain large enough l we have

$$\int_{\mathbb{R}^n} \Delta^l_{u(\cdot)}(x)\nu^u(dx) = \epsilon_0 > 0. \tag{10.51}$$

An application of Ito's Lemma to the twice continuously differentiable function $\phi(\cdot)$ yields

$$\begin{aligned} E[c_T(u(\cdot))] &= \mu T + \phi(x_0) - E^u_x[\phi(x_T)] + E^u_x\Big[\int_0^T \Delta_u(x_t)\,dt\Big] \\ &\geq \mu T + \phi(x_0) - E^u_x[\phi(x_T)] + E^u_x\Big[\int_0^T \Delta^l_u(x_t)\,dt\Big]. \end{aligned} \tag{10.52}$$

Since $\nu^u(dy)$ is the equilibrium measure and $\Delta^l_u(\cdot)$ is bounded and measurable, it follows from (10.51) that

$$\lim_{T\to\infty} E^u_x[\Delta^l_u(x_T)] = \epsilon_0. \tag{10.53}$$

Since $J(u(\cdot))$ is finite only when

$$\liminf_{T\to\infty} E^u_x[\phi(x_T)] < \infty$$

it follows from (10.52) and (10.53) that

$$J(u(\cdot)) \geq \mu + \epsilon_0$$

proving that $J(u(\cdot)) > J(u_0(\cdot))$ and concluding the proof of the theorem. ∎

Example 10.1 We describe here a class of systems for which there exists a solution to (10.7) which belongs to $\mathcal{A}$.

Let $(x, u) \to b(x, u)$ be a homogeneous function, namely

$$b(\beta x, \beta u) = \beta b(x, u)$$

for every $\beta > 0$, $x \in \mathbb{R}^n$ and $u \in \mathbb{R}^m$. We assume that the diffusion matrix σ is the identity matrix so that $a_{ij} = \delta_{ij}$. There is a constraint set $\Gamma \subset \mathbb{R}^m$ so that $u_t \in \Gamma$ for every $t \geq 0$. We assume that there exist p, a_1, $a_2 > 0$ satisfying

$$a_1\|x\|^p \leq k(x, u) \leq a_2\|x\|^p \tag{10.54}$$

for all large $\|x\|$ and every $u \in \Gamma$. We assume that $0 \in \mathring{\Gamma}$ (the inside of Γ), and for some c_1, $c_2 > 0$

$$x'b(x, u) \leq c_2 - c_1\|x\|^2 \text{ for every } x, \tag{10.55}$$

which is equivalent to

$$\max_{\|x\|=1} \{x'b(x, 0)\} < 0.$$

If (10.55) holds, then to every admissible control $u(\cdot)$ there corresponds a positively recurrent diffusion which has an equilibrium measure $\nu^u(dx)$ such that

$$\int_{R^d} \|x\|^r \nu^u(dx) < \infty \text{ for every } r \geq 0.$$

In fact we have the following result (see Lemma 3.6 in [182]):

Lemma 10.4 *Let* (10.54) *hold. Then for every* $r \geq 0$ *and every* $\epsilon > 0$ *there is an* $R > 0$ *such that*

$$\int_{\|x\|>R} \|x\|^r \nu^u(dx) < \epsilon$$

for every admissible control.

We will show next that under the same conditions as those of Theorem 10.9, there exists an ϵ-overtaking optimal control, for every $\epsilon > 0$. In this result, however, we restrict the controls to be Markov controls. Recall that a control $u(\cdot)$ is a Markov control if there is a measurable function $g(\cdot) : \mathbb{R}^n \to \mathbb{R}^m$ such that $g(x) \in U(x)$ for every $x \in \mathbb{R}^n$, and $u(t) = g(x(t))$.

Theorem 10.10 *Assume that* (10.7) *has a solution* $(\phi(\cdot), \mu)$ *such that* $\phi(\cdot)$ *is twice contiuously differentiable, bounded on* $\mathbb{R}^n$ *and belongs to* $\mathcal{A}$. *Then for every* $\epsilon > 0$ *and every initial state* x_0 *there exists an* ϵ*-overtaking optimal control in the class of all Markov controls.*

Proof: It follows from Theorem 10.9 that if $u_1(\cdot) \in \mathcal{U}_0$ and $u_2(\cdot) \notin \mathcal{U}_0$ then $J(u_1(\cdot)) < J(u_2(\cdot))$, and in particular $u_1(\cdot)$ overtakes $u_2(\cdot)$ ($\mathcal{U}_0$ is as in (10.50)). Therefore it is enough to prove existence of an ϵ-overtaking optimal control while considering only controls in $\mathcal{U}_0$. Then $\Delta_u(x) \equiv 0$ and (10.52) implies that

$$Ec_T(u(\cdot)) = \mu T + \phi(x_0) - E^u_{x_0}\phi(x_T). \tag{10.56}$$

We define the functional

$$\psi : \mathcal{U}_0 \to \mathbb{R},\ \psi(u(\cdot)) = \int_{\mathbb{R}^n} \phi(y)\nu^u(dy)$$

which satisfies the relation

$$|\psi(u)| \leq C_1 \int_{\mathbb{R}^n} h(y)\nu^u(dy) + C_2 \leq C_1 \int_{\mathbb{R}^n} k(y, u(y))\nu^u(dy) + C_2$$

and we claim that this implies $|\psi(u)| \leq C_3|\mu| + C_4$ for some constants C_3, $C_4 > 0$, thus $\psi(\cdot)$ is a bounded functional on $\mathcal{U}_0$. To prove this claim it is enough to show that

$$\int_{\mathbb{R}^n} k(x, u(x))\nu^u(dx) \leq \frac{a_2}{a_1}\mu + C_5 \tag{10.57}$$

where a_1, a_2 are the same as in (10.54) and $C_5 > 0$ is some constant. We claim that the relation (10.57) is implied by Assumption B(i) and

$$\liminf_{t\to\infty} E^u\left[\|x_t\|^p\right] \geq \int_{\mathbb{R}^n} \|x\|^p \nu^u(dx) \tag{10.58}$$

as follows: We note that

$$\int_{\mathbb{R}^n} k(x, u(x))\nu^u(dx) \leq a_2 \int_{\mathbb{R}^n} \|x\|^p \nu^u(dx) + \beta_1 \leq a_2(\liminf_{t\to\infty} E^u\|x_t\|^p) + \beta_1$$

by (10.54), so that

$$\int_{\mathbb{R}^n} k(x, u(x))\nu^u(dx) \leq a_2 \left[\frac{1}{a_1} \liminf_{t\to\infty} E^u[k(x_t, u(x_t))] + \beta_2\right] + \beta_1$$

using (10.54), which implies (10.57) since

$$\liminf_{t\to\infty} E^u[k(x_t, u_t)] \leq \mu.$$

Given an $\epsilon > 0$ there thus exists a control $u_\epsilon(\cdot) \in \mathcal{U}_0$ such that

$$\psi(u_\epsilon(\cdot)) > \psi(u(\cdot)) - \frac{1}{2}\epsilon \text{ for every } u(\cdot) \in \mathcal{U}_0. \tag{10.59}$$

It follows that

$$\lim_{T\to\infty} E^u_{x_0}[\phi(x_T)] = \psi(u(\cdot)), \quad \lim_{T\to\infty} E^{u_\epsilon}_{x_0}[\phi(x_T)] = \psi(u_\epsilon(\cdot))$$

and in view of (10.59) this implies that for some $T_\epsilon > 0$,

$$E[c_T(u_\epsilon(\cdot))] < E[c_T(u(\cdot))] + \epsilon \quad \text{for all } T > T_\epsilon,$$

concluding the proof of the theorem. ∎

10.5 On almost-sure overtaking optimality

The question which we consider in this section is the existence of controls which are almost-surely optimal. This, in contrast to the usual approach which looks for minimizers of the expected value of the cost. It is quite clear that, except for very special situations, there will be no almost-sure (a.s.) optimal controls for finite-horizon problems. As is demonstrated in Example 10.2 below, even under a relaxed definition of almost-sure optimality, it is in general the case that such optimal controls do not exist.

The situation is different for infinite-horizon control systems. In Borkar [22] the a.s. minimization of the cost per unit of time is obtained for controlled Markov chains on infinite-horizon. We are interested here in a more refined notion of a.s. optimality. For a control $\mathbf{u}$ let $t \rightarrow c_t^{\mathbf{u}}(\omega)$ be the cost-process. We say that a control $\mathbf{u}^*$ is *a.s. overtaking optimal* if for every control $\mathbf{u}$ there exists a random variable $\tau^{\mathbf{u}}$ such that

$$c_t^{\mathbf{u}^*}(\omega) < c_t^{\mathbf{u}}(\omega) \text{ for every } t > \tau^{\mathbf{u}}(\omega), \text{ a.s.}$$

One possible approach to the problem is as follows. We are interested in a situation where $Ec_t^{\mathbf{u}}(\omega) \rightarrow \infty$ as $t \rightarrow \infty$, and where $c_t^{\mathbf{u}}$ can be expressed in the following way

$$c_t^{\mathbf{u}}(\omega) = \lambda^{\mathbf{u}} t + A_t^{\mathbf{u}} + M_t^{\mathbf{u}}$$

where $\lambda^{\mathbf{u}}$ is a constant number, $A_t^{\mathbf{u}}$ is a nonnegative nondecreasing process and $M_t^{\mathbf{u}}$ is a martingale. Suppose that $\mathbf{u}^*$ is such that $A_t^{\mathbf{u}^*} \equiv 0$ and $\lambda^{\mathbf{u}^*} < \lambda^{\mathbf{u}}$ for every $\mathbf{u}$ different from $\mathbf{u}^*$. Then the a.s. overtaking optimality of $\mathbf{u}^*$ will follow once we have shown that $\frac{1}{t} M_t^{\mathbf{u}} \rightarrow 0$ a.s. as $t \rightarrow \infty$, for every control $\mathbf{u}$. Notice that in order to apply this argument we need *a unique* control $\mathbf{u}^*$ with a minimal value $\lambda^{\mathbf{u}^*}$. We demonstrate this approach by applying it to the LQG (Linear Quadratic Gaussian) regulation and tracking problem, where we show that the linear feedback control which has a minimal long-run average ecpected cost is a.s. overtaking optimal.

Example 10.2 *The finite horizon case: no a.s. overtaking controls*
We describe here an example which demonstrates that, in general, one should not expect the existence of a.s. optimal controls for finite horizon problems.

We consider a probability space $(\Omega, \mathcal{F}, P)$ with a filtration $\{\mathcal{F}_t\}_{t \geq 0}$ and an $\{\mathcal{F}_t\}$-adapted n-dimensional Brownian motion $\{\beta_t\}_{t \geq 0}$. We consider a control system whose dynamics is described by the stochastic differential equation

$$dx_t = b(x_t, u_t)\, dt + \sigma(x_t) d\beta_t, \ x_0 \text{ is given.} \tag{10.60}$$

In (10.60) $x \in \mathbb{R}^n$ and $u \in \mathbb{R}^m$ are the state and the control variables respectively. To this equation corresponds the infinitesimal generator $\mathcal{L}^u$ which, for twice continuously differentiable functions $x \rightarrow f(x)$, is given by

$$\mathcal{L}^u f(x) = \sum_{i=1}^{n} b_i(x, u) \frac{\partial f}{\partial x_i} + \frac{1}{2} \sum_{i,j=1}^{n} a_{ij}(x) \frac{\partial^2 f}{\partial x_i \partial x_j}$$

where the matrix (a_{ij}) is the matrix $(\sigma\sigma')$. Let $T > 0$ be fixed and let the control problem be defined on the $[0, T]$ time interval. The diffusion coefficient $a(\cdot)$ is assumed to be Lipschitz continuous and bounded on $\mathbb{R}^n$, and such that there exists a constant $\lambda_0 > 0$ such that

$$y'a(x)y \geq \lambda_0 |y|^2$$

for every $x, y \in \mathbb{R}^n$. The function $(x,u) \to b(x,u)$ is continuous and bounded on $\mathbb{R}^n \times \mathbb{R}^m$. *A control* $u(\cdot)$ is a measurable function $x \to u(x)$ from $\mathbb{R}^n$ into $K \subset \mathbb{R}^m$, where K is a compact set. With a control $\mathbf{u}$ there is associated a random cost C_u which is given by

$$C_u(\omega) = \phi(x_T) \tag{10.61}$$

where x_t is the solution of (10.60), and $\phi(\cdot)$ is a real valued continuous function on $\mathbb{R}^n$. Clearly it is hopeless to find a control $\mathbf{u}^*$ such that, for every $\mathbf{u}$, $C_{u^*}(\omega) \leq C_u(\omega)$ *a.s.* Rather than that we are looking for a control $\mathbf{u}^*$ which will be a.s. optimal according to the following definition:

Definition 10.3 *We say that a control* $\mathbf{u}^*$ *is almost-surely optimal if for every other control* $\mathbf{u}$ *there exists a random variable* $\psi_u : \Omega \to \mathbb{R}$ *such that*

(i) ψ_u *and* C_u *have the same distribution laws.*

(ii) $C_{u^*} \leq \psi_u$ a.s.

For a real random variable X we denote by F_X the corresponding distribution function, namely

$$F_X(s) = P(X \leq s), \ -\infty < s < \infty.$$

Let X and Y be two real random variables. If there exists a real random variable Z such that

(i) Y and Z have the same distribution laws, and
(ii) $X(\omega) \leq Z(\omega)$ a.s.,

then clearly

$$F_Y(s) \leq F_X(s) \text{ for every } -\infty < s < \infty.$$

Let $\phi(\cdot)$ be the function in the cost expression (10.61) and define the level sets

$$G_\alpha = \{x \in \mathbb{R}^n : \phi(x) \leq \alpha\}.$$

For a control $\mathbf{u}$ let $P(t,x)$ $0 \leq t \leq T$, $x \in \mathbb{R}^n$ be defined by

$$P(t,x) = P\{x_T \in G_\alpha | x_t = x\}.$$

Then $(t,x) \to P(t,x)$ satisfies the Kolmogorov backward equation

$$\begin{cases} \dfrac{\partial P}{\partial t} + \sum_{i=1}^{n} \dfrac{\partial P}{\partial x_i} b_i(x,u) + \dfrac{1}{2} \sum_{i,j=1}^{n} a_{ij} \dfrac{\partial^2 P}{\partial x_i \partial x_j} = 0 \\ P(T,x) = \begin{cases} 1 & \text{if } x \in G_\alpha \\ 0 & \text{if } x \notin G_\alpha. \end{cases} \end{cases} \tag{10.62}$$

Let $Q(t,x)$ be the solution to the following quasilinear equation

$$\begin{cases} \dfrac{\partial Q}{\partial t} + \min\limits_u \sum\limits_{i=1}^{n} \dfrac{\partial Q}{\partial x_i} b_i(x,u) + \dfrac{1}{2} \sum\limits_{i,j=1}^{n} a_{ij} \dfrac{\partial^2 Q}{\partial x_i \partial x_j} = 0 \\ Q(T,x) = \begin{cases} 1 & \text{if } x \in G_\alpha \\ 0 & \text{if } x \notin G_\alpha . \end{cases} \end{cases} \tag{10.63}$$

It then follows from the maximum principle that $Q(t,x) \geq P(t,x)$ for every solution $P(\cdot,\cdot)$ of (10.63) and for every control $u(\cdot)$. Moreover, the equality $Q(t,x) = P(t,x)$ will hold for every $0 \leq t \leq T$ and $x \in \mathbb{R}^n$ only if $u(\cdot)$ realizes the equality in the minimization in (10.63) for almost every t and x. It thus follows that if $\mathbf{u}^*$ is almost-surely optimal according to Definition 10.3, then it must realize the equality in (10.63), and this *for every level set* G_α. However, unless the function $\phi(\cdot)$ and the infinitesimal generator

$$\mathcal{L}^u = \sum_{i=1}^{n} b_i(x,u) \frac{\partial}{\partial x_i} + \frac{1}{2} \sum_{i,j=1}^{n} a_{ij}(x) \frac{\partial^2}{\partial x_i \partial x_j}$$

are related in a very special way then there would not exist a control $\mathbf{u}^*$ which simultaneously realizes the minimum in (10.63) for every level set.

The infinite-horizon LQG problem

We consider the linear control system

$$dx_t = (Ax_t + Bu_t)\,dt + G d\beta_t \tag{10.64}$$

where $x_t \in \mathbb{R}^n$ and $u_t \in \mathbb{R}^m$ are the state and the control variables, respectively, and $\{\beta_t\}_{t\geq 0}$ is a p-dimensional Brownian motion. The matrices A, B and G are constant and of appropriate dimensions. The processes are defined on a probability space $(\Omega, \mathcal{F}, P)$ and $\{\beta_t\}_{t\geq 0}$ is a Brownian motion with respect to a filtration $\{\mathcal{F}_t\}_{t\geq 0}$ on this probability space.

Along with (10.64) a trajectory

$$\Gamma : [0,\infty) \to \mathbb{R}^n \tag{10.65}$$

is given and assumed to be periodic with a period T_0, i.e.

$$\Gamma(t + T_0) = \Gamma(t) \text{ for all } t \geq 0.$$

We assume that $\Gamma(\cdot)$ is piecewise continuous.

We consider Markov controls which are functions of the current state, namely there is a continuous function $\varphi(\cdot,\cdot) : [0,\infty) \times \mathbb{R}^n \to \mathbb{R}^m$ such that for some $K > 0$

$$\begin{cases} \|\varphi(t,x) - \varphi(t,y)\| \leq K\|x - y\| \text{ for all } x,y \in \mathbb{R}^n \text{ and } t \geq 0 \\ \varphi(t + T_0, x) = \varphi(t,x) \text{ for every } x \in \mathbb{R}^n \text{ and } t \geq 0 \\ u_t = \varphi(t, x_t). \end{cases} \tag{10.66}$$

Definition 10.4 *Let* $\mathbf{u}_t$ *be a Markov control such that* (10.66) *is satisfied and* $\mathbf{x}_t$ *be the solution of* (10.64) *corresponding to* $\mathbf{u}_t$. *Then we say that* $\mathbf{u}_t$ *is an admissible control if the function*

$$t \to E\left[\|x_t\|^2\right] \tag{10.67}$$

is bounded on $[0, \infty)$.

With an admissible control $\mathbf{u}$ we associate the cost-flow process

$$c_T(\mathbf{u}(\cdot)) = \int_0^T \left\{\|x_t - \Gamma(t)\|_Q^2 + \|u_t\|_R^2\right\} dt \tag{10.68}$$

where R is a constant positive-definite $m \times m$ matrix, Q is a constant positive-semidefinite $n \times n$ matrix and we denote $\|x\|_Q^2 = x'Qx$ and $\|u\|_R^2 = u'Ru$ for $x \in \mathbb{R}^n$ and $u \in \mathbb{R}^m$.

We will employ the overtaking optimality criterion as presented in Defintion 10.2. As remarked in Section 10.2 we get a stronger version of overtaking optimality due to the fact that there exists a *unique* admissible control which minimizes the long-run expected average cost, namely minimizes the functional

$$\mathbf{u} \to \liminf_{T\to\infty} \frac{1}{T} E c_T(\mathbf{u}).$$

It is well known that this unique optimal control is expressed by the feedback law

$$u_t^* = -R^{-1}B'[Kx_t^* - g(t)] \tag{10.69}$$

where K is the positive-definite symmetric solution of the algebraic Riccati equation

$$KA + A'K - KBR^{-1}B'K + Q = 0 \tag{10.70}$$

and $g(\cdot)$ is the periodic function

$$g(t) = \int_t^\infty e^{(s-t)F'} Q\Gamma(s) ds \tag{10.71}$$

where $F = A - BR^{-1}B'K$.

Let $\mathbf{u}$ be an admissible control and let $\mathbf{x}_t$ be the corresponding solution of (10.64). Then $\mathbf{x}.$ is a diffusion process in $\mathbb{R}^n$ and there is associated with it a transition probability function $P(0, t, x, dy)$. Let

$$Q(x, dy) = P(0, T_0, x, dy)$$

(T_0 is the period of $\Gamma(\cdot)$ in (10.65)) and let $\{\xi_k\}_{k=0}^\infty$ be a discrete time Markov process in $\mathbb{R}^n$ whose one-step transition probability is $Q(x, dy)$. Since the distribution of $\{\xi_k\}_{k=0}^\infty$ is identical to that of $\{x_{kT_0}\}_{k=0}^\infty$, and since by (10.67) $\{E[\|x_{kT_0}\|^2]\}_{k=0}^\infty$ is a bounded sequence, it follows that $\{\xi_k\}_{k=0}^{\infty}$ is positevely recurrent and has an equilibrium measure, say $\nu(dx)$. Moreover, since $\{E[\|x_{kT_0}\|^2]\}_{k=0}^{\infty}$ is a bounded sequence it follows that

$$\int_{\mathbb{R}^n} \|x\|^2 \nu(dx) < \infty. \tag{10.72}$$

(This is so since the existence of a bound b such that $E[\|x_{kT_0}\|^2] < b$ for all $k \geq 0$ implies that $\int_{B(0,R)} \|x\|^2 \nu(dx) \leq b$ for every ball $B(0,R)$ of radius R centered at the origin in $\mathbb{R}^n$. Then (10.72) follows by letting $R \uparrow \infty$ and using the monotone convergence Theorem.)

We will need the following corollary of the stong law of large numbers for solutions of (10.64) corresponding to admissible controls. We denote by E_ν the expectation operation when the initial distribution of x_0 is $\nu(dx)$.

Theorem 10.11 *Let $\mathbf{u}_t$ be an admissible control, $\mathbf{x}_t$ the corresponding solution of (10.64) and $\nu(dx)$ be as above, the equilibrium measure of $\{x_{kT_0}\}_{k=0}^{\infty}$. Let $f : [0,\infty) \times \mathbb{R}^n \to \mathbb{R}$ be such that*

$$f(t+T_0, x) = f(t,x) \text{ for every } t \geq 0 \text{ and } x \in \mathbb{R}^n,$$

and

$$0 \leq f(t,x) \leq C_1|x|^2 + C_2 \text{ for some } C_1,\ C_2 > 0 \text{ and all } t \geq 0 \text{ and } x \in \mathbb{R}^n. \tag{10.73}$$

Then

$$\liminf_{T\to\infty} \frac{1}{T} \int_0^T f(t,x_t)\, dt = \frac{1}{T_0} E_\nu\Big[\int_0^{T_0} f(t,x_t)\, dt\Big] \ a.s. \tag{10.74}$$

Proof: For every integer $N \geq 1$ we have, by the periodicity of $t \to f(t,x)$

$$\frac{1}{NT_0} \int_0^{NT_0} f(t,x_t) = \frac{1}{T_0} \int_0^{T_0} \Big[\frac{1}{N} \sum_{k=0}^{N-1} f(t, x_{t+kT_0})\Big]\, dt. \tag{10.75}$$

In view of (10.72) and (10.73) it follows from the strong law of large numbers that

$$\frac{1}{N} \sum_{k=0}^{N-1} f(t, x_{t+kT_0}) \to E_\nu[f(t,x_t)] \text{ as } N \to \infty \ a.s.$$

for every $0 \leq t \leq T_0$. It follows fron (10.75) and Fatou's Lemma that for almost every ω

$$\liminf_{N\to\infty} \frac{1}{NT_0} \int_0^{NT_0} f(t,x_t)\, dt \geq \frac{1}{T_0} \int_0^{T_0} E_\nu[f(t,x_t)]\, dt$$

which implies, by the nonnegativity of $f(\cdot,\cdot)$ that

$$\liminf_{T\to\infty} \frac{1}{T} \int_0^T f(t,x_t)\, dt \geq \frac{1}{T_0} E_\nu\Big[\int_0^{T_0} f(t,x_t)\, dt\Big] \ a.s. \tag{10.76}$$

We will show next that

$$\frac{1}{T}\int_0^T f(t,x_t)\,dt \to \frac{1}{T_0}E_\nu\Big[\int_0^{T_0} f(t,x_t)\,dt\Big] \quad \text{in probability} \tag{10.77}$$

and this, combined with (10.76), implies (10.74).

For every $M > 0$ we define

$$f_M(t,x) = \begin{cases} f(t,x) & \text{if } \|x\| \le M \\ 0 & \text{if } \|x\| > M. \end{cases}$$

Then $f_M(\cdot,\cdot)$ is a bounded, nonnegative function and, by the periodicity property $f_M(t+T_0,x) = f_M(t,x)$ we have

$$\frac{1}{NT_0}\int_0^{NT_0} f(t,x_t)\,dt = \frac{1}{T_0}\int_0^{T_0}\Big[\frac{1}{N}\sum_{k=0}^{N-1} f_M(t,x_{t+kT_0})\Big]\,dt. \tag{10.78}$$

It follows from the strong law of large numbers that

$$\frac{1}{N}\sum_{k=0}^{N-1} f_M(t,x_{t+kT_0}) \to E_\nu[f_M(t,x_t)] \text{ a.s.}$$

for every $0 \le t \le T_0$. Therefore, for almost every ω

$$\frac{1}{N}\sum_{k=0}^{N-1} f_M(t,x_{t+kT_0}) \to E_\nu[f_M(t,x_t)]$$

for almost every $0 \le t \le T_0$ in the Lebesgue measure. It then follows from (10.78) and the bounded convergence Theorem that

$$\frac{1}{NT_0}\int_0^{NT_0} f_M(t,x_t)\,dt \to \frac{1}{T_0}E_\nu\Big[\int_0^{T_0} f_M(t,x_t)\,dt\Big] \text{ a.s. as } N \to \infty$$

and hence

$$\frac{1}{T}\int_0^T f_M(t,x_t)\,dt \to \frac{1}{T_0}E_\nu\Big[\int_0^{T_0} f_M(t,x_t)\,dt\Big] \text{ a.s. as } T \to \infty. \tag{10.79}$$

We have thus established for a bounded function $f(\cdot,\cdot)$ a property which is stronger than that in (10.74). In particular it follows from (10.79) that

$$\frac{1}{T}\int_0^T f_M(t,x_t)\,dt \to \frac{1}{T_0}E_\nu\Big[\int_0^{T_0} f_M(t,x_t)\,dt\Big] \quad \text{in probability.} \tag{10.80}$$

For every $T > 0$ and $M > 0$ we have that

$$E\Big[\Big|\frac{1}{T}\int_0^T f(t,x_t)\,dt - \frac{1}{T}\int_0^T f_M(t,x_t)\,dt\Big|\Big] = \frac{1}{T}\int_D\int f(t,x_t)\,dtP(d\omega)$$

$$\leq \frac{1}{T} \int_D \int [C_1 |x_t(\omega)|^2 + C_2] P(d\omega)\, dt,$$

where D is the set

$$D = \{(t,\omega) \in [0,\infty) : |x_t(\omega)| > M,\ 0 \leq t \leq T\}.$$

It follows from (10.72) and (10.73) that

$$\int_D \|x_t\|^2 P(d\omega) \to 0 \text{ as } M \to \infty$$

uniformly in t and therefore there are $\delta_M > 0$, $\delta_M \to 0$ as $M \to \infty$, such that

$$E\Big[\Big|\frac{1}{T}\int_0^T f(t,x_t)\, dt - \frac{1}{T}\int_0^T f_M(t,x_t)\, dt\Big|\Big] \leq \delta_M \text{ for every } T > 0. \tag{10.81}$$

Moreover, we have by the monotone convergence Theorem that

$$E_\nu\Big[\int_0^{T_0} f_M(t,x_t)\, dt\Big] \to E_\nu\Big[\int_0^{T_0} f(t,x_t)\, dt\Big] \text{ as } M \to \infty. \tag{10.82}$$

The following argument shows that (10.80), (10.81) and (10.82) imply (10.77). Let $\epsilon > 0$ be given and M be sufficiently large so that, by (10.82)

$$E_\nu\Big[\frac{1}{T_0}\int_0^{T_0} f(t,x_t)\, dt\Big] - E_\nu\Big[\frac{1}{T_0}\int_0^{T_0} f_M(t,x_t)\, dt\Big] < \frac{\epsilon}{3}.$$

For such an M we have that

$$P\Big\{\Big|\frac{1}{T}\int_0^T f(t,x_t)\, dt - E_\nu\Big[\frac{1}{T_0}\int_0^{T_0} f(t,x_t)\, dt\Big]\Big| > \epsilon\Big\}$$

$$\leq P\Big\{\Big|\frac{1}{T}\int_0^T f(t,x_t)\, dt - \frac{1}{T}\int_0^T f_M(t,x_t)\, dt\Big| > \frac{\epsilon}{3}\Big\}$$

$$+ P\Big\{\Big|\frac{1}{T}\int_0^T f_M(t,x_t)\, dt - E_\nu\Big[\frac{1}{T_0}\int_0^{T_0} f_M(t,x_t)\, dt\Big]\Big| > \frac{\epsilon}{3}\Big\}. \tag{10.83}$$

Suppose that M is large enough so that the first term in the right hand side of (10.83) is (by (10.81) and the Chebychev's inequality) smaller than $\epsilon/2$, for every $T > 0$. For such an M the second term in the right-hand-side of (10.83) will be smaller than $\epsilon/2$ for all large T, by (10.80). This proves (10.77) and completes the proof of the theorem. ∎

We will establish now the a.s. overtaking optimality of u^* in (10.69).

Theorem 10.12 *The admissible control*

$$u^*(t,x) = -R^{-1}B'[Kx_t - g(t)]$$

is a.s. overtaking optimal.

Proof: Let $\mathbf{x}^*$ be the solution of (10.64) corresponding to $\mathbf{u}^*$ and to an initial value x_0. It is proved in Section 10.3 (in the proof of Theorem 10.5) that $t \to E\left[\|x_t^*\|^2\right]$ is a bounded function on $[0,\infty)$, thus $\mathbf{u}^*$ is indeed an admissible control.

Let $\mathbf{u}$ be an admissible control with a corresponding solution x_t, and let $\nu^u(dx)$ be the equilibrium measure of $\{x_{kT_0}\}_{k=0}^{\infty}$. We claim that for every $f : [0,\infty) \times \mathbb{R}^n \to \mathbb{R}$ which satisfies the conditions of Theorem 10.11 the following holds

$$\frac{1}{T}\int_0^T E[f(t,x_t)]\,dt \to \frac{1}{T_0}\int_0^{T_0} E_{\nu^u}[f(t,x_t)]\,dt \text{ as } T \to \infty. \tag{10.84}$$

This is so since by (10.80) and the bounded convergence Theorem we have for $f_M(\cdot,\cdot)$

$$\frac{1}{T}\int_0^T E[f_M(t,x_t)]\,dt \to \frac{1}{T_0}\int_0^{T_0} E_{\nu^u}[f_M(t,x_t)]\,dt \text{ as } T \to \infty \tag{10.85}$$

for every $M \geq 1$. The validity of (10.84) now follows from (10.81), (10.82) and (10.85). By (10.66) the feedback function $\varphi_u(\cdot,\cdot)$ satisfies

$$|\varphi_u(t,x)| \leq C_1|x| + C_2$$

for some C_1, and $C_2 > 0$ and $x \in \mathbb{R}^n$. Therefore we have by (10.84)

$$\frac{1}{T}E\left[\int_0^T \left\{\|x_t - \Gamma(t)\|_Q^2 + \|u_t\|_R^2\right\} dt\right] \to E_{\nu^u}\left[\int_0^{T_0} \left\{\|x_t - \Gamma(t)\|_Q^2 + \|\varphi_u(t,x_t)\|_R^2\right\} dt\right] \tag{10.86}$$

as $T \to \infty$, and this for every initial value x_0. We denote by λ^u the right-hand-side of (10.86). By (10.74) we have that (recall (10.68))

$$\liminf_{T\to\infty} \frac{c_T(\mathbf{u})}{T} = \lambda^u \text{ a.s.} \tag{10.87}$$

It was established in Section 10.3 that $\mathbf{u}^*$ is the *unique minimizer* of the mapping

$$\mathbf{u} \to \liminf_{T\to\infty} E\left[\frac{c_T(\mathbf{u})}{T}\right]$$

and in fact we will prove next that for $\mathbf{u}^*$ the following holds

$$\lim_{T\to\infty} \frac{c_T(\mathbf{u}^*)}{T} = \lambda^{u^*} \text{ a.s.} \tag{10.88}$$

It turns out that

$$\lambda^u > \lambda^{u^*} \text{ whenever } \mathbf{u} \neq \mathbf{u}^*. \tag{10.89}$$

The almost-sure optimality of $u^*(\cdot)$ is implied by (10.87), (10.88) and (10.89) as follows. Let $\mathbf{u} \neq \mathbf{u}^*$ and $\lambda^u = \lambda^{u^*} + 3\epsilon$ for an $\epsilon > 0$. There is a set Ω_0, $P(\Omega_0) = 1$, such that for every $\omega \in \Omega_0$ there is a $T(\omega) > 0$ so that $c_T(\mathbf{u}) > (\lambda^u - \epsilon)T$ and $c_T(\mathbf{u}^*) < (\lambda^{u^*} + \epsilon)T$

for every $T > T(\omega)$. Therefore $c_T(\mathbf{u}) - c_T(\mathbf{u}^*) > \epsilon T$ for every $\omega \in \Omega_0$ and every $T > T(\omega)$, which proves the a.s. optimality of u^*.

It still remains to prove (10.88), namely that the limit of $\dfrac{c_T(\mathbf{u}^*(\cdot))}{T}$ exists as $T \to \infty$. We have

$$c_T(u^*) = \int_0^T g(t)'BR^{-1}B'g(t) - 2\int_0^T \Gamma(t)'Qx^*(t)\,dt$$

$$+ 2\int_0^T x^*(t)'KGd\beta(t) + x_0'Kx_0 - x^*(T)'Kx(T). \tag{10.90}$$

We will show that the limits for the first three terms exist a.s., and they sum up to λ^{u^*}. It turns out that, by (10.87), the last term (which is nonpositive) does not contribute anything to the limit expression.

Considering the first term in (10.90) we have

$$\frac{1}{T}\int_0^T g(t)'BR^{-1}B'g(t)\,dt \to \frac{1}{T_0}\int_0^{T_0} g(t)'BR^{-1}B'g(t)\,dt$$

since $g(\cdot)$ is a deterministic, periodic function.

For the solution $x^*(\cdot)$ we have $x^*(t) = X^*(t) + \theta(t)$ where $X^*(\cdot)$ is a deterministic trajectory which converges to a periodic trajectory $\gamma(\cdot)$, and $\theta(\cdot)$ is given by

$$\theta(t) = \int_0^t e^{F'(t-s)}Gd\beta(s). \tag{10.91}$$

This follows from the discussion in Section 10.3.

Considering the second term in (10.90) we have two contributions corresponding to the two terms $X^*(\cdot)$ and $\theta(\cdot)$. The first is

$$\frac{1}{T}\int_0^T \Gamma(t)'QX^*(t)\,dt \to \frac{1}{T_0}\int_0^{T_0} \Gamma(t)'Q\gamma(t)\,dt$$

and the second is

$$\frac{1}{T}\int_0^T \Gamma(t)'Q\theta(t)\,dt = \frac{1}{T}\int_0^T \left[\int_s^T \Gamma(t)'Qe^{F'(t-s)}G\,dt\right] d\beta_s, \tag{10.92}$$

using (10.91). We denote the deterministic integrand in the right hand side of (10.92) by $h(t,s)$ and we have

$$\begin{cases} \displaystyle\int_s^T |h(t,s)|\,dt \le B_0 & \text{for every } 0 \le s \le T < \infty \\ \displaystyle\int_0^t |h(t,s)|^2 ds \le B_0 & \text{for every } t > 0 \end{cases} \tag{10.93}$$

for a certain $B_0 > 0$, as a consequence of the stability of F and the boundedness of $\Gamma(\cdot)$. We denote the expression in (10.92) by $(1/T)M_T$ so that

$$M_T = \int_0^T \Big[\int_s^T h(t,s)\,dt\Big]\,d\beta_s.$$

We will show that $(1/n)M_n \to 0$ a.s. as $n \to \infty$ (n is an integer). For every $T > 0$ we have

$$E\Big[|M_T|^2\Big] = \int_0^T \left|\int_s^T h(t,s)\,dt\right|^2 ds \le BT$$

for some $B > 0$, using (10.93). Therefore

$$E\Big[\Big|\frac{1}{k^{3/2}} M_{k^{3/2}}\Big|^2\Big] \le \frac{B}{k^{3/2}}$$

for every integer $k \geq 1$, and it follows from the Chebychev's inequality that

$$P\Big[\Big|\frac{1}{k^{3/2}} M_{k^{3/2}}\Big| > k^{-1/8}\Big] \le \frac{B}{k^{5/4}}.$$

It follows from the Borel-Cantelli Lemma that

$$\frac{1}{k^{3/2}} M_{k^{3/2}} \to 0 \text{ a.s. as } k \to \infty \tag{10.94}$$

To deduce that $(1/n)M_n \to 0$ a.s. we represent every integer n as $n = k^{3/2} + p$ for some integer k, where p satisfies $0 \le p \le C\sqrt{k}$, for some fixed constant C. We compute

$$E\Big[|M_n - M_{k^{3/2}}|^2\Big] = \int_0^{k^{3/2}+p} \left|\int_{\max(k^{3/2},s)}^{k^{3/2}+p} h(t,s)\,dt\right|^2 ds$$

$$\le p \int_{k^{3/2}}^{k^{3/2}+p} \int_0^t |h(t,s)|^2 ds\,dt \le B_0 p^2 \tag{10.95}$$

using (10.93) and the Cauchy-Schwartz inequality. Now the following standard argument implies that $(1/n)M_n \to 0$ a.s. as $n \to \infty$. We observe that

$$\sum_{1 \le p \le C\sqrt{k}} p^2 \le C_1 k^{3/2} \tag{10.96}$$

for some constant $C_1 > 0$, and thus it follows from (10.95), (10.96) and the Borel-Cantelli Lemma that for every $\epsilon > 0$ the following holds a.s.: $(1/n)|M_n - M_{k^{3/2}}| < \epsilon$ for all large enough n and k such that $0 \le n - k^{3/2} \le C\sqrt{k}$. Combined with (10.94) this implies that $1/n)M_n \to 0$ a.s. as $n \to \infty$.

Considering the third term in (10.90) we denote

$$N_T = \int_0^T x^*(t)' K G\,d\beta(t)$$

and since $E\left[\|x^*(t)\|^2\right] \leq B_1$ for some $B_1 > 0$ and all $t > 0$, it follows, arguing as above, that $(1/k^2)N_{k^2} \to 0$ a.s. as $k \to \infty$. Then the bound on $E\left[\|x^*(t)\|^2\right]$ enables to deduce (by the same argument that was described above) that $(1/n)N_n \to 0$ a.s. as $n \to \infty$.

We thus have proved that the limit

$$\lim_{n\to\infty} \frac{1}{n} \left[c_n(\mathbf{u}^*) + x^*(n)'Kx^*(n)\right]$$

exists and it is easy to see that it is almost-surely equal to λ^*. This implies, as remarked in the sentence which follows (10.90), that

$$\limsup_{n\to\infty} \frac{1}{n} c_n(\mathbf{u}^*) \leq \lambda^{u^*} \text{ a.s.}$$

Together with (10.87) (applied to $\mathbf{u}^*$) this yields

$$\lim_{n\to\infty} \frac{1}{n} c_n(\mathbf{u}^*) = \lambda^{u^*} \text{ a.s.}$$

The positivity of the cost integrand implies that $\lim_{T\to\infty} \frac{1}{T} c_T(\mathbf{u}^*) = \lambda^{u^*}$ a.s. and concludes the proof of the theorem. ∎

Chapter 11

Maximum Principle and Turnpike Properties for Systems with Random Modal Jumps

11.1 Introduction

This chapter deals with a class of stochastic control systems where the random disturbances are stochastic jump processes taking values in a finite set E. Each value $i \in E$ can be interpreted as a "mode" of the dynamic system. Systems with modal disturbances have been studied in the control literature for a long time in particular in the case of linear dynamics and quadratic performance criterion (e.g. [182]). Here we demonstrate that these processes may be viewed as a sequence of control problems with random stopping time. Each of these problems can be rewritten as a deterministic infinite horizon control system with state and control dependent discount rate. Exploiting this property we first extend the infinite horizon Maximum Principle and some turnpike properties from Chapter 6 to the class of control problems with random stopping time. We conclude by applying these turnpike properties to the case of control systems with random modal jumps.

11.2 Optimal control under random stopping time

Consider a control system of the form

$$\left.\begin{aligned} \dot{x}(t) &= f(x(t), u(t)) \text{ a.e. } t \geq 0 \\ x(0) &= x_0\,,\ x(t) \in X \text{ for } t \geq 0 \\ u(t) &= U \text{ a.e. } t \geq 0 \end{aligned}\right\} \tag{11.1}$$

As before, $x_0 \in \mathbb{R}^n$ is a given fixed initial state, $x(\cdot) : [0,\infty) \to \mathbb{R}^n$ is a trajectory, $u(\cdot) : [0,\infty) \to \mathbb{R}^m$ is a control and $f(\cdot,\cdot) : \mathbb{R}^n \times \mathbb{R}^m \to \mathbb{R}^n$ is continuous with continuous first partial derivatives with respect to x and satisfies the growth condition

$$\|f(x,u)\| \leq K + M\|x\| \tag{11.2}$$

for all $(x,u) \in X \times U$, where K and M are constants, and $X \subset \mathbb{R}^n$ and $U \subset \mathbb{R}^m$ are given closed sets.

Let (Ω, B, P_u) be a probability space and let T be a random stopping time for the above system. The probability P_u depends on the control, $u(\cdot)$, as follows:

> For an admissible pair $(x(\cdot), u(\cdot))$ of (11.1), we assume that for each $t \geq 0$ and dt, we have

$$P_u[T \in (t, t+dt) | T \geq t] = q(x(t), u(t)) + o(dt). \tag{11.3}$$

Here we assume that $q(\cdot,\cdot) : \mathbb{R}^n \times \mathbb{R}^m \to [0,\infty)$ is continuous with continuous first partial derivatives in x and $\frac{o(dt)}{dt} \to 0$ uniformly in (x,u) when $dt \to 0$. The reward, or payoff, of the above system is defined as the conditional expectation

$$J(x_0, u(\cdot)) = E_u\left[\int_0^T e^{-\rho t} f_0(\cdot)(x(t), u(t))\, dt + e^{-\rho T}\Phi(x(T))\bigg|\, x_0\right] \tag{11.4}$$

in which $f_0(\cdot,\cdot) : \mathbb{R}^n \times \mathbb{R}^m \to (-\infty, 0]$ and $\Phi(\cdot) : \mathbb{R}^n \to (-\infty, 0]$ are continuous with continuous first partial derivatives in x and $\rho \geq 0$ is a constant discount rate. The stochastic optimal control problem is now briefly written

$$\max\{J(x_0, u(\cdot)) | (x(\cdot), u(\cdot)) \text{ admissible}\}. \tag{11.5}$$

The above optimal control problem can, under appropriate assumptions, be reformulated as an equivalent infinite horizon deterministic optimal control problem. In this way the tools developed in the previous chapters can be adapted to this problem as well. To see this reformulation consider the elementary probability of the time interval $(t, t+dt)$ for the stopping time T, for a given admissible pair $(x(\cdot), u(\cdot))$. From (11.3), this probability is given by

$$q(x(t), u(t)) \exp(-\int_0^t q(x(s), u(s))\, ds) dt.$$

Thus the reward function (11.4) can be written as

$$\begin{aligned} J(x_0, u(\cdot)) \;=\; & \int_0^\infty \left[\int_0^t \left\{e^{-\rho s} f_0(x(s), u(s))\, ds + e^{-\rho t}\Phi(x(t))\right\}\right. \\ & \left. q(x(t), u(t)) \exp\left(-\int_0^t q(x(s), u(s))\, ds\right)\right] dt. \end{aligned} \tag{11.6}$$

We now let

$$\mathcal{U}(t) = \int_0^t e^{-\rho s} f_0(x(s), u(s))\, ds,$$

$$\mathcal{V}(t) = -\exp\left[\int_0^t q(x(s), u(s))\, ds\right]$$

and assume that

$$\lim_{t\to\infty} \mathcal{U}(t)\mathcal{V}(t) = 0. \tag{11.7}$$

This permits us to integrate by parts in (11.6) to get

$$J(x_0, u(\cdot)) = \int_0^\infty \Big[\{ f_0(x(t), u(t)) + \Phi(x(t)) q(x(t), u(t)) \} \exp(-\rho t - \int_0^t q(x(s), u(s))\, ds) \Big]\, dt \tag{11.8}$$

Consequently, we see that the random stopping time control problem is equivalent to a deterministic infinite horizon optimal control problem.

Remark 11.1 *Condition* (11.7) *is implied by the following two conditions:*

$$\int_0^\infty e^{-\rho t} f_0(x(t), u(t))\, dt > -\infty \tag{11.9}$$

$$\int_0^\infty q(x(t), u(t))\, dt = \infty \tag{11.10}$$

Condition (11.9) *is standard when* $\rho > 0$*, whereas* (11.10) *means that the stopping time occurs almost surely before infinity. We remark however that* (11.7) *may hold even when* $\rho = 0$.

The infinite horizon optimal control problem described above differs from those considered earlier in that the discount factor, given by $\exp(-\rho t - \int_0^t q(x(s), u(s))\, ds)$, depends on the past history of the admissible pair $(x(\cdot), u(\cdot))$. Models in economic growth exhibiting an identical behavior have been studied by several researchers. Specifically we note the works of Carlson [44], Becker, et. al. [17], Epstein [68], Epstein and Hynes [69], and Nairay [145]. The motivation to consider discount factors of this form arose to address the concerns made by Ramsey (and others) in his 1928 paper when he questioned the use of discounting on ethical grounds, as well as the remarks made by Fisher [77] in 1930 who suggested that the discount rate should depend on the past history of the model.

To study these new models most of the researchers listed above place the optimal control model in *standard form* by introducing a new state variable. Specifically, for an admissible pair $(x(\cdot), u(\cdot))$ we introduce the new variable $x^o \in [0, \infty)$ and the auxiliary state equation

$$\begin{aligned} \dot{x}^o(t) &= q(x(t), u(t)) \\ x^o(0) &= 0. \end{aligned} \tag{11.11}$$

With this new state variable the infinite horizon optimal control problem can be written as

maximize $J(x_0, u(\cdot)) = \int_0^\infty \left[e^{-\rho t - x^o(t)}\{f_0(x(t),u(t)) + \Phi(x(t))q(x(t),u(t))\}\right] dt$
subject to

$$\begin{aligned}
\dot{x}^o(t) &= q(x(t),u(t)) \text{ a.e. } t \geq 0 \\
\dot{x}(t) &= f(x(t),u(t)) \text{ a.e. } t \geq 0 \\
x^o(0) &= 0 \\
x(0) &= x_0 \\
x(t) &\in X \text{ for } t \geq 0 \\
u(t) &\in U \text{ a.e. } t \geq 0.
\end{aligned} \tag{11.12}$$

We now have an infinite horizon control problem to which we may try to apply the results developed in previous chapters. Unfortunately many of these results are not directly applicable since, even under the usual strong concavity conditions, the model as formulated above does not satisfy the required concavity conditions. The reason for this is the presence of the state and control dependent discount factor. Indeed to insure these conditions are met one must typically assume that the function

$$(x,u) \to -\{f_0(x,u) + \Phi(x)q(x,u)\}$$

is log-convex. That is, the natural logarithm of this function must be a convex function, a condition which is not generally satisfied.

On the other hand, those results which do not require convexity requirements may be exploited. In particular we close this section by examining the maximum principle.

Theorem 11.1 *If $(x^*(\cdot), u^*(\cdot))$ is a strongly optimal solution for the random stopping time problem, then there exist two constants, $\alpha \geq 0$ and $\beta \in \mathbb{R}$, and a locally absolutely continuous function $\lambda(\cdot) : [0,\infty) \to \mathbb{R}^n$ such that for the Hamiltonian,*

$$\hat{H}(x,u,\alpha,\beta,\lambda,\mathcal{E}) = \alpha\,[f_0(x,u) + (\Phi(x) - \mathcal{E})\,q(x,u)] + \beta q(x,u) - +\lambda' f(x,u)$$

one has

$$\begin{aligned}
\hat{H}(x^*(t),u^*(t),\alpha,\beta e^{\rho t + x^{o*}(t)},\lambda(t),\mathcal{E}(t)) &= \max_{u \in U}\left\{\mathcal{H}(x^*(t),u,\alpha,\beta e^{\rho t + x^{o*}(t)},\lambda(t),\mathcal{E}(t))\right\} \\
\dot{\lambda}(t) &= \nabla_x \mathcal{H}(x^*(t),u^*(t),\alpha,\beta e^{\rho t + x^{o*}(t)},\lambda(t),\mathcal{E}(t)) \\
&\quad + \lambda(t)'(\rho + q(x^*(t),u^*(t)))
\end{aligned}$$

$$(\alpha, \beta - \alpha\mathcal{E}(0), \lambda(0)) \neq 0$$

where

$$\mathcal{E}(t) = \int_t^\infty e^{-\rho(s-t)} \{f_0(x^*(s), u^*(t)) + \Phi(x^*(s))q(x^*(s), u^*(t))\} \exp\left(\int_s^t q(x^*(\tau), u^*(\tau))\, d\tau\right) ds$$

Proof: Appealing to the Maximum Principle, Theorem 2.3, there exists a constant $\alpha \geq 0$ and an absolutely continuous function $(\mu^o, \mu)(\cdot) : [0, \infty) \to \mathbb{R}^{1+n}$ such that for the Hamiltonian,

$$H(x^o, x, u, t, \alpha, \mu^o, \mu) = \alpha e^{-\rho t - x^o} [f_0(x, u) + \Phi(x)q(x, u)] + \mu^o q(x, u) + \mu' f(x, u)$$

one has

$$H(x^{o*}(t), x^*(t), u^*(t), t, \alpha, \mu^o(t), \mu(t)) = \max_{u \in U} [H(x^{o*}(t), x^*(t), u, t, \alpha, \mu^o(t), \mu(t))]$$

$$\dot{\mu}^o(t) = -\frac{\partial}{\partial x^o} H(x^{o*}(t), x^*(t), u^*(t), t, \alpha, \mu^o(t), \mu(t))$$

$$\dot{\mu}(t) = -\nabla_x H(x^{o*}(t), x^*(t), u^*(t), t, \alpha, \mu^o(t), \mu(t))$$

$$(\alpha, \mu^o(0), \mu(0)) \neq 0$$

where $x^{o*}(\cdot)$ denotes the auxiliary state variable evolution associated with the optimal pair $(x^*(\cdot), u^*(\cdot))$. From the differential equation for $\mu^o(\cdot)$ and the strong optimality of $(x^*(\cdot), u^*(\cdot))$ it follows that there exists a constant $\beta \in \mathbb{R}$ such that

$$\begin{aligned}\mu^o(t) &= \beta - \alpha \int_t^\infty e^{-\rho s - x^{o*}(s)} [f_0(x^*(s), u^*(t)) + \Phi(x^*(s))q(x^*(s), u^*(t))]\, ds \\ &= e^{-\rho t - x^{o*}(t)} \left(\beta e^{\rho t + x^{o*}(t)} - \mathcal{E}(t)\right).\end{aligned}$$

Further identifying $\lambda(\cdot)$ through the formula

$$\lambda(t) = \mu(t) \exp\left(\rho t + \int_0^t q(x^*(s), u^*(s))\, ds\right)$$

one can translate the above conditions into the desired conclusions. ∎

Remark 11.2 *The function $\mathcal{E}(\cdot)$ given above is equivalently expressed as*

$$\mathcal{E}(t) = \max\left[E_u\left\{\int_t^T e^{-\rho(s-t)} f_0(x(s), u(s))\, ds + e^{-\rho(T-t)}\Phi(x(T))\, |T \geq t,\ x(t) = x^*(t)\right\}\right],$$

i.e., the maximum expected reward-to-go. This expression, which is related to the adjoint variable $\mu^o(t)$ (recall $\mu^o(t) = \beta - \alpha\mathcal{E}(t)$) anticipates the future accumulated discounted utility of the system along the optimal trajectory.

11.3 Turnpike properties

11.3.1 A global result

We consider the infinite horizon optimal control problem described by the control system (11.1) in conjunction with the objective functional (11.8). For brevity and to conform with the notations used in earlier chapters we assume that $\rho \equiv 0$ and that $\Phi(x) \equiv 0$ for all $x \in \mathbb{R}^n$. Thus the objective functional we consider is more conveniently written as

$$J(x_0, u(\cdot)) = \int_0^\infty f_0(x(t), u(t)) \exp\Big(-\int_0^t q(x(s), u(s))\, ds\Big)\, dt. \tag{11.13}$$

The consideration of the turnpike property for these models requires special attention. One reason for this is that the requisite concavity conditions are not in general satisfied for these models. In addition, since the *discount rate*, $q(x,u)$, is strictly positive, the *standard formulation* given in (11.12) has no steady states and so even if the concavity requirements were met, earlier results would not be applicable. Therefore to develop an analogous theory for the random stopping time problem (or more precisely the equivalent optimal control problem) we must carefully define what we mean by the *Turnpike property.* We begin by defining the Hamiltonian

$$\mathbf{H}(x^o, x, p^o, p) = \max_{u \in U} \left\{ f_0(x,u) e^{-x^o} + p^o q(x,u) + p' f(x,u) \right\}, \tag{11.14}$$

and assume that $\mathbf{H}$ is differentiable. We further assume that, as in Sections 6.1–6.2 , the maximum principle for an extremal trajectory $(x^o(\cdot), x(\cdot))$ and its associated costate trajectory is equivalent to the Hamiltonian system

$$\begin{aligned}
\dot{x}^o(t) &= \frac{\partial}{\partial p^o} \mathbf{H}(x^o(t), x(t), p^o(t), p(t)) \\
\dot{x}(t) &= \frac{\partial}{\partial p} \mathbf{H}(x^o(t), x(t), p^o(t), p(t)) \\
\dot{p}^o(t) &= -\frac{\partial}{\partial x^o} \mathbf{H}(x^o(t), x(t), p^o(t), p(t)) \\
\dot{p}(t) &= -\frac{\partial}{\partial x} \mathbf{H}(x^o(t), x(t), p^o(t), p(t))
\end{aligned} \tag{11.15}$$

for almost all $t \geq 0$. To simplify this $2n+2$ dimensional system slightly we observe from (11.14) that

$$\mathbf{H}(x^o, x, p^o, p) = e^{-x^o} \mathcal{H}(x, e^{x^o} p^o, e^{x^o} p) \tag{11.16}$$

in which $\mathcal{H}(\cdot,\cdot,\cdot) : \mathbb{R}^n \times \mathbb{R} \times \mathbb{R}^n \to \mathbb{R}$ is defined by

$$\mathcal{H}(x, \pi^o, \pi) = \max_{u \in U} \left\{ f_0(x,u) + \pi^o q(x,u) + \pi' f(x,u) \right\}. \tag{11.17}$$

With this notation the Hamiltonian system becomes

$$\begin{aligned}
\dot{x}^o(t) &= \frac{\partial}{\partial \pi^o}\mathcal{H}(x(t), e^{x^o(t)}p^o(t), e^{x^o(t)}p(t)) \\
\dot{x}(t) &= \frac{\partial}{\partial \pi}\mathcal{H}(x(t), e^{x^o(t)}p^o(t), e^{x^o(t)}p(t)) \\
\dot{p}^o(t) &= e^{-x^o(t)}\mathcal{H}(x(t), e^{x^o(t)}p^o(t), e^{x^o(t)}p(t)) \\
&\quad - \frac{\partial}{\partial \pi^o}\mathcal{H}(x(t), e^{x^o(t)}p^o(t), e^{x^o(t)}p(t))p^o(t) \\
&\quad - \frac{\partial}{\partial \pi}\mathcal{H}(x(t), e^{x^o(t)}p^o(t), e^{x^o(t)}p(t))'p(t) \\
\dot{p}(t) &= -e^{-x^o(t)}\frac{\partial}{\partial x}\mathcal{H}(x(t), e^{x^o(t)}p^o(t), e^{x^o(t)}p(t)).
\end{aligned}$$

We now simplify this system further by using the transformation

$$\begin{pmatrix} \pi^o(t) \\ \pi(t) \end{pmatrix} = e^{x^o(t)} \begin{pmatrix} p^o(t) \\ p(t) \end{pmatrix}. \tag{11.18}$$

With these new variables we obtain the following $2n+1$ system of equations

$$\begin{aligned}
\dot{x}(t) &= \frac{\partial}{\partial \pi}\mathcal{H}(x(t), \pi^o(t), \pi(t)) \\
\dot{\pi}^o(t) &= \mathcal{H}(x(t), \pi^o(t), \pi(t)) - \frac{\partial}{\partial \pi}\mathcal{H}(x(t), \pi^o(t), \pi(t))'\pi(t) \\
\dot{\pi}(t) &= \frac{\partial}{\partial \pi^o}\mathcal{H}(x(t), \pi^o(t), \pi(t))\pi(t) - \frac{\partial}{\partial x}\mathcal{H}(x(t), \pi^o(t), \pi(t))
\end{aligned} \tag{11.19}$$

The remaining function, $x^o(\cdot)$, (since $x^o(0) = 0$) is completely determined by (11.19) through the formula

$$x^o(t) = \int_0^t \frac{\partial}{\partial \pi^o}\mathcal{H}(x(s), \pi^o(s), \pi(s))\, ds. \tag{11.20}$$

Paralleling the results of Section 6.2 we restrict our considerations to those trajectories of the Modified Hamiltonian System (11.19) satisfying the *transversality condition*

$$\lim_{t\to\infty} e^{-x^o(t)}p(t)'x(t) = 0. \tag{11.21}$$

A steady state for this system is a triple $(\bar{x}, \bar{\pi}^o, \bar{\pi})$ satisfying the equations

$$\begin{aligned}
0 &= \frac{\partial}{\partial \pi}\mathcal{H}(\bar{x}, \bar{\pi}^o, \bar{\pi}) \\
0 &= \mathcal{H}(\bar{x}, \bar{\pi}^o, \bar{\pi}) - \frac{\partial}{\partial \pi}\mathcal{H}(\bar{x}, \bar{\pi}^o, \bar{\pi})'\bar{\pi} \\
0 &= \frac{\partial}{\partial \pi^o}\mathcal{H}(\bar{x}, \bar{\pi}^o, \bar{\pi})\bar{\pi} - \frac{\partial}{\partial x}\mathcal{H}(\bar{x}, \bar{\pi}^o, \bar{\pi})
\end{aligned} \tag{11.22}$$

To state our result we now make the following assumption.

Assumption 11.1 *There exists a unique stationary point* $(\bar{x}, \bar{\pi}^o, \bar{\pi})$ *such that the following two conditions hold:*

For $x \neq \bar{x}$

$$(\pi - \bar{\pi})'\nabla_\pi \mathcal{H}(x, \pi^o, \pi) - \nabla_x \mathcal{H}(x, \pi^o, \pi)'(x - \bar{x}) + \frac{\partial}{\partial \pi^o}\mathcal{H}(x, \pi^o, \pi)\,[\bar{\pi}'(x - \bar{x})] > 0 \tag{11.23}$$

for all $(x, \pi^o, \pi) \in \mathbb{R}^n \times \mathbb{R} \times \mathbb{R}^n$.

For every $\epsilon > 0$ *there exists* $\delta > 0$ *such that* $\|x - \bar{x}\| > \epsilon$ *implies*

$$(\pi - \bar{\pi})'\nabla_\pi \mathcal{H}(x, \pi^o, \pi) - \nabla_x \mathcal{H}(x, \pi^o, \pi)'(x - \bar{x}) + \frac{\partial}{\partial \pi^o}\mathcal{H}(x, \pi^o, \pi)\,[\bar{\pi}'(x - \bar{x})] > \delta - \frac{\partial}{\partial \pi^o}\mathcal{H}(x, \pi^o, \pi)\,[(\pi - \bar{\pi})'(x - \bar{x})] \tag{11.24}$$

for all $(x, \pi^o, \pi) \in \mathbb{R}^n \times \mathbb{R} \times \mathbb{R}^n$.

Remark 11.3 *We observe that condition* (11.24) *coincides with its analogue* (6.21) *whenever the discount rate* $q(x, u)$ *is identically constant (i.e.,* $q(x, u) \equiv \rho$*) since in this case it happens that*

$$\frac{\partial}{\partial \pi^o}\mathcal{H}(x, \pi^o, \pi) \equiv \rho.$$

The other condition (11.23) *is stronger than the concavity in* x *and convexity in* π *of the function*

$$\mathcal{H}(x, \pi^o, \pi) - \frac{\partial}{\partial \pi^o}\mathcal{H}(\bar{x}, \bar{\pi}^o, \bar{\pi})\,[\bar{\pi}'x].$$

We now state an analogue of Theorem 6.1.

Theorem 11.2 *In addition to Assumption* (11.1) *suppose that there exists a positive number* $r > 0$ *such that*

$$\frac{\partial}{\partial \pi^o}\mathcal{H}(x, \pi^o, \pi) \geq r$$

for all $(x, \pi^o, \pi) \in \mathbb{R}^n \times \mathbb{R} \times \mathbb{R}^n$. *If* $(x(\cdot)\,, \pi^o(\cdot)\,, \pi(\cdot)\,)$ *is a solution to the Modified Hamiltonian system* (11.19) *satisfying the asymptotic boundary condition* (11.21) *is such that* $(x(\cdot)\,, \pi(\cdot)\,)$ *is uniformly bounded, then*

$$\lim_{t \to \infty} x(t) = \bar{x}.$$

Proof: Let $(x(\cdot)\,, \pi^o(\cdot)\,, \pi(\cdot)\,)$ be as indicated above and consider the function $V : [0, \infty) \to \mathbb{R}$ defined by

$$V(t) = -(x(t) - \bar{x})'(\pi(t) - \bar{\pi}).$$

For almost all $t \geq 0$ we have, as a consequence of (11.24),

$$\begin{aligned} \dot{V}(t) &= -\dot{x}(t)'(\pi(t)-\bar{\pi}) - (x(t)-\bar{x})'\dot{\pi}(t) \\ &= -\nabla_\pi \mathcal{H}(x(t),\pi^o(t),\pi(t))'(\pi(t)-\bar{\pi}) \\ &\quad + (x(t)-\bar{x})'\nabla_x \mathcal{H}(x(t),\pi^o(t),\pi(t)) \\ &\quad - \frac{\partial}{\partial \pi^o}\mathcal{H}(x(t),\pi^o(t),\pi(t))\,(x(t)-\bar{x})'\,\pi(t) \\ &\leq 0. \end{aligned}$$

Further the condition (11.23) combined with the definition of $x^o(\cdot)$ given in (11.20) gives us

$$\begin{aligned} \frac{d}{dt}e^{-x^o(t)}V(t) &= e^{-x^o(t)}\dot{V}(t) - e^{-x^o(t)}V(t)\dot{x}^o(t) \\ &= -e^{-x^o(t)}\Big[(\pi-\bar{\pi})'\nabla_\pi \mathcal{H}(x,\pi^o,\pi) - \nabla_x \mathcal{H}(x,\pi^o,\pi)'(x-\bar{x}) \\ &\quad + \frac{\partial}{\partial \pi^o}\mathcal{H}(x,\pi^o,\pi)[\bar{\pi}'(x-\bar{x})]\Big] \\ &\leq 0. \end{aligned}$$

Moreover, the asymptotic boundary condition and the boundedness of $(x(\cdot)\,,\pi(\cdot)\,)$ for $t \geq 0$ gives us

$$\begin{aligned} \lim_{t\to\infty} e^{-x^o(t)}V(t) &= \lim_{t\to\infty}\left\{-e^{-x^o(t)}\left[(x(t)-\bar{x})'(\pi(t)-\bar{\pi})\right]\right\} \\ &= \lim_{t\to\infty}\left\{-e^{-x^o(t)}\left[x(t)'\pi(t)+\bar{x}'\bar{\pi}-\bar{x}'\pi(t)-x(t)'\bar{\pi}\right]\right\} \\ &\geq \lim_{t\to\infty}\left\{-e^{-x^o(t)}\left[x(t)'\pi(t)+\bar{x}'\bar{\pi}\right]\right\} = 0, \end{aligned}$$

where the last inequality follows from the fact that $x^o(t) \geq rt$ for all $t \geq 0$. The remainder of the proof now follows exactly as in Theorem 6.1.

Remark 11.4 *As remarked previously, the above result parallels Theorem* 6.2. *Unfortunately the conditions given in Assumption* (11.1) *are more difficult to satisfy than in the case of constant rate discounting. Indeed, even if one supposes that the Hamiltonian*

$$\mathcal{H}(x,\pi^o,\pi) = \max_{u\in U}\{f_0(x,u)+\pi^o q(x,u)+\pi' f(x,u)\}$$

is α-concave in x and β-convex in (π^o,π) it is not enough to insure the conditions needed in the above result. To see this we remark that these concavity–convexity conditions only imply

$$\begin{aligned} &(\pi-\bar{\pi})'\nabla_\pi \mathcal{H}(x,\pi^o,\pi) - \nabla_x \mathcal{H}(x,\pi^o,\pi)'(x-\bar{x}) + \frac{\partial}{\partial \pi^o}\mathcal{H}(x,\pi^o,\pi)\,[\bar{\pi}'(x-\bar{x})] \\ &\geq \alpha\|x-\bar{x}\|^2 + \beta\left[\|\pi-\bar{\pi}\|^2 + (\pi^o-\bar{\pi}^o)^2\right] \\ &+ \left[\frac{\partial}{\partial \pi^o}\mathcal{H}(\bar{x},\bar{\pi}^o,\bar{\pi}) - \frac{\partial}{\partial \pi^o}\mathcal{H}(x,\pi^o,\pi)\right]\,[\pi^o-\bar{\pi}^o-\bar{\pi}'(x-\bar{x})] \end{aligned}$$

Thus to insure the condition (11.23) *requires the right hand side to be nonnegative. Similar difficulties are encountered in trying to insure* (11.24) *holds as well. What one does see however is that if the quantity*

$$\left[\frac{\partial}{\partial \pi^o}\mathcal{H}(\bar{x},\bar{\pi}^o,\bar{\pi}) - \frac{\partial}{\partial \pi^o}\mathcal{H}(x,\pi^o,\pi)\right]$$

is sufficiently small, then the condition (11.23) *is likely to hold. Similarly, if one additionally has*

$$\left[\frac{\partial \mathcal{H}}{\partial \pi^o}(x,\pi^o,\pi)\right]^2 < 4\alpha\beta$$

holding, then (11.24) *may hold as well. While these observations do not constitute a proof, they do provide some insight.*

11.3.2 A local result

We now consider the local stability of the modified Hamiltonian system given in (11.19). To consider this system we assume that $\mathcal{H}(\cdot,\cdot,\cdot) : \mathbb{R}^n \times \mathbb{R} \times \mathbb{R}^n \to \mathbb{R}$, given by (11.17), is twice continuously differentiable and that $(\bar{x},\bar{\pi}^o,\bar{\pi})$ is an isolated steady state. We now linearize the system (11.19) in the vicinity of $(\bar{x},\bar{\pi}^o,\bar{\pi})$ and obtain the first variational system

$$\begin{aligned} \dot{\delta x} &= A_1\,\delta x + B_1\,\delta\pi^o + C_1\,\delta\pi \\ \dot{\delta\pi^o} &= A_2\,\delta x + B_2\,\delta\pi^o + C_2\,\delta\pi \\ \dot{\delta\pi} &= A_3\,\delta x + B_3\,\delta\pi^o + C_3\,\delta\pi \end{aligned} \tag{11.25}$$

where the matrices A_j, B_j, C_j, $j = 1,2,3$ are defined by the first order Taylor development of the RHS of (11.19) (e.g., $A_1 = \nabla_x[\nabla_\pi \mathcal{H}(\cdot,\cdot,\cdot)]$ evaluated at $(\bar{x},\bar{\pi}^o,\bar{\pi})$). The following proposition is a classical result of stability analysis.

Proposition 11.1 *If the linear system* (11.25) *admits n eigenvalues with negative real parts and $n+1$ eigenvalues with positive real parts, then there will locally exist an n-dimensional stable manifold in the extended phase space $\mathbb{R}^n \times \mathbb{R} \times \mathbb{R}^n$. Given an initial state x_0, in the vicinity of $\bar{x}$, there will exist initial values π_0^o and π_0 such that the phase space trajectory converges to $(\bar{x},\bar{\pi}^o,\bar{\pi})$.*

Remark 11.5 *If the linearized system* (11.25) *admits this eigenvalue structure, the extremal steady state will be a local attractor for the optimal trajectories.*

11.3.3 An example: Economics of innovation

Consider a firm which maintains a research and development (R&D) department in order to be ready to exploit a possible "technological breakthrough" (e.g. computerized integrated manufacturing). Let x be a state variable which measures the "size" of the R&D department. The state equation is

$$\frac{d}{dt}x(t) \quad = \quad u(t) - \gamma x(t), \quad t \geq 0, \tag{11.26}$$

$$x(0) \quad = \quad x_0, \quad \text{given initial state,} \tag{11.27}$$

where the control u represents the investment effort in the R&D department and γ is the depreciation rate of the R&D facility. The current reward function is given by

$$f_0(x,u) = a + bx + cu^2, \quad a,b,c < 0, \tag{11.28}$$

where $-a$ represents the current constant production cost of the firm (i.e. determined by the current technology used), $-b$ and $-c$ are the maintenance and development costs of R&D respectively. The time of occurence of the technological breakthrough is modeled as a random stopping time with intensity

$$q(x,u) = \alpha x. \tag{11.29}$$

The breakthrough generates a new constant production cost. Assume that the larger the R&D department is at the time of breakthrough the lower this new cost will be. This permits us to represent the ability of a "prepared" firm to seize the opportunity offered by a new technology. This leads us to consider the anticipated total discounted (negative) reward after the stopping time as given by the function

$$\Phi(x) = \frac{a}{\rho}\exp(-x) = \int_0^\infty \exp(-\rho t) a \exp(-x)\, dt. \tag{11.30}$$

If the firm maximizes the expected discounted reward over an infinite time horizon, (11.26)-(11.30) define a control system with a random stopping time.

The modified Hamiltonian system (11.19) for this example becomes

$$\begin{aligned}
\dot{x} &= -\frac{\pi}{2c} - \gamma x \\
\dot{\pi}^o &= (\rho + \alpha x)\pi^o + a + bx + \frac{\alpha a}{\rho}e^{-x}x + \frac{\pi^2}{4c} \\
\dot{\pi} &= (\rho + \alpha x + \gamma)\pi - \alpha\pi^o - (b + \frac{\alpha a}{\rho}e^{-x}(1-x)).
\end{aligned}$$

An extremal steady state is a stationary point of this system. It has been shown in [24] that such a point is uniquely defined if the following conditions hold

$$a\gamma < 2\alpha, \; 2c\gamma\rho < a(a+b).$$

One verifies immediately that these conditions imply

$$\bar{x} > 0,\ \bar{\pi} < 0,\ \bar{\pi}^o < 0.$$

The linearized system in the vicinity of the extremal steady state $(\bar{x}, \bar{\pi}, \bar{\pi}^o)$ is given by

$$\begin{aligned}
\dot{\delta x} &= -\gamma\,\delta x - \frac{1}{2c}\,\delta\pi \\
\dot{\delta\pi^o} &= (\alpha\bar{\pi}^o + b + \frac{\alpha a}{\rho}e^{-\bar{x}} - \frac{\alpha a}{\rho}\bar{x}e^{-\bar{x}})\,\delta x + \frac{\bar{\pi}}{2c}\,\delta\pi + (\rho + \alpha\bar{x})\,\delta\pi^o \\
\dot{\delta\pi} &= (\alpha\bar{\pi} + \frac{2\alpha a}{\rho}e^{-\bar{x}} - \frac{\alpha a}{\rho}\bar{x}e^{-\bar{x}})\,\delta x + (\rho + \alpha\bar{x} + \gamma)\,\delta\pi - \alpha\,\delta\pi^o.
\end{aligned}$$

Applying the Routh criterion to this system one gets the following sufficient conditions for the existence of a saddle point with one single negative real part eigenvalue

$$\begin{aligned}
\gamma\rho c &> \frac{\alpha}{2(\gamma+\rho)}(b + \frac{\alpha a}{\rho} + 2a) \\
\gamma\rho c &> \frac{a}{6}(\frac{\alpha}{\rho} - 1).
\end{aligned}$$

A similar local analysis for the case where $q(\cdot,\cdot)$ depends linearly on u only is developed in [105].

11.4 Piecewise Deterministic Control Systems

We now consider a class of control systems subject to random disturbances having the form of jump processes. These systems have been called Piecewise Deterministic Markov Systems by Davis [61] since the system behaves in a deterministic way between two successive random jump times. We prefer to call them Piecewise Deterministic Control Systems (*PDCS*). Models of this type have recently been utilized in the planning of manufacturing systems (see e.g., Olsder and Suri [146], Gershwin and Kimemia [83], Gershwin and Maimon [84], and Akella and Kumar [2]). In these models the random jumps arise as a result of failures within production cells. In all of the above mentioned works the authors have viewed these disturbance processes as homogeneous Markov chains with constant jump rates. A similar model with nonconstant jump rates has been offered in Boukas and Haurie [23] where the failure rates depend on the ages of the machines in each production cell. Furthermore, they incorporate preventive maintenance by the introduction of a controlled jump rate. A complete dynamic programming analysis of a specific model of this type appears in Fleming et al. [79]. They deal with production planning under random switching demands. This work was the first to establish a turnpike property for a *PCDS*.

By following the approach of Davis [61], [60] and Vermes [178] we show that the discrete event dynamic programming conditions for the associated Markov decision process are linked to the class of control problems with random stopping time studied in the previous sections. We are then able to extend the turnpike property considered by Fleming et al. [79] to a whole class of convex systems with constant jump rates.

11.4.1 PDCS dynamics

In this section we introduce the structure of the *PDCS* and provide several motivating examples arising from production planning in advanced manufacturing systems (e.g., flexible manufacturing cells). To begin we consider a class of systems described by a family of state equations

$$\frac{d}{dt}x(t) = f^i(x(t), u(t)),\, i \in E, \quad t \geq 0 \tag{11.31}$$

$$x(0) = x_0, \qquad \text{given initial state} \tag{11.32}$$

where E is a finite set, $x \in \mathbb{R}^n$ is the state variable and $u \in \mathbb{R}^m$ is the control variable. The functions $f^i : \mathbb{R}^n \times \mathbb{R}^m \mapsto \mathbb{R}^n$ are assumed to be continuous with respect to x and u and satisfy the growth condition (11.2).

For each $i \in E$ let $U^i : \mathbb{R}^n \mapsto 2^{\mathbb{R}^m}$ be an upper semi-continuous point to set mapping. A control is a measurable function $u(\cdot) : [0, \infty) \mapsto \mathbb{R}^m$. A pair $a = (u(\cdot), x(\cdot)) : [0, \infty) \mapsto \mathbb{R}^m \times \mathbb{R}^n$ is admissible at (x_0) if $x(\cdot)$ is the unique trajectory, solution of (11.31)–(11.32), generated by $u(\cdot)$, and if a.e. on $[0, \infty)$, $u(t) \in U^i(x(t))$. We denote by $\Gamma(i, x_0)$ the set of all pairs $a = (u(\cdot), x(\cdot))$ admissible at (i, x_0).

Let $(\Omega, \mathcal{B})$ be a measurable space and $(\xi(t) : t \geq 0)$ a stochastic process with value in E. Given (i, x_0) and an admissible pair $a = (u(\cdot), x(\cdot)) \in \Gamma(i, x_0)$, let a probability measure P_a be defined on $(\Omega, \mathcal{B})$ such that the following holds

$$P_a[\xi(t + dt) = j \mid \xi(t) = i] = q_{ij}(x(t), u(t))dt + o(dt) \tag{11.33}$$

where the $q_{ij} : \mathbb{R}^n \times \mathbb{R}^m \mapsto \mathbb{R}$ are continuous functions satisfying

$$\sum_{j \in E} q_{ij} = 0, \qquad q_{ij} \geq 0 \quad \text{if } i \neq j, \qquad \sum_{j \in E - \{i\}} q_{ij} > 0,$$

and, uniformly in x and u, $\frac{o(dt)}{dt} \to 0$ when $dt \to 0$.

The system evolves in the following way: at time $t^\circ = 0$, $\xi(0) = i$ is observed as well as $x(0) = x_0$. An admissible pair $a = (u(\cdot), x(\cdot)) \in \Gamma(i, x_0)$ is picked. If the process ξ jumps to state j at time τ, then one defines $\xi^1 = j$, $x^1 = x(\tau)$, $t^1 = t^\circ + \tau$. One operates a time translation of value τ, and one redefines $\xi(0) = j$, $x(0) = x^1$. Then a new admissible pair is picked in $\Gamma(j, x^1)$; one defines similarly the successive jump times $t^2, \dots, t^n, \dots$, and visited states $(\xi^2, x^2), \dots, (\xi^n, x^n), \dots$, etc.

We consider a transition reward defined as follows

$$g\left((i,x),\, u(\cdot)\,,x(\cdot)\,\right) = E_a\left[\int_0^T e^{-\rho t} f_0^i\,(x(t),\, u(t))\, dt \mid \xi(0) = i,\; x(0) = x\right] \tag{11.34}$$

where $\{\, f_0^i(\cdot,\cdot) : \mathbb{R}^n \times \mathbb{R}^m \mapsto \mathbb{R},\ i \in E\,\}$ is a class of continuous functions, ρ is a constant positive discount rate, and T is the stopping time corresponding to the first jump of the ξ process, given $\xi(0) = i$, $x(0) = x$ and the admissible pair $a = (u(\cdot)\,, x(\cdot)\,) \in \Gamma(i, x_0)$.

Putting together all the successive transitions, the system can be formally described by the state equations

$$\frac{d}{dt}\,x(t) = f^{\xi(t)}(x(t),\, u(t))\,, \quad t \geq 0, \tag{11.35}$$

$$u(t) \in U^{\xi(t)}(x(t)), \tag{11.36}$$

$$x(0) = x_0, \tag{11.37}$$

$$\xi(0) = \xi_0, \tag{11.38}$$

where $\xi(\cdot)$ is a jump process with values in E and jump rates $q_{ij}(x, u)$. This system could be controlled via a feedback law $u(\xi, x)\,: E \times \mathbb{R}^n \mapsto U^{\xi}(x)$, associating a control value u with the observed state (ξ, x). This feedback law generates the reward functional

$$E_{u(\cdot,\cdot)}\left[\int_0^\infty e^{-\rho t} f_0^{\xi(t)}\,(x(t),\, u(t))\, dt \mid \xi(0) = \xi_0,\; x(0) = x_0\right]. \tag{11.39}$$

This is not always a convenient definition as some regularity conditions must be imposed on $u(\cdot,\cdot)$ (e.g. see [154]). A more precise approach, which uses the formalism of Markov Decision Processes, is proposed in Section 11.4.2. However the above formulation gives an intuitive feeling about the stochastic control structure of this class of systems.

The following examples show how these systems have been associated with manufacturing flow control models or related problems.

Example 11.1 Gershwin and Kimemia [83] have considered the following system

$$\frac{d}{dt}\,x(t) = u(t) - d,$$

$$u(t) \in U^{\xi(t)},$$

where $U^{\xi(t)}$ is a convex set, with the cost rate functions

$$f_0^{\xi}\,(x,\, u) = \sum_{l=1}^{n} [\alpha_l \max\{0, x_l\} + \beta_l \max\{0, -x_l\}].$$

The ξ process is a Markov chain with constant jump rates $q_{ij}(x,u) \equiv q_{ij}$. It could describe a flexible manufacturing cell where ξ represents the set of operational machines, $x = (x_l)_{l=1,\dots,n}$ is the surplus level for n different parts, $d = (d_l)_{l=1,l\dots,n}$ is the parts demand rate vector, $u = (u_l)_{l=1,\dots,n}$ is the production rates for different parts, $-\alpha_l$ is the unit holding cost rate and $-\beta_l$ is the unit back-order cost rate for the surplus of part l. The piecewise deterministic structure comes from the machine failure and repair process (described by the constant jump rates q_{ij}). The control sets U^{ξ} depend on the set of operational machines.

Example 11.2 Fleming, Sethi and Soner [79] have considered the following system

$$\frac{d}{dt}x(t) = u(t) - d^{\xi(t)},$$

$$u(t) \in U,$$

where U is a convex set, the ξ process is a Markov chain with the constant jump rates $q_{ij}(x,u) \equiv q_{ij}$ and one considers a concave reward rate function $f_0(x,u)$. This corresponds to the case where the production system has to satisfy a demand d^{ξ} which is randomly jumping from one value d^i to another d^j. These demand variations may be caused by some perturbed downstream operating conditions in an hierarchical production system.

Example 11.3 Boukas and Haurie [23] have studied the following structure

$$\frac{d}{dt}x(t) = f^{\xi(t)}(x(t),u(t)),$$

$$u(t) \in U^{\xi(t)},$$

$$f_0^{\xi}(x,u) = \sum_{l=1}^{p} [\alpha_l \max\{0,x_l\} + \beta_l \max\{0,-x_l\}],$$

where the ξ process has state and control dependent jump rates $q_{ij}(x,u)$. In this model, the state vector x is composed of the parts surplus levels and of the machines ages, ξ represents the set of operational machines. The control vector u is composed of the parts production rates and of the preventive maintenance jump rates (e.g. when a machine is stopped for some maintenance operations). The jump rates $q_{ij}(x,u)$ describe the changes in machines' operational states. The failure rate is age dependent; the preventive maintenance rate is controlled.

A variant of this model could be the following, $x = (y, z)$ where y is the parts surplus level, z is the "state of wear" of the machines, $u = (v, w)$ where v is the parts production rates and w is the machine maintenance effort. The controlled system would then have the following structure

$$\begin{aligned} \frac{d}{dt} y(t) &= v(t) - d, \\ \frac{d}{dt} z(t) &= f(z, v, w), \\ v(t) \in V^{\xi(t)} \quad &, \quad w(t) \in W^{\xi(t)}, \end{aligned}$$

where $f(z, v, w)$ is the wear accumulation function. The reward rate functions $f_0^{\xi}(x, u)$ are concave in (x, u) and the failure rates which enter into the computation of $q_{ij}(z, w)$ are increasing in z, decreasing in w.

11.4.2 Reformulation as a Markov Renewal Decision Process

We can view *PDCS* as hybrid systems in which part of the state variable (i.e., ξ) is discrete while the remaining state variables (i.e., x) are continuous. The discrete variable jumps at random times. Between two successive jumps one has a deterministic system which can be open-loop controlled. By observing the system after each of these jump times, and choosing an open-loop control to be effective until the next jump these models can be viewed as a discrete event stochastic system in the class of the so called Markov Renewal Decision Processes (*MRDP*). These ideas will be developed below.

The system described by (11.31)–(11.34) is a Discounted Markov Renewal Decision Process ($MRDP$) with Borel state and action spaces defined by the six-tuple

$$(S,\ A,\ \Gamma,\ Q, \gamma,\ \rho),$$

where each element is described below. This class of systems has recently been studied in great detail by Bertsekas and Shreve [19]. This formalism has been extended in [102] and used in [114] for the analysis of optimal replacement in deteriorating systems.

First we define the state set $S = E \times \mathbb{R}^n$ with its class of Borel sets $\mathcal{B}(S)$. The system will be observed immediately after each jump. Now define the action set $A = \{a = (u(\cdot)\,, x(\cdot)\,)\}$, where $u(\cdot)$ is a control and $x(\cdot)$ is a trajectory generated by $u(\cdot)$. There are restrictions on the possible actions, we thus consider the mapping $\Gamma : S \mapsto 2^A$ defined previously and we impose that the action be picked in $\Gamma(s)$ if the system is observed in state s. The mapping Γ is called the *action constraint map*.

The dynamics of the sequential process in the state space S is defined by a transition probability kernel $Q(\cdot \mid s,\ a)$ which is the conditional probability measure on $(S \times [0,\ \infty)\,,\ \mathcal{B}(S \times [0,\ \infty)))$, given the observed state s and the chosen action a, and induced by the transition rates (11.33). This kernel defines the family of conditional probability laws governing the delay to the next jump, and the state s' reached right after the next jump, given the current state s and action a.

Finally we consider the transition reward $\gamma(s, a)$ defined by

$$\gamma(s, a) = g\left((i, x), u(\cdot), x(\cdot)\right) \tag{11.40}$$

where the RHS, defined in (11.34), gives the expected cost before the next jump occurs, if currently $s = (i, x)$, and $a = (u(\cdot), x(\cdot))$. This completes the definition of the data characterizing the $MRDP$ under study.

Define a policy as a mapping

$$\pi : S \mapsto A \text{ such that } \pi(s) \in \Gamma(s).$$

A policy π is admissible if it induces a probability measure over the product space

$$\prod_{n=0}^{\infty} \{[0, \infty) \times S \times A\}$$

such that the process $\{t^n, s^n, a^n : n = 0, \ldots, \infty\}$, where t^n is the time of the nth jump of the ξ process, $s^n = (\xi(t^n), x(t^n))$, and $a^n = \pi(s^n)$, is well defined, and the total discounted expected cost given the initial state s^o

$$V_\pi(s^o) = E_\pi\left[\sum_{n=0}^{\infty} e^{-\rho t^n} \gamma(s^n, a^n)\right] \tag{11.41}$$

exists. An optimal policy is an admissible policy π^* such that, for any initial state s_0

$$V_{\pi^*}(s^o) = \sup\{V_\pi(s^o) : \pi \text{ admissible}\}. \tag{11.42}$$

In the remainder of this section we write down the dynamic programming equations characterizing the solution of this Markov Renewal Decision process. We are using the Denardo contracting operator formalism [62].

Let $V(\cdot,\cdot) = (V(j,\cdot))_{j\in E}$, $V(j,\cdot) : \mathbb{R}^n \mapsto \mathbb{R}$, be a function continuous in x. Then we define, for any $s = (i, x)$ and any admissible action a, the *local return function*

$$\begin{aligned} h(s, a, V(\cdot,\cdot)) \quad = \quad & E_a\Big[\int_0^T e^{-\rho t} f_0^i(x(t), u(t))\, dt + \\ & e^{-\rho T}\, V(\xi(T), x(T)) \,|\, \xi(0) = i,\; x(0) = x\Big]. \end{aligned} \tag{11.43}$$

Now, we introduce the operators $\mathcal{T}_\pi$, for a given policy π, and $\mathcal{T}^*$ acting on $V(\cdot,\cdot)$, defined by

$$\mathcal{T}_\pi(V(\cdot,\cdot))(s) \quad = \quad h(s, \pi(s), V(\cdot,\cdot)) \tag{11.44}$$

$$\mathcal{T}^*(V(\cdot,\cdot))(s) \quad = \quad \sup_{a\in\Gamma(s)} h(s, a, V(\cdot,\cdot)). \tag{11.45}$$

In (11.44) $\mathcal{T}_\pi(V(\cdot,\cdot))(s)$ is a continuous function of x. In (11.45), the RHS corresponds to a control problem where the randomness is essentially concentrated in the stopping time T. This is precisely the random stopping time control problem discussed in Sections 2 and 3.

11.5 Global turnpike property for constant jump rates

When the jump rates q_{ij} are constant (this corresponds to the models developed in Rishel [153], [154], Gershwin et al. [83], Fleming, Sethi and Soner [79]), one can use Rockafellar's [156] or Brock and Scheinkman's results [30] and show that under strict convexity conditions optimal trajectories globally converge toward optimal steady states.

Following the procedure already used in earlier chapters we define for all $(x,v) \in \mathbb{R}^n \times \mathbb{R}^n$, for all $i \in E$,

$$F^i(x,v) = \{u \in U^i(x) : v = f^i(x,u)\},$$

(one assumes $U^i(x) \neq$), as well as the reward function with value in $\mathbb{R} \cup \{-\infty\}$

$$\mathcal{L}^i(x,v) = \begin{cases} \sup\{f_0^i(x,u) : u \in F^i(x,v)\}, & \text{if } F^i(x,v) \neq \emptyset, \\ -\infty, & \text{if } F^i(x,v) = \emptyset. \end{cases}$$

The stochastic control problem defined in the RHS of (11.45) is then equivalent to the following problem of Lagrange (after proceeding as in Section 11.2)

$$\max\{\int_0^\infty e^{-(\rho+\sum_{j\in E-\{i\}} q_{ij})t}\,[\mathcal{L}^i(x(t),\dot{x}(t)) + \textstyle\sum_{j\in E-\{i\}} q_{ij}\,V(j,x(t))]\,dt \mid x(\cdot) \in AC_{loc}\ s.t.\ x(0)=x_0\}. \tag{11.46}$$

Define, $\forall\, i \in E$, the Hamiltonian

$$\mathcal{H}^i(x,\lambda) = \sup_v\{\mathcal{L}^i(x,v) + \lambda^T v\} + \sum_{j\in E-\{i\}} q_{ij}\,V(j,x), \tag{11.47}$$

it has been shown by Rockafellar [156] that, if $\mathcal{H}^i$ is concave in x and convex in λ, a solution of (11.46) is also a solution to the modified Hamiltonian system

$$(-\dot{\lambda}(t) + (\rho + \sum_{j\in E-\{i\}} q_{ij})\lambda(t), \dot{x}(t)) \in \partial\mathcal{H}^i(x(t),\lambda(t)), \tag{11.48}$$

where the set $\partial\mathcal{H}^i(x,\lambda)$ consists of the subgradients of $\mathcal{H}^i$ at (x,λ),

Proposition 11.2 *Assume that there exists a unique stationary point* $(\bar{x}^i, \bar{\lambda}^i)$ *for the hamiltonian system* (11.48), *i.e. s.t.*

$$((\rho + \sum_{j\in E-\{i\}} q_{ij})\bar{\lambda}^i, 0) \in \partial\mathcal{H}^i(\bar{x}^i,\bar{\lambda}^i),$$

and let $\mathcal{S}^i \subset \mathbb{R}^n \times \mathbb{R}^n$ *be a neighborhood of* $(\bar{x}^i, \bar{\lambda}^i)$ *on which* $\mathcal{H}^i(\cdot,\cdot)$ *is finite, α-convex in* x*, β-concave in* λ*, with*

$$(\rho + \sum_{j\in E-\{i\}} q_{ij})^2 < 4\alpha\beta. \tag{11.49}$$

Then, every solution $(x(\cdot), \lambda(\cdot))$ *of the modified Hamiltonian system* (11.48) *such that* $(x(0), \lambda(0)) \in \mathcal{S}^i$ *and*

$$\lim_{t\to\infty} e^{-(\rho+\sum_{j\in E-\{i\}} q_{ij})t} \|x(t)\| \|\lambda(t)\| \neq \infty,$$

must converge toward $(\bar{x}^i, \bar{\lambda}^i)$.

Proof: It is a direct adaptation of Rockafellar's results [156]. ∎

This local result can be extended to the whole space $\mathbb{R}^n \times \mathbb{R}^n$ using Rockafellar's procedure [156]. We further refer the reader to Chapter 6 for other results that can be easily modified to the case considered here. The steady state $\bar{x}^i$ is the turnpike associated with the value i for the jump process $\xi(\cdot)$. If, for each $i \in E$, the steepness condition (11.49) holds, then there will exist a set of attractors $\{\bar{x}^i,\, i \in E\}$ for the trajectory. The current value of the jump process $\xi(t)$ will indicate the active attractor.

Remark 11.6 *One notices that the concavity–convexity property of* $\mathcal{H}^i$ *is a consequence of the joint concavity in* $(x, \dot{x})$ *of the integrand*

$$\mathcal{L}^i(x, \dot{x}) + \sum_{j\in E-\{i\}} q_{ij} V(j, x)$$

in (11.46)*. Actually it is only a consequence of the concavity in* $(x, \dot{x})$ *of* $\mathcal{L}^i(x, \dot{x})$*, for any* $i \in E$*. Indeed, using the notations developed in the previous section, the fixed-point condition for the operator* $\mathcal{T}^*$ *can also be rewritten*

$$V(i,x) = \sup E_{x(\cdot)} \Big[\int_0^T e^{-\rho t} \mathcal{L}^i(x(t), \dot{x}(t)) dt \tag{11.50}$$
$$+ e^{-\rho T}\, V(x(T), \xi(T)) \,|\, x(\cdot) \in AC_{loc} \text{ s.t. } x(0) = x \Big], \quad i \in E.$$

It is immediate to check that $\mathcal{T}^*$ *is contracting and, under the concavity assumption on* $\mathcal{L}^i$*, closed on the class of functions* $V(j,x)$*,* $j \in E$*, which are concave in* x*. This implies the concavity–convexity of the hamiltonians* $\mathcal{H}^i$*.*

Example 11.4 Consider the following variant of the system studied by Fleming, Sethi and Soner [79],

maximize

$$E_u\Big[\int_0^\infty e^{-\rho t}[L^{\xi(t)}(x(t), u(t))]\, dt\Big] \tag{11.51}$$

subject to

$$\frac{d}{dt}x(t) = Ax(t) + u(t) - d^{\xi(t)} \tag{11.52}$$

$$x(0) = x_0 \tag{11.53}$$

$$\xi(0) = i \tag{11.54}$$

where x, $u \in \mathbb{R}^n$, $\xi \in E$, where E is a finite set, and A is an $n \times n$ matrix. One has to consider the family of $DCPITH$'s

maximize

$$\int_0^\infty e^{-(\rho+\sum_{j\in E-\{i\}} q_{ij})t} [L^i(x(t), u(t)) + \sum_{j\in E-\{i\}} q_{ij} V(j, x(t))]\, dt \tag{11.55}$$

subject to

$$\frac{d}{dt}x(t) = Ax(t) + u(t) - d^i \tag{11.56}$$

$$x(0) = x_0. \tag{11.57}$$

Assume the functions $L^i(x, u)$ are strictly concave in both arguments. As the state equations are linear in x and u, the associated functions $\mathcal{L}^i(x, v)$ are also strictly concave and finite for all (x, v). As noted above this implies that the hamiltonians $\mathcal{H}^i$ are concave in x, convex in λ.

Consider the set of optimal steady states $\bar{x}^i$, defined by the conditions

$$((\rho + \sum_{j\in E-\{i\}} q_{ij})\bar{\lambda}^i, 0) \in \partial\mathcal{H}^i(\bar{x}^i, \bar{\lambda}^i),\ i \in E$$

with

$$\mathcal{H}^i(x, \lambda) = \sup_{u\in\mathbb{R}^n} \{f_0^i(x, u) + \lambda^T u\} + \sum_{j\in E-\{i\}} q_{ij} V(j, x) + \lambda^T Ax - \lambda^T d^i. \tag{11.58}$$

The global asymptotic stability of these points, for optimal trajectories is obtained if the hamiltonians $\mathcal{H}^i$ are $\alpha-$concave in x, $\beta-$convex in λ, with

$$(\rho + \sum_{j\in E-\{i\}} q_{ij})^2 < 4\alpha\beta.$$

Since the convexity of the Bellman functions $V(j, x)$ is assured for each j (see Remark 5.1), this property is implied by the $\alpha-$concavity in x, $\beta-$convexity in λ of the functions

$$\sup_{u\in\mathbb{R}^n} \{f_0^i(x, u) + \lambda^T u\}. \tag{11.59}$$

If we further assume state and control separation in L^i, i.e. $f_0^i(x, u) = \tilde{\ell}^i(x) + \tilde{k}^i(u)$, with the required regularity, the optimal control is defined by

$$\frac{\partial \tilde{k}^i}{\partial u_j}(u) = \frac{\partial \tilde{k}^i}{\partial u_j}(u_j) = -\lambda_j. \tag{11.60}$$

One gets

$$u^*(\lambda_j) = (\frac{\partial \tilde{k}^i}{\partial u_j})^{-1}(-\lambda_j) = G_j(\lambda_j). \tag{11.61}$$

By direct calculations, one obtains

$$\frac{\partial^2}{\partial x_j \partial x_k}\mathcal{H}^i(x,\lambda) = \frac{\partial^2}{\partial x_j \partial x_k}\bar{\ell}^i(x) \tag{11.62}$$

and

$$\frac{\partial^2}{\partial \lambda_j^2}\mathcal{H}^i(x,\lambda) = -\frac{1}{\frac{\partial^2}{\partial u_j^2}\tilde{k}^i(G_j(\lambda_j))} = -[\frac{\partial^2}{\partial u_j^2}\tilde{k}^i(G_j(\lambda_j))]^{-1}. \tag{11.63}$$

The definite positiveness of the matrices

$$\begin{bmatrix} \bar{\ell}^i_{x_1x_1} - \alpha & \bar{\ell}^i_{x_1x_2} & \dots & \bar{\ell}^i_{x_1x_n} \\ \bar{\ell}^i_{x_2x_1} & \bar{\ell}^i_{x_2x_2} - \alpha & \dots & \bar{\ell}^i_{x_2x_n} \\ \vdots & \vdots & \dots & \vdots \\ \bar{\ell}^i_{x_nx_1} & \bar{\ell}^i_{x_nx_2} & \dots & \bar{\ell}^i_{x_nx_n} - \alpha \end{bmatrix}$$

and

$$\begin{bmatrix} [\tilde{k}^i_{u_1u_1}]^{-1} - \beta & 0 & \dots & 0 \\ 0 & [\tilde{k}^i_{u_2u_2}]^{-1} - \beta & \dots & 0 \\ \vdots & \vdots & \dots & \vdots \\ 0 & 0 & \dots & [\tilde{k}^i_{u_nu_n}]^{-1} - \beta \end{bmatrix}$$

is a sufficient condition to get $\alpha-$concavity in x, $\beta-$convexity in λ for $\mathcal{H}^i(x,\lambda)$.

Bibliography

[1] ALBOUY, A. and A., BRETON, Interprétation Economique du Principe du Maximum, *Revue Française d'Informatique et de Recherche Operationnelle*, **14**, 1968.

[2] AKELLA R. A. and P. R. KUMAR, Optimal Control of Production Rate in Failure Prone Manufacturing system, *IEEE Transactions on Automatic Control*, **31**, pp. 116–126, 1986.

[3] ANGELL, T.S., On Controllability for Non-linear Hereditary Systems: A Fixed Point Approach, *Nonlinear Analysis*, **4**, pp. 529-545, 1980.

[4] ATHANS, M. and P. FALB, *Optimal Control*, McGraw Hill, 1966.

[5] ANDERSON, B.D.O. and J.B. MOORE, *Linear Optimal Control*, Englewood Cliffs, NJ, Prentice Hall, 1971.

[6] ARTSTEIN, Z. and A. LEIZAROWITZ, Tracking Periodic Signals with the Overtaking Criterion, *IEEE Transactions on Automatic Control*, **30**, pp. 1122–1126, 1985.

[7] ARROW, K. J. and M. KURZ, *Public Investment, Rate of Return, and Optimal Fiscal Policy*, Johns Hopkins Press, Baltimore, Maryland, 1970.

[8] ASTROM, K.J., *Introduction to Stochastic Control Theory*, Academic Press, New York, 1970.

[9] ATSUMI, H., Neoclassical Growth and the Efficient Program of Capital Accumulation, *Review of Economic Studies*, **30**, pp. 127–136, 1963.

[10] BALAKRISHNAN, A. V., *Applied Functional Analysis*, Applications of Mathematics Vol. 3, Springer-Verlag of New York, 1979.

[11] BALAKRISHNAN, A.V. *Stochastic Differential Systems I*, Springer-Verlag, 1978.

[12] BALDER, E.J., An Existence Result for Optimal Economic Growth, *Journal of Mathematical Analysis and Applications*, **95**, pp. 195-213, 1983.

[13] BARBU, V., Boundary Control Problems with Convex Cost Criterion, *SIAM Journal on Control and Optimization*, **18**, pp. 227-243, 1980.

[14] BARBU, V. and G. DAPRATO, *Hamilton-Jacobi Equations in a Hilbert Space*, Research Notes in Mathematics, Vol. 86, Pitman Publishing Inc., Marshfield, Massachussetts, 1983.

[15] BATES, G.R., Lower Closure and Existence Theorems for Optimal Control Problems with Infinite Horizon, *Journal of Optimization Theory and Applications*, **24**, pp. 639-649, 1978.

[16] BAUM, R.F., Existence Theorems for Lagrange Control Problems with Unbounded Time Domain, *Journal of Optimization Theory and Applications*, **19**, pp. 89-116, 1976.

[17] BECKER, R.A., J.H. BOYD and B.Y. SUNG, Recursive Utility and Optimal Capital Accumulation, *Journal of Economic Theory*, **47**, pp. 76–100, 1990.

[18] BELLMAN, R. and R. BUCY, Asymptotic Control Theory, *SIAM Journal on Control*, **2**, pp. 11–18, 1964.

[19] BERTSEKAS D. P. and S.E. SHREVE, *Stochastic Optimal Control : The Discrete Time Case*, Academic Press, 1978

[20] BLAQUIÈRE, A. and G. LEITMANN, On the Geometry of Optimal Processes, in G. Leitmann edit., *Topics in Optimization*, Academic Press 1967.

[21] BORKAR V.S. and M.K. GHOSH, Ergodic Control of Multidimensional Diffusions II: Adaptive Control. Preprint.

[22] BORKAR V.S., On Minimum Cost per Unit Time Control of Markov Chains, *SIAM Journal on Control and Optimization* **22**, pp. 965-978.

[23] BOUKAS E.K. and A. HAURIE, Planning Production and Preventive Maintenance in an FMS: a Stochastic Control Approach, *IEEE Transactions On Automatic Control*, **35**, pp. 1029–1031, 1990.

[24] BOUKAS E.K., A. HAURIE and PH. MICHEL, An Optimal Control Problem with a Random Stopping Time, *Journal of optimization theory and applications*, **64**, pp. 471–480, 1990.

[25] BRIDGELAND, T.F., On the Existence of Optimal Feedback Controls, *SIAM Journal on Control*, Ser. A, **1**, pp. 261-74, 1963.

[26] BROCK, W.A., The Global Asymptotic Stability of Optimal Control: A Survey of Recent Results, *University of Chicago Working Paper*, Feb. 1976.

[27] BROCK, W. A., On Existence of Weakly Maximal Programmes in a Multisector Economy, *Review of Economic Studies*, **37**, pp. 275–280, 1970.

[28] BROCK, W.A. and A. HAURIE, On Existence of Overtaking Optimal Trajectories Over an Infinite Time Horizon, *Mathematics of Operations Research*, **1**, pp. 337–346, 1976.

[29] BROCK, W.A. and J. SHEINKMAN, Global Asymptotic Stability of Optimal Control Systems with Application to Economic Analysis in J.D. Pitchford and S.J. Turnovsky eds., *Application of Control Theory to Economics*, North Holland, 1978.

[30] BROCK, W.A. and J. SHEINKMAN, Global Asymptotic Stability of Optimal Control Systems with Applications to the Theory of Economic Growth, *Journal of Economic Theory*, **12**, pp. 164–190, 1976.

[31] BRYSON, A. and Y.C. HO, *Applied Optimal Control*, Blaisdell 1969.

[32] BUCY, R.S. and P.D. JOSEPH, *Filtering for Stochastic Process with Applications to Guidance*, Wiley-Interscience, 1968.

[33] BUTKOVSKIY, A. B., *Distributed Control Systems*, American Elsevier, 1969.

[34] CARATHÉODORY, C., *Calculus of Variations and Partial Differential Equations*, Chelsea, New York, 1982.

[35] CASS, D., Optimum Growth in an Aggregative Model of Capital Accumulation, *Review of Economic Studies*, **32**, pp. 233–240, 1965.

[36] CASS, D., Optimal Growth in an Aggregative Model of Capital Accumulation, A Turnpike Theorem, *Econometrica*, **34**, pp. 833–850, 1965.

[37] CASS, D. and K. SHELL, The Structure and Stability of Competitive Dynamical Systems, *Journal of Economic Theory*, **12**, pp. 30–70, 1976.

[38] CARLSON, D.A., The Existence of Catching-up Optimal Solutions for a Class of Infinite Horizon Optimal Control Problems with Time Delay, *SIAM Journal on Control and Optimization*, **28**, pp. 402-422, 1990.

[39] CARLSON, D.A., Uniformly Overtaking and Weakly Overtaking Optimal Solutions in Infinite-Horizon Optimal Control: When Optimal Solutions are Agreeable, *Journal of Optimization Theory and Applications*, **64**, pp. 55–69, 1990.

[40] CARLSON, D.A., The Controllability of Infinite Horizon Optimal Control Problems, *Nonlinear Analysis, Theory, Method, and Applications*, **11**, pp. 437-453, 1987.

[41] CARLSON,D.A., A Carathéodory-Hamilton-Jacobi Theory for Infinite Horizon Optimal Control Problems, *Journal of Optimization Theory and Applications*, **48**, pp. 265-87, 1986.

[42] CARLSON, D.A., On the Existence of Catching up Optimal Solutions for Lagrange Problems Defined on Unbounded Intervals, *Journal of Optimization Theory and Applications*, **49**, pp. 207-225, 1986.

[43] CARLSON, D.A., Carathéodory-Hamilton-Jacobi Theory for Nonsmooth Problems of Lagrange and Problems of Optimal Control, *Journal of Mathematical Analysis and Applications*, **162**, pp. 259-273, 1985.

[44] CARLSON, D.A., Asymptotic Stability for Optimal Trajectories of Infinite Horizon Optimal Control Models with State and Control Dependent Discounting, preprint.

[45] CARLSON, D. A., A. JABRANE and A. HAURIE, Existence of Overtaking Solutions to Infinite Dimensional Control Problems on Unbounded Time Intervals, *SIAM Journal on Control and Optimization*, **25**, pp. 1517-1541, 1987.

[46] CESARI, L., *Optimization-Theory and Applications*, Applications of Mathematics, Vol. 17, Springer-Verlag, New York, 1983.

[47] CESARI, L., *Optimization - Theory and Applications: Problems with Ordinary Differential Equations*, Applications of Mathematics Series, Vol. 17, Springer-Verlag of New York, New York, 1983.

[48] CESARI, L., Existence Theorems for Weak and Usual Optimal Solutions in Lagrange Problems with Unilateral Constraints, I and II, *Transactions of American Mathematical Society*, **124**, pp. 369–429, 1966.

[49] CESARI, L., J.R. LAPALM and T. NISHIURA, Remarks on Some Existence Theorems for Optimal Control, *Journal of Optimization Theory and Applications*, **31**, pp. 397-416, 1969.

[50] CESARI, L. and M.B. SURYANARAYANA, On Recent Existence Theorems in the Theory of Optimization, *Journal of Optimization Theory and Applications*, **31**, pp. 397-416, 1980.

[51] CHUKWU, E.N., Null Controllability in Function Space of Nonlinear Retarded Systems with Limited Control, *Journal of Mathematical Analysis and Applications*, **103**, pp. 198-210, 1984.

[52] CHUKWU, E.N., On the Null-controllability of Nonlinear Delay Systems with Restrained Controls, *Journal of Mathematical Analysis and Applications*, **76** pp. 283-296, 1980.

[53] CINQUINI, S., Sopra l'Esistenza dell'Estromo Assuloto per Integrali Estesi a Intervalli Infiniti, *Rendiconti della Accademia Nazionale dei Lincei*, Ser. 8, **32**, pp. 320-325, 845-851, 1962.

[54] CINQUINI, S., Una Nuova Estensione dei Moderni Methodi del Calculo delle Variazioni, *Annali delle Scuola Normale Superiore de Pisa*, Ser. 2, **9**, pp. 258-261, 1940.

[55] CLARK, C., *Mathematical Bioeconomics: The Optimal Management of Renewable Resources*, John Wiley and Sons, New York, 1976.

[56] CLARKE, F.H., *Optimization and Nonsmooth Analysis*, John Wiley and Sons, Inc., 1983.

[57] CLIFF, E.M. and T.L. VINCENT, An Optimal Policy for a Fish Harvest, *Journal of Optimization Theory and Applications*, **12**, pp. 485–496, 1973.

[58] CRANDALL, M.G. and P-L. LIONS, Viscosity Solutions of Hamilton-Jacobi Equations, *Transactions of the American Mathematical Society*, **277**, pp. 1-42, 1983.

[59] DARBO, G., L'estremo assuloto per gli Integrali su Intervallo Infinito, *Rendiconti de Seminaro Mathematico dell'Universita de Padova*, **22**, pp. 319-416, 1953.

[60] DAVIS, M.H A., Control of Piecewise-Deterministic Processes Via Discrete-Time Dynamic Programming, *Proceedings of the 3rd Bad Honnef Symposium on Stochastic Differential Systems*, Springer Verlag, Berlin, 1985.

[61] DAVIS, M.H.A., Piecewise-deterministic Markov Processes : A General Class of Non-diffusion Stochastic Models, *Journal of Royal Statistical Society*, **46**, pp. 353–388, 1984.

[62] DENARDO, E.V., Contractions Mappings in the Theory Underlying Dynamic Programming, *SIAM Review*, **9**, pp. 165–177, 1967.

[63] DERZO, N. and S.P. SETHI, Distributed Parameter Systems Approach to the Optimal Cattle Ranching Problem, *Optimal Control Applications and Methods*, **1**, pp. 3–10, 1980.

[64] DORFMAN, R., An Economic Interpretation of Optimal Control Theory, *American Economic Review*, **59**, pp. 817–831, 1969.

[65] DUNFORD, N. and J.T. SCHWARTZ, *Linear Operators, Part I: General Theory*, John Wiley and Sons, Inc., New York, New York, 1958.

[66] EIROLA, T. and V. KAITALA, On the Existence of Overtaking Trajectories, *Journal of Optimization Theory and Applications*, **49**, pp. 227–237, 1986.

[67] EKELAND, I. and R. TEMAM, *Convex Analysis and Variational Problem*, North Holland, Amsterdam, 1976.

[68] EPSTEIN, L.G., A Simple Dynamic General Equilibrium Model, *Journal of Economic Theory*, **41**, (1987), pp. 68–95.

[69] EPSTEIN, L.G. and J.A. HYNES, The Rate of Time Preference and Dynamic Economic Analysis, *Journal of Political Economy*, **91**, pp. 611–635, 1983.

[70] FAEDO, S., Il Calcolo delle Variazioni per gli Integrali su Intervallo Infiniti, *Commentationes, Pontificia Academia Scientiarum*, **8**, pp. 319–421, 1944.

[71] FAEDO, S., Il Calcolo delle Variazioni per gli Integrali su Intervallo Infinity, *Rendiconti di Mathematica Applicata*, **8**, pp. 94–125, 1949.

[72] FAEDO, S., Il Calcolo della Variazioni per gli Integrali Estesi a Intervalli Infiniti, *Annali della Scuola Normale Superiore di Pisa*, **7**, pp. 91–132, 1953.

[73] FATTORINI, H.O., Boundary Control Systems, *SIAM Journal on Control and Optimization*, **6**, pp. 349-385, 1968.

[74] FEDERER, H., *Geometric Measure Theory*, Springer-Verlag, Berlin, 1969.

[75] FEINSTEIN, C.D. and D.G. LUENBERGER, Analysis of the Asymptotic Behavior of Optimal Control Trajectories: The Implicit Programming Problem, *SIAM Journal of Control and Optimization*, **19**, pp. 561–585, Sept. 1981.

[76] FEINSTEIN, C.D. and S.S. OREN, A "Funnel" Turnpike Theorem for Optimal Growth Problems with Discounting, *Journal of Economic Dynamics and Control*, **9**, pp. 25–39, 1985.

[77] FISHER, I., *Theory of Interest*, MacMillan Publishing Co., New York, New York, (1930).

[78] FLEMING, W.H. and R.W. RISHEL, *Deterministic and Stochastic Optimal Control*, Springer-Verlag, New York, 1975.

[79] FLEMING W., S.P. SETHI and H.M. SONER, An Optimal Stochastic Production Planning Problem with Randomly Fluctuating Demand, *SIAM Journal on Control and optimization*, **25**, pp. 1494-1502, 1987.

[80] FRIEDMAN, A., *Stochastic Differential Equations and Applications, vol. I.* Academic Press, New York, 1974.

[81] GALE, D., On Optimal Development in a Multi-Sector Economy, *Review of Economic Studies*, **34**, pp. 1–19, 1967.

[82] GANDOLFO, G., *Mathematical methods and models in economic dynamics*, North-Holland, London, 1971.

[83] GERSHWIN S.B. and J. KIMEMIA, An Algorithm for Computer Control of a Flexible Manufacturing System, *IIE Transactions*, **15**, pp. 353–362, 1983.

[84] GERSHWIN S.B. and O.Z. MAIMON, Dynamic Scheduling and Routing for Flexible Manufacturing Systems that have Unreliable Machines, *Operations Research*, **36**, pp. 279–292, 1988.

[85] GETZ, W.M., On Harvesting Two Competing Populations, *Journal of Optimization Theory and Applications*, **28**, pp. 585-602, 1979.

[86] GILBERT, F.G., Optimal Periodic Control: A General Theory of Necessary Conditions, *SIAM Journal on Control and Optimization*, **15**, pp. 717–746, 1977.

[87] GILBERT, W. and W. SMITH, Optimal Exploitation of a Multispecies Community, *Mathematical Bioscience*, **33**, pp. 121–132, 1977.

[88] GOH, B.S., G. LEITMANN and T.L. VINCENT, Optimal Control of a Prey-Predator System, *Mathematical Bioscience*, **19**, pp. 263–286, 1974.

[89] HALKIN, H., Necessary Conditions for Optimal Control Problems with Infinite Horizon, *Econometrica*, **42**, pp. 267–273, 1974.

[90] HALKIN, H., Extremal Properties of Biconvex Contingent Equations, in L. Weiss ed., *Ordinary Differential Equations*, Academic Press 1971.

[91] HALKIN, H., Mathematical Foundations of System Optimization, in G. Leitmann edit., *Topics in Optimization*, Academic Press 1967.

[92] HAMMOND, P.J. and J.A. MIRRLEES, Agreeable Plans, *Models of Economic Growth*, Edited by J.A. Mirrlees and N.H. Stern, Macmillan, New York, New York, pp. 283–299, 1990.

[93] HAMMOND, P.J. and J. KENNAN, Uniformly Optimal Infinite-Horizon Plans, *International Economic Review*, **20**, pp. 283–296, 1979.

[94] HARTL, R.F., Optimal Dynamic Advertising Policies for Hereditary Processes, *Journal of Optimization Theory and Applications*, **43**, pp. 51–72, 1984.

[95] HARTL, R.F. and S.P. SETHI, Optimal Control of a Class of Systems with Continuous Lags: Dynamic Programming Approach and Economic Interpretations, *Journal of Optimization Theory and Applications*, **43**, pp. 73–88, 1984.

[96] HARTMAN, P., *Ordinary Differential Equations*, Wiley, 1964.

[97] HAURIE, A., Stability and Optimal Exploitations over an Infinite Time Horizon of Interacting Populations, *Optimal Control Applications and Methods*, **3**, pp. 241-256, 1982.

[98] HAURIE, A., Existence and Global Asymptotic stability of Optimal Trajectories for a class of Infinite-Horizon, Nonconvex Systems, *Journal of Optimization Theory and Applications*, **31**, pp. 515-33, 1980.

[99] HAURIE, A., Optimal Control on a Infinite Time Horizon: The Turnpike Approach, *Journal of Mathematical Economics*, **3**, pp. 81–102, 1976.

[100] HAURIE, A. and N.M. HUNG, Turnpike Properties for the Optimal Use of a Natural Resource, *Review of Economic Studies*, **44**, pp. 329–336, 1976.

[101] HAURIE, A. and N.M. HUNG, Further Aspects of the Turnpike Theory in Continuous Time with Applications, *Journal of Dynamical Systems Management and Control*, **98**, pp. 1–6, 1976.

[102] HAURIE A. and P. L'ECUYER, Approximation and Bounds in Discrete Event Dynamic Programming, *IEEE Transactions on Automatic Control*, **31**, pp. 227–235, 1986.

[103] HAURIE, A. and SETHI, S., Decision and Forecast Horizons, Agreeable Plans, and the Maximum Principle, *Operations Research Letters*, **3**, pp. 261–265, 1984.

[104] HAURIE, A., S.P. SETHI and R.F. HARTL, Optimal Control of Age-structured Population Model with Applications to Social Services Planning, *Large Scale Systems*, **6**, pp. 133-158, 1984.

[105] HAURIE A. and CH. VAN DELFT, Turnpike Properties for a Class of Piecewise Deterministic Systems Arising in Manufacturing Flow Control, *Annals of Operations Research*, to appear 1991.

[106] HERMES AND LASALLE, *Functional Analysis and Time Optimal Control*, Academic Press, 1969.

[107] ISARD, W. and P. LIOSSATOS, *Spatial dynamics and Space-Time Development*, North Holland, Amsterdam, 1979.

[108] JABRANE, A., *Modélisation des Problémes Spatio-Temporels: Extension des Propriétés du Turnpike à des Systèmes à Paramètres Répartis*, Département d'Informatique et de Recherche Opérationnelle, Université de Montréal, 1987.

[109] KALECKI, M., A Macrodynamic Theory of Business Cycles, *Econometrica*, **3**, pp. 327-344, 1935.

[110] KALMAN, R.E., Contribution to the Theory of Optimal Control, *Bol. Soc. Mat. Mex.*, **5**, pp. 102–119, 1960.

[111] KOOPMAN, T.C., On the Concept of Optimal Economic Growth, In the Economic Approach to Development Planning, *Pontificiae Academiae Scientarum Scripia Veria*, **28**, pp. 225–287, North-Holland, Amsterdam, 1965.

[112] KURATOWSKI, K, and C. RYLL-NARDZEWSKI, A General Theorem on Selectors, Bulletin Academie Polonaise Science, **13**, pp. 397–403, 1965.

[113] KURZ, M., Optimal Paths of Capital Accumulation Under the Minimum Time Objective, *Econometrica*, **33**, pp. 42–66, 1965.

[114] L'ECUYER P. and A. HAURIE, Discrete Event Dynamic Programming with Simultaneous Events, *Mathematics of Operations Research*, **13**, pp. 152–163, 1988.

[115] LEE, E.B. and L. MARKUS, *Foundations of Optimal Control Theory*, John Wiley and Sons, Inc., New York, New York, 1967.

[116] LEITMANN, G., *The Calculus of Variations and Optimal Control*, Plenum Press 1981.

[117] LEITMANN, G. and H. STALFORD, A Sufficiency Theory for Optimal Control, *Journal of Optimization Theory and Applications*, **8**, 1971.

[118] LEIZAROWITZ, A., Optimal Control for Diffusions in R^d-A min-max max-min Formula for the Minimal Cost-growth Rate. *Journal of Mathematical Analysis and Applications*, 1990.

[119] LEIZAROWITZ, A., Optimal Trajectories of Infinite Horizon Deterministic Control Systems, *Applied Mathematics and Optimization*, **19**, pp. 11–32, 1989.

[120] LEIZAROWITZ, A., Infinite-horizon Stochastic Regulation and Tracking with the Overtaking Criterion. *Stochastics,* **22**, pp. 117-150, 1987.

[121] LEIZAROWITZ, A., Convergence of Viable Solutions of Differential Inclusions with Convex, Compact Graphs, *SIAM Journal on Control and Optimization*, **23**, pp. 514–522, 1985.

[122] LEIZAROWITZ, A., Existence of Overtaking Optimal Trajectories for Problems with Convex Integrands, *Mathematics of Operations Research*, **10**, pp. 450–461, 1985.

[123] LEIZAROWITZ, A., Existence of Overtaking Optimal Trajectories with Convex Integrands, *Mathematics of Operations Research*, **10**, pp. 450–461, 1985.

[124] LEIZAROWITZ, A., Infinite Horizon Autonomous Systems with Unbounded Cost, *Applied Mathematics and Optimization*, **13**, pp. 19–43, 1985.

[125] LEONTIEF, W., Lags and the Stability of Dynamic Systems, *Econometrica*, **29** pp. 659–669, 1961.

[126] LEVHARI, D. and N. LIVIATAN, Stability in the Saddlepoint Sense, *Journal of Economic Theory*, pp. 88–93, 1972.

[127] LIONS, P-L., *Generalized Solutions to Hamilton-Jacobi Equations*, Res. Notes in Math., Vol. 69, Pitman Publishing Inc., Boston, Mass. 1982.

[128] LIONS, J. L., *Optimal Control of Systems Governed by Partial Differential Equations*, Springer-Verlag, New York, New York, 1971.

[129] MAFFEZZONI, C., Hamilton-Jacobi Theory for Periodic Control Problems, *Journal of Optimization Theory and Applications*, **14**, pp. 21–29, 1974.

[130] MAGILL, M., Some New Results on the Local Stability of the Process of Capital Accumulation, *Journal of Economic Theory*, **15**, pp. 174–210, 1977.

[131] MAGILL, M., *On a General Economic Theory of Motion*, Springer-Verlag, New York, 1970.

[132] MAGILL, J.P., Infinite Horizon Programs, *Econometrica*, **49**, pp. 679–711, 1981.

[133] MAGILL, J.P., On a Class of Variational Problems Arising in Mathematical Economics, *Journal of Mathematical Analysis and Applications*, **82**, pp. 66–74, 1981.

[134] MANGASARIAN, O.L., Sufficient Conditions for Optimal Control of Nonlinear Systems, *SIAM Journal on Control*, **4**, pp. 139–152, 1966.

[135] MANGASARIAN, O.L., *Nonlinear Programming*, McGraw Hill, 1969.

[136] MARCUS, M. and V.J. MIZEL, *Limiting Equations for Problems Involving Long Range Memory*, Memoirs of the American Mathematical Society, Vol. 43, 1983.

[137] MARUYAMA, T., A Variational Problem in Relating to the Theory of Optimal Economic Growth, *Proceeding of the Japanese Academy*, **57**, Ser. A, pp. 387–386, 1981.

[138] MARUYAMA, T., Optimal Economic Growth with Infinite Planning Horizon, *Proceeding of the Japanese Academy*, **57**, Ser. A, pp. 469–472, 1981.

[139] MCKENZIE, L.W., Turnpike Theory, *Econometrica*, **44**, pp. 841–866, 1976.

[140] MCKENZIE, L.W., Accumulation Programs of Maximum Utility and the Von Neumann Facet, *Value, Capital, and Growth*, J.N. Wolfe ed., Aldine, Chicago, 1968.

[141] MCKENZIE, L. W., Turnpike Theorems for a Generalized Leontief Model, *Econometrica*, **31**, pp. 165–180, 1963.

[142] MICHEL, P., Une Démonstration Élémentaire du Principe du Maximum de Pontryagin, *Bulletin the Mathématique Économiques*, **14**, pp. 9–23, 1977.

[143] MIELE, A., Flight Mechanics and Variational Problems of a Linear Type, *Journal of Aero-Space Sciences*, **25**, pp. 581–590, 1958.

[144] MIELE, A., Extremization of Linear Integrals by Green's Theorem, *Optimization Techniques*, Edited by G. Leitmann, Academic Press, 1962.

[145] NAIRAY, A., Asymptotic Behavior and Optimal Properties of a Consumption-Investment Model with Variable Time Preference, *Journal of Economic Dynamics and Control*, **7**, pp. 283–313, 1984.

[146] OLSDER G.J. and R. SURI, Time Optimal Control of Parts-routing in a Manufacturing System with Failure Prone Machines, *IEEE International Conference on Decision and Control*, Albuquerque, New Mexico, 1980.

[147] PAUWELS, W., Optimal Dynamic Advertising Policies in the Presence of Continuously Distributed Time Lag, *Journal of Optimization Theory and Applications*, **22**, pp. 79–89, 1977.

[148] PETERSON, D.W., A Sufficient Maximum Principle, *IEEE Transaction on Automatic Control*, pp. 85–86, Feb. 1971.

[149] PETERSON, D.W. and J.H. ZALKIND, A Review of Direct Sufficient Conditions in Optimal Control Theory, *International Journal on Control*, **28**, 1978.

[150] PONTRYAGIN, L.S., V.G. BOLTYANSKII, R.V. GAMKRELIDZE and E.F. MISHCHENKO, *The Mathematical Theory of Optimal Processes*, Wiley-Interscience, New York, New York, 1962.

[151] RADNER, R., Paths of Economic Growth that are Optimal with Regard Only to Final State: A Turnpike Theorem, *Review of Economic Studies*, **28**, pp. 98–104, 1961.

[152] RAMSEY, F., A Mathematical Theory of Saving, *Economic Journal*, **38**, pp. 543–549, 1928.

[153] RISHEL, R., Control of Systems with Jump Markov Disturbances, *IEEE Transactions on Automatic Control*, pp. 241–244, 1975.

[154] RISHEL, R., Dynamic Programming and Minimum Principles For Systems With Jump Markov Disturbances, *SIAM Journal on Control and Optimization*, **13**, pp. 338–371, 1975.

[155] ROCKFELLAR, R.T., Convex Processes and Hamiltonian Dynamical Systems, in *Convex Analysis and Mathematical Economics*, Ed. J. Krein, Lecture Notes on Economy and Mathematical Systems, **168**, 122–136, 1979.

[156] ROCKAFELLAR, R.T., Saddle Points of Hamiltonian Systems in Convex Lagrange Problems Having Nonzero Discount Rate, *Journal of Economic Theory*, **12**, pp. 71–113, 1976.

[157] ROCKAFELLAR, R.T., Integral Functionals, Normal Integrands, and Measurable Selections, in *Nonlinear Operators and the Calculus of Variations*, Bruxelles, 1975, J.P. Gessez, E.J. Lami Dozo, J. Mawhin, and L. Waelbroeck, eds. Lec. Notes in Math., Vol. 543, Springer-Verlag, Berlin, pp. 157-207, 1976.

[158] ROCKAFELLAR, R.T., Existence Theorems for General Control Problems of Bolza and Lagrange, *Advances in Mathematics*, **15**, pp. 312–333, 1975.

[159] ROCKAFELLAR, R.T., Saddle Points of Hamiltonian Systems in Convex Problems of Lagrange, *Journal of Optimization Theory and Applications*, **12**, pp. 367–399, 1973.

[160] ROCKAFELLAR, R.T., *Convex Analysis*, Princtone University Press, New Jersey, 1970.

[161] ROCKAFELLAR, R.T., Conjugate Convex Functions in Optimal Control and the Calculus of Variations, *Journal of Mathematical Analysis and Applications*, **32**, pp. 174–222, 1970.

[162] ROSEN, R., *Dynamical System Theory in Biology*, Wiley-Interscience, 1970.

[163] SAMUELSON, P.A., The General Saddlepoint Property of Optimal Control Motions, *Journal of Economic Theory*, **5**, pp. 102–120, 1972.

[164] SAMUELSON, P.A., A Catenary Turnpike Theorem Involving Consumption and the Golden Rule, *American Economic Review*, **55**, pp. 486–496, 1965.

[165] SANDERS, J., A Note on Optimal Fish Harvest, *Journal of Optimization Theory and Applications*, **24**, pp. 361–369, 1978.

[166] SEIERSTAD, A. and K. SYDSÆTER, Sufficient Conditions in Optimal Control Theory, *International Economic Review*, **18**, 1977.

[167] SETHI, S.P., Optimal Control of the Vidale-Wolfe Advertising Model, *Operations Research*, **22**, pp. 998–1023, 1973.

[168] SETHI, S.P. and G.L. THOMPSON, *Optimal Control Theory: Management Science Applications*, Martinus Nijhoff Pub. Co., Boston, MA, 1981.

[169] SHELL, K., The Theory of Hamiltonian Dynamic Systems, in J.D. Groke ed. *The Theory and Application of Differential Games*, D. Reidel, 1975.

[170] SNOW, D.R., Carathéodory-Hamilton-Jacobi Theory in *Optimal Control, Journal of Mathematical Analysis and Applications*, **16**, pp. 99–118, 1967.

[171] STOLERU, L.G., An Optimal Policy for Economic Growth, *Econometrica*, **33**, pp. 321–348, 1965.

[172] TAKEKUMA, S, A Sensitivity Analysis in Optimal Economic Growth, *Journal of Mathematical Economics*, **7**, pp. 193–208, 1980.

[173] THISSE, J. and D. WEISERBS, Existence and Uniqueness of an Optimal Control with an Infinite Horizon, *Cahiers du CERO*, Vol. 21, pp. 17–22, 1979.

[174] TOMAN, M.A., Optimal Control with Unbounded Horizon, *Journal of Economic Dynamics and Control*, **9**, pp. 291–316, 1986.

[175] UNDERWOOD, R.G. and D. YOUNG, Null Controllability of Nonlinear Functional Differential Equations, *SIAM Journal on Control and Optimization*, **17** pp. 753–772, 1979.

[176] UZAWA, H., Optimal Growth in a Two-sector Model of Capital Accumulation, *Review of Economic Studies*, **31**, pp. 1-24, 1964.

[177] VARAIYA, P., On the Trajectories of a Differential System, in Balakrishnan, Neustadt, eds. *Mathematical Theory of Control*, Academic Press, 1967.

[178] VERMES, D., Optimal Control of Piecewise Deterministic Markov Process, *Stochastics*, **14**, pp. 165–208, 1985.

[179] VINCENT, T.L., Pest Management Programs via Optimal Cntrol Theory, *Biometrics*, **31**, pp. 1–10, 1975.

[180] VON WEIZÄCKER, C.C., Existence of Optimal Programs of Accumulation for an Infinite Time Horizon, *Review of Economic Studies*, **32**, pp. 85–104, 1965.

[181] WILLEMS, J.C., The L.Q.G. Problem. In *Stochastic Systems: The Mathematics of Filtering and Identification and Applications*, M. Hazewinkel and J.C. Willems (eds.), Reidel Publishing Company, pp. 29–44, 1980.

[182] WONHAM, W.M., Random Differential Equations in Control Theory in *Probabilistic Methods in Applied Mathematics*, Barucha, A.T. and Reid (eds.), Academic Press, pp. 131–217, 1970.

[183] YANO, M., A Note on the Existence of an Optimal Capital Accumulation in the Continuous-Time Horizon, *Journal of Economic Theory*, **27**, pp. 421–429, 1981.

Index

FSC
www.fsc.org
MIX
Papier aus verantwortungsvollen Quellen
Paper from responsible sources
FSC® C105338

If you have any concerns about our products,
you can contact us on
ProductSafety@springernature.com

In case Publisher is established outside the EU,
the EU authorized representative is:
Springer Nature Customer Service Center GmbH
Europaplatz 3, 69115 Heidelberg, Germany

Printed by Libri Plureos GmbH
in Hamburg, Germany